AN INTRODUCTION TO

SECOND ORDER PARTIAL DIFFERENTIAL EQUATIONS

Classical and Variational Solutions

AN INTRODUCTION TO SECOND ORDER PARTIAL DIFFERENTIAL EQUATIONS

Classical and Variational Solutions

Doina Cioranescu

Université Pierre et Marie Curie (Paris 6), France

Patrizia Donato

Université de Rouen, France

Marian P. Roque

University of the Philippines Diliman, Philippines

World Scientific

NEW JERSEY • LONDON • SINGAPORE • BEIJING • SHANGHAI • HONG KONG • TAIPEI • CHENNAI • TOKYO

Published by

World Scientific Publishing Co. Pte. Ltd.
5 Toh Tuck Link, Singapore 596224
USA office: 27 Warren Street, Suite 401-402, Hackensack, NJ 07601
UK office: 57 Shelton Street, Covent Garden, London WC2H 9HE

Library of Congress Cataloging-in-Publication Data
Names: Cioranescu, D. (Doina), author. | Donato, Patrizia, author. | Roque, Marian P., author.
Title: An introduction to second order partial differential equations :
classical and variational solutions / by Doina Cioranescu (Université Pierre et Marie Curie (Paris 6), France), Patrizia Donato (Université de Rouen, France),
Marian P. Roque (University of the Philippines, Diliman, Philippines).
Other titles: Second order partial differential equations | Partial differential equations
Description: New Jersey : World Scientific, 2017. | Includes bibliographical references and index.
Identifiers: LCCN 2017042805 | ISBN 9789813229174 (hardcover : alk. paper)
Subjects: LCSH: Differential equations, Partial--Study and teaching (Higher) |
Differential equations, Partial--Study and teaching (Graduate)
Classification: LCC QA377 .C5627 2017 | DDC 515/.353--dc23
LC record available at https://lccn.loc.gov/2017042805

British Library Cataloguing-in-Publication Data
A catalogue record for this book is available from the British Library.

Printed in Singapore

To all those who will read this book. May it give them some insight in the fascinating field of partial differential equations. To Ion and our son Jean-Michel with love.

Doina Cioranescu

To the peacebuilder Daisaku Ikeda and mentor of my life, with deepest gratitude. To my husband Eric and sons Maikel and Johendry, persons of great value, with all my love.

Patrizia Donato

To my parents, Martin and Consolacion, brothers Marlon and Marius, husband Rolly, and daughter Tin-Tin, with love and gratitude for having supported all my endeavors. Unlike partial differential equations, their love requires no conditions.

Marian P. Roque

Preface

The study of partial differential equations (PDEs) is at the crossroads of mathematical analysis, measure theory, topology, differential geometry, scientific computing, and many other branches of mathematics. Modeling physical phenomena, partial differential equations are fascinating topics because of their increasing presence in treating real physical processes. In recent years, PDEs have become essential tools when modeling in fields such as materials science, fluid mechanics, quantum mechanics, mathematical finance, biology and biomedicine, and environmental sciences.

The aim of this book is to introduce this important subject to graduate and post-graduate students in Mathematics with sufficient background in advanced calculus, real analysis, and functional analysis, as well as to young researchers in the field. Indeed, as explained below, various reading levels are possible, and because of that we also hope it will be useful to colleagues teaching PDEs at different levels.

The book is essentially devoted to second order linear partial differential equations and consists of two parts that are each self-contained. Part I (Chapters 1 to 5) gives a comprehensive overview of classical PDEs, that is, equations which admit smooth (strong) solutions. Part II (Chapters 6 to 10) deals with variational PDEs, where weak solutions are considered. They are defined via a weak (variational) formulation of the equations and are searched in suitable function spaces (here in particular, Sobolev spaces). These spaces, being the essential tools in the treatment of variational PDEs, are introduced and extensively detailed.

The first chapter answers the basic question: What is a PDE? In this chapter, we give the basic definitions encountered in the study of PDEs, the general notation used throughout the book and a list of some classical partial differential equations. Chapter 2 presents the classification of PDEs and their canonical forms. Characteristic curves and some existence theorems are also given. The classical examples of Laplace, heat, and wave equations are introduced in Chapters 3, 4, and 5, respectively. Part I is aimed to be an introductory presentation of the subject, it is why we choose not to include too many details but to state only the main methods and results, with proofs for some theorems.

For the study of weak solutions, a review of L^p-spaces and distributions is necessary. This is done in Chapter 6 where we also prove some less classical results, which are specifically needed in the succeeding chapters. A comprehensive and detailed discussion of Sobolev spaces and Sobolev continuous and compact embeddings is presented in Chapters 7 and 8, respectively. Examples of variational elliptic problems with different boundary conditions are discussed in Chapter 9. Finally, variational parabolic and hyperbolic problems are studied in Chapter 10.

The Sobolev spaces theory provides the foundation for variational PDEs. For young mathematicians, understanding the proofs can be comforting and undoubtedly formative. We decided to prove almost all the important results we stated, for completeness and for a deeper understanding of the theory underlying variational PDEs. While writing this book, we observed that some of the proofs we wanted to present in these chapters are not usually found in a PDE book, since they involve advanced topics in functional analysis. We hope that their presence in this book will be appreciated by the reader. We also give several references for further reading in every chapter.

For Chapters 7 to 10, we have chosen to explain proofs in detail to enable young researchers to do an independent study on the topics covered. We also completed the presentation by an important number of examples, in particular for the most delicate definitions.

According to the authors' experience, the material covered here is more than sufficient for a one-semester graduate course and may be extended to two or three semesters, depending on the level of the students. For an introductory course, we suggest a detailed discussion of Chapters 1 and 2 followed by solutions of the Laplace, heat, and wave equations in $\mathbb{R}^2$ from Chapters 3, 4, and 5. Definitions and theorems (without proofs) from Chapters 6, 7 and 8, together with the Lax-Milgram Theorem, can then be given before presenting some examples of variational elliptic problems from

Chapter 9. Sections and proofs skipped in the first course, can be discussed in detail in a second course. A more advanced course could contain the proofs of the most delicate results from Chapter 7 as well as those of the Sobolev embedding theorems. Then one could present the results of the eigenvalue problems from Chapter 9, and apply them to the study of the variational evolution problems done in Chapter 10.

Finally, let us mention that this book is based on the long experience of the authors as researchers and teachers in the field of PDEs, teaching both in their home universities and in research schools abroad. And as significant, the book is founded on the scientific collaborations and deep friendship between the authors, which have been enriched through the years. The first two authors have a scientific collaboration for about thirty years, and the collaboration with the third author started more than twelve years ago, when they gave a graduate PDE course for three years at the Institute of Mathematics, UP Diliman under the European Asia Link IMAMIS program. In particular, the lecture notes of this course were published in 2012, under the name "Introduction to classical and variational partial differential equations", by the University of the Philippines Press.

The present book is a more developed and detailed version of these lecture notes.

Doina Cioranescu
Patrizia Donato
Marian P. Roque

Contents

List of Symbols

PART I

Classical Partial Differential Equations

Chapter 1

What is a Partial Differential Equation?

The first chapter contains generalities about partial differential equations (PDEs) as well as a list of the most popular PDEs in the literature, starting with the classical Laplace equation, the heat and wave equations, and ending with the Maxwell equations. A general existence and uniqueness theorem is given for analytic data.

Throughout the book, we only consider real functions and spaces, unless otherwise specified. The space $\mathbb{R}^N$ is equipped with the Lebesgue measure dx. To simplify the notation, x belongs either to $\mathbb{R}$ or to $\mathbb{R}^N (N > 1)$, and in this case, one has to recall that $x = (x_1, \ldots, x_N)$. It is assumed that the reader is familiar with the basic properties of Lebesgue measure and integration theory. For a detailed discussion on these topics, we refer the reader to [3], [4], [33], and [34].

1.1 Introduction

A partial differential equation is a relation containing one or more partial derivatives of an unknown function depending on two or several independent variables. If at least one of these partial derivatives is of order m, and if there are no derivatives of order higher than m, the PDE is said to be of order m.

To be more explicit, let us introduce the notion of multi-index of order $|\alpha|$ that will be used very often in this book.

Definition 1.1. *The N-tuple* α

$$\alpha = (\alpha_1, \ldots, \alpha_N),$$

is called multi-index of order $|\alpha| = \alpha_1 + \ldots + \alpha_N$ *if* α_i *is a nonnegative integer for every* $i \in \{0, 1, \ldots, N\}$.

In the following, we will use the notation

$$\partial^\alpha = \frac{\partial^{|\alpha|}}{\partial x_1^{\alpha_1} \dots \partial x_N^{\alpha_N}}, \tag{1.1}$$

where ∂^α is the identity for $\alpha = 0$. For $k \in \mathbb{N}$, the set of all partial derivatives of order k of a function u is denoted $D^k(u)$, that is,

$$D^k(u) = \{\partial^\alpha(u), \quad |\alpha| = k\}.$$

Definition 1.2. *Let $\mathcal{O}$ be an open set in $\mathbb{R}^N$. A partial differential equation (PDE) of order m is an equation of the form*

$$F\left(x, u, D^1(u), D^2(u), \dots, D^m(u)\right) = f(x), \tag{1.2}$$

for $x = (x_1, x_2, \dots, x_N) \in \mathcal{O}$, where f is a given function.

The mapping

$$u \mapsto F\left(x, u, D^1(u), D^2(u), \dots, D^m(u)\right),$$

is called a differential operator.

A solution of (1.2) is a function $u = u(x)$ whose partial derivatives in (1.2) are well-defined in $\mathcal{O}$ and which satisfies equation (1.2) at each point x of $\mathcal{O}$. If $N = 1$, then (1.2) is simply an ordinary differential equation of order m.

When one has to solve the partial differential equation (1.2) with a nonzero right-hand side, that is when $f \not\equiv 0$, one says that (1.2) is an inhomogeneous PDE, while if $f \equiv 0$, (1.2) is a homogeneous PDE.

One can also consider systems of partial differential equations. In this case, one has several expressions of the form (1.2) containing one or several unknown functions as well as their derivatives.

In the study of partial differential equations, the notion of linearity plays an important role. The PDE given by (1.2) is called linear if F is linear with respect to the unknown u and all its derivatives.

The linearity of F implies that it can be written as

$$F\big(x, u, D^1(u), D^2(u), \dots, D^m(u)\big) = \sum_{|\alpha| \le m} a_\alpha(x)\partial^\alpha(u), \tag{1.3}$$

where the functions $a_\alpha = a_\alpha(x)$ (for $|\alpha| \le m$) are called the coefficients of the differential operator.

In the case of a homogeneous linear ordinary differential equation, any linear combination of its solutions is still a solution of the same equation. This is known as the principle of superposition. For homogeneous linear partial differential equations, the same principle holds. Indeed,

Theorem 1.3 (Principle of Superposition). *If $u_1, u_2, \ldots, u_k$ are solutions of (1.2) with $f \equiv 0$ and F is linear, then $U = c_1u_1 + c_2u_2 + \ldots + c_ku_k$ is also a solution of (1.2) for any constants $c_1, c_2, \ldots, c_k$. That is,*

$$F\left(x, U, D^1(U), D^2(U), \ldots, D^m(U)\right) = 0 \quad \text{for } x \in \mathcal{O}.$$

The proof of this result follows simply from the fact that the differentiation operator is linear.

If F is linear only with respect to the derivatives of order m, i.e., if

$$\begin{aligned} &F\left(x, u, D^1(u), D^2(u), \ldots, D^m(u)\right) \\ &\quad = \sum_{|\alpha|=m} a_\alpha(x)\partial^\alpha(u) + \overline{F}\left(x, u, D^1(u), D^2(u), \ldots, D^{m-1}(u)\right), \end{aligned}$$

then equation (1.2) is called semilinear.

If F can be written with more general coefficients in the following form:

$$\begin{aligned} &F\left(x, u, D^1(u), D^2(u), \ldots, D^m(u)\right) \\ &\quad = \sum_{|\alpha|=m} a_\alpha\left(x, u, D^1(u), D^2(u), \ldots, D^{m-1}(u)\right)\partial^\alpha(u) \\ &\qquad + \overline{F}\left(x, u, D^1(u), D^2(u), \ldots, D^{m-1}(u)\right), \end{aligned}$$

then (1.2) is called quasilinear.

Obviously, the superposition principle is not valid in these cases. Nevertheless, the fact that a PDE can be classified as semilinear or quasilinear, renders its study easier.

Example 1.4. We now give some examples of second order PDEs. In the following, c is a constant, t is the time variable, and x is the usual Cartesian coordinate.

(1) $\dfrac{\partial u}{\partial t} = c^2\dfrac{\partial^2 u}{\partial x^2}$ One-dimensional heat equation

(2) $\dfrac{\partial u}{\partial t} = c^2\left(\dfrac{\partial^2 u}{\partial x_1^2} + \dfrac{\partial^2 u}{\partial x_2^2}\right)$ Two-dimensional heat equation

(3) $\dfrac{\partial^2 u}{\partial t^2} = c^2\dfrac{\partial^2 u}{\partial x^2}$ One-dimensional wave equation

(4) $\dfrac{\partial^2 u}{\partial t^2} = c^2\left(\dfrac{\partial^2 u}{\partial x_1^2} + \dfrac{\partial^2 u}{\partial x_2^2}\right)$ Two-dimensional wave equation

(5) $\dfrac{\partial^2 u}{\partial x_1^2} + \dfrac{\partial^2 u}{\partial x_2^2} = 0$ Two-dimensional Laplace equation

(6) $\dfrac{\partial^2 u}{\partial x_1^2} + \dfrac{\partial^2 u}{\partial x_2^2} = f(x_1, x_2)$ Two-dimensional Poisson equation

(7) $\dfrac{\partial^2 u}{\partial x_1^2} + \dfrac{\partial^2 u}{\partial x_2^2} + \dfrac{\partial^2 u}{\partial x_3^2} = 0$ Three-dimensional Laplace equation

(8) $\dfrac{\partial^2 u}{\partial x_1^2} + \dfrac{\partial^2 u}{\partial x_2^2} + \dfrac{\partial u}{\partial x_1} = cu^3$

(9) $u\dfrac{\partial^2 u}{\partial x_1^2} + \dfrac{\partial u}{\partial x_1}\dfrac{\partial^2 u^2}{\partial x_2} = f(x_1, x_2).$ ◇

The first seven equations are linear, the last two are nonlinear. Equations (6) and (9) with $f \neq 0$ are inhomogeneous while all the others are homogeneous. Equations (8) and (9) are nonlinear, since the terms u^3 and $u\dfrac{\partial^2 u}{\partial x_2^2}$ do not satisfy a linear relation. Notice that (8) is semilinear and that (9) is quasilinear.

A general method to solve linear problems is the separation of variables. It consists of searching for a solution expressed as a product of functions, each depending only on one variable. For instance, in equation (1) or (3) above, we look for a solution u of the form

$$u(x,t) = \mathcal{F}(x)\mathcal{G}(t).$$

Substituting this in equation (1) or (3) gives two ordinary differential equations involving $\mathcal{F}$ and $\mathcal{G}$ separately. We shall use this method to solve the Laplace equation (Chapter 3), the heat equation (Chapter 4), and the wave equation (Chapter 5) in the one and two-dimensional cases and in some simple geometric domains (essentially discs or rectangles).

We shall be considering only linear equations in this book, that is of the form (1.3). However, let us point out that many physical situations are modeled by nonlinear PDEs (for example, the Navier–Stokes equations describing fluid flows, which are discussed in the next section). In most cases, their study presents severe mathematical difficulties. To get rid of these difficulties, one is led to sacrifice the precision of the mathematical models, either by neglecting the nonlinear perturbations (if they are small), or by linearizing the equation in the neighborhood of a known solution. This reinforces again the interest of linear partial differential equations.

Before going further, we introduce some notation that will be needed throughout the book.

1.2 General notation

(1) As a rule, we use the following notation:

- $\mathbb{N}$ for the set of natural numbers, $\mathbb{N} = 1, 2, \ldots,$
- $\mathbb{N}_0 = \mathbb{N} \cup \{0\}$,
- $\mathbb{R}$ for the set of the real numbers,
- $\mathbb{R}^+ = \{x \in \mathbb{R},\ x > 0\}$, $\quad \mathbb{R}_0^+ = \{x \in \mathbb{R},\ x \geq 0\}$,
- $(x, y)_{\mathbb{R}^N}$ for the scalar product in $\mathbb{R}^N$ given by

$$(x, y)_{\mathbb{R}^N} = \sum_{i=1}^{N} x_i y_i,$$

- $|x|$ for the Euclidean norm in $\mathbb{R}^N$ given by

$$|x| = \Big(\sum_{i=1}^{N} x_i^2\Big)^{1/2},$$

- $\mathcal{O}$ for an open subset of $\mathbb{R}^N$,
- Ω for a bounded open subset of $\mathbb{R}^N$,
- $|\omega|$ for the Lebesgue measure of a measurable set $\omega \subset \mathbb{R}^N$.

(2) A property is said to hold almost everywhere (a.e.) in a set X if it holds on X except on a measurable subset of X with zero Lebesgue measure.

(3) The space of continuous functions on $\mathcal{O} \subset \mathbb{R}^N$ is denoted $C^0(\mathcal{O})$.

(4) Let u be a function defined on $\mathbb{R}^N$. If all its partial derivatives up to order $k \in \mathbb{N}$ are continuous on $\mathcal{O} \subset \mathbb{R}^N$, we say that u is of class $C^k(\mathcal{O})$ (or u belongs to $C^k(\mathcal{O})$).

(5) If k is a positive integer, for any open set $\mathcal{O}$ in $\mathbb{R}^N$, the Banach space $C_b^k(\mathcal{O})$ is defined as follows:

$$C_b^k(\mathcal{O}) = \Big\{u \,\Big|\, \partial^\alpha u \in C^0(\mathcal{O}) \text{ and bounded } \forall \alpha \in \mathbb{N}_0^N,\ |\alpha| \leq k\Big\}, \tag{1.4}$$

with the norm

$$\|u\|_{C_b^k(\mathcal{O})} = \sum_{|\alpha| \leq k} \sup_{x \in \mathcal{O}} \left| \frac{\partial^\alpha u(x)}{\partial x_1^{\alpha_1} \ldots \partial x_N^{\alpha_N}} \right|.$$

For the definition of a Banach space, see Definition 6.1 in Chapter 6.

(6) We denote by $\overline{\mathcal{O}}$ the closure of $\mathcal{O}$, that is to say, $\overline{\mathcal{O}} = \mathcal{O} \cup \partial\mathcal{O}$, where $\partial\mathcal{O}$ is the boundary of $\mathcal{O}$. We say $u \in C^k\big(\overline{\mathcal{O}}\big)$ whenever u is the restriction to $\overline{\mathcal{O}}$ of a function in $C^k(\mathcal{O}_1)$, for some open set $\mathcal{O}_1$ containing $\overline{\mathcal{O}}$ (this means in particular that they are also defined on $\partial\mathcal{O}$). In the case of a bounded open set, the space $C^k(\overline{\Omega})$ is a Banach space for the norm

$$\|u\|_{C^k(\overline{\Omega})} = \sum_{|\alpha|\leq k} \max_{x\in\overline{\Omega}} \left| \frac{\partial^\alpha u(x)}{\partial x_1^{\alpha_1}\dots\partial x_N^{\alpha_N}} \right|. \tag{1.5}$$

(7) We say that $u \in C^\infty(\mathcal{O})$ if $u \in C^k(\mathcal{O})$, for every $k \in \mathbb{N}$. We say that $u \in C^\infty(\overline{\mathcal{O}})$ if $u \in C^k(\overline{\mathcal{O}})$, for every $k \in \mathbb{N}$.

(8) The Laplacian Δ is the following partial differential operator:

$$\Delta = \frac{\partial^2}{\partial x_1^2} + \dots + \frac{\partial^2}{\partial x_N^2}.$$

(9) If u is a scalar function defined on $\mathbb{R}^N$, then the gradient of u, denoted grad u or ∇u, is the vector function

$$\text{grad } u = \nabla u = \left(\frac{\partial u}{\partial x_1}, \dots, \frac{\partial u}{\partial x_N}\right). \tag{1.6}$$

(10) The normal derivative of u on a curve (if $N = 2$) or on a surface (if $N \geq 3$) γ is given by

$$\frac{\partial u}{\partial n} = (\nabla u, \vec{n}) = \sum_{i=1}^{N} \frac{\partial u}{\partial x_i} n_i, \tag{1.7}$$

where $\vec{n}$ is the unit normal vector to γ.

Recall that for $i = 1, \dots, N$, the component n_i of $\vec{n}$ is the cosine of the angle between $\vec{n}$ and the x_i-axis. As a rule, when taking the normal derivative of u on the boundary $\partial\omega$ of an open set $\omega \subset \mathbb{R}^N$, $\vec{n}$ will denote the outward unit normal to $\partial\omega$. To simplify, and if no ambiguity occurs, we denote this vector by n.

(11) The divergence of u, denoted $\operatorname{div} u$, is the scalar function defined as

$$\operatorname{div} u = \frac{\partial u}{\partial x_1} + \dots + \frac{\partial u}{\partial x_N}.$$

(12) Let $\vec{v} = (v_1, \dots, v_N)$ be a vector function defined on $\mathbb{R}^N$. The gradient of $\vec{v}$, denoted grad $\vec{v} = \nabla\vec{v}$ is the following square matrix:

$$\nabla\vec{v} = \left(\frac{\partial v_i}{\partial x_j}\right)_{(i,j=1,\dots,N)},$$

while the scalar function div $\vec{v}$ is defined by

$$\text{div } \vec{v} = \frac{\partial v_1}{\partial x_1} + \dots + \frac{\partial v_N}{\partial x_N},$$

and if no ambiguity occurs, we skip the vectorial notation and write simply ∇v and div v, instead of $\nabla\vec{v}$ and div $\vec{v}$, respectively.

1.3 Boundary and initial conditions

In general, a partial differential equation can have more than one or even an infinite number of solutions. For example, the functions

$$u = x_1^2 - x_2^2, \quad u = e^{x_1} \cos x_2, \quad u = \ln(x_1^2 + x_2^2),$$

which are very different in nature, are all solutions of the two-dimensional Laplace equation given in Example 1.4(5).

We shall see later that the uniqueness of a solution of a physical problem is obtained by using some additional conditions contained in the model. For instance, if the problem is posed in a bounded open set Ω of $\mathbb{R}^N$, a condition assigning a given value on its boundary $\partial\Omega$ is called a boundary condition.

The most classical boundary conditions are the following:

(1) The Dirichlet boundary condition: a given value is imposed for the solution u on $\partial\Omega$,

(2) The Neumann boundary condition: the normal derivative of u has to take a given value on $\partial\Omega$,

(3) The Robin boundary condition: it involves both u and its normal derivative in an expression written on $\partial\Omega$.

Another type of conditions are the initial ones. They can be imposed when the time t is a variable in the problem: one assigns a value to u and/or to derivatives of u with respect to t at the initial time $t = 0$.

For the wave equations (3) and (4), one may prescribe two initial conditions: the initial displacement and the initial velocity. For the heat equations (1) and (2), one may prescribe the initial temperature.

Example 1.5. Consider equations (5) and (6) from Example 1.4. The most frequently used boundary conditions for these equations in a bounded open set $\Omega \subset \mathbb{R}^2$ whose boundary is denoted $\partial\Omega$, are

- $u(x) = 0 \quad \text{for } x \in \partial\Omega,$ (Homogeneous Dirichlet condition)
- $u(x) = \varphi(x) \quad \text{for } x \in \partial\Omega,$ (Nonhomogeneous Dirichlet condition)
- $\dfrac{\partial u}{\partial n}(x) = 0 \quad \text{for } x \in \partial\Omega,$ (Homogeneous Neumann condition)
- $\dfrac{\partial u}{\partial n}(x) = \varphi(x) \quad \text{for } x \in \partial\Omega,$ (Nonhomogeneous Neumann condition)
- $\dfrac{\partial u}{\partial n}(x) + h(x)u(x) = g(x) \quad \text{for } x \in \partial\Omega,$ (Robin condition)

where φ, h and g are given functions defined on $\partial\Omega$. ◇

Example 1.6. Consider the heat equation (2) from Example 1.4 for $x \in \Omega \subset \mathbb{R}^2$ and $t \in (0,T)$ for some $T > 0$. As in the preceding example, we can take as boundary condition one of those listed above, in particular,

$$\text{(i)} \quad u(x,t) = \varphi(x,t) \qquad \text{for } x \in \partial\Omega \text{ and } t \in (0,T),$$

$$\text{(ii)} \quad \frac{\partial u}{\partial n}(x,t) = \varphi(t) \qquad \text{for } x \in \partial\Omega \text{ and } t \in (0,T),$$

$$\text{(iii)} \quad \left[\frac{\partial u}{\partial n} + hu\right](x,t) = g(x,t) \qquad \text{for } x \in \partial\Omega \text{ and } t \in (0,T),$$

where φ, h, and g are given functions on $\partial\Omega \times (0,T)$.

Condition (i) signifies that the boundary is maintained at a given temperature φ. Condition (ii) means that the flux of the heat through the boundary is prescribed. Finally, condition (iii) translates to the fact that through the boundary, there is an exchange of heat with the environment that has a given temperature.

As for the initial condition, the most natural one is to impose an initial temperature

$$u(x,0) = \phi(x) \quad \text{for } x \in \Omega. \qquad \diamond$$

Example 1.7. When treating the wave equation (3) from Example 1.4, it is usual (and natural) to specify the initial position and velocity of all the points of the vibrating string (see next section for its modeling), that is,

$$u(x,0) = \phi(x) \quad \text{for } x \in \Omega,$$

$$\frac{\partial u}{\partial t}(x,0) = \psi(x) \quad \text{for } x \in \Omega,$$

to which one has to add a boundary condition (it can be one of those mentioned in Example 1.5 above). ◇

1.4 Some classical partial differential equations

We can encounter partial differential equations in an extremely wide range of fields: acoustics, quantum mechanics, elasticity, electromagnetism, thermodynamics, fluid mechanics, geophysics, to name a few. They model, for instance, heat transfer, vibrations, wave propagations, fluid flows, and elastic or plastic deformations or displacements.

The most known partial differential equations are the Laplace equation, the heat equation, and the wave equation. We shall show later how the heat and the wave equations are derived from the balance laws.

1.4.1 *Laplace and Poisson equations*

The homogeneous Laplace equation has the following form:

$$\Delta u = 0,$$

where $u = u(x),\ x \in \mathbb{R}^N$ and Δ is the Laplacian defined in Section 1.2. The corresponding nonhomogeneous equation

$$\Delta u = f, \tag{1.8}$$

where $f = f(x)$ is a given function, is called the Poisson equation.

The Laplace and Poisson equations model a wide range of physical phenomena. For instance, the temperature in a homogeneous medium in a stationary region satisfies the Laplace equation which in this case, is called the stationary thermal diffusion equation. The same equation also describes the displacements of a tight membrane. In both situations, if an external force is given (an external source of heat, in the first one), obviously, one is concerned with the Poisson equation. These equations also appear in electrical problems. The potential of an electrostatic field satisfies the Poisson equation with a function f proportional to the density of the charge (without such a charge, it is the Laplace equation which has to be considered).

The most frequently used boundary conditions for these equations in a bounded open set $\Omega \subset \mathbb{R}^N$ are either the Dirichlet condition, the Neumann condition, or the Robin condition as described in Section 1.3. In the first case, the problem to be solved is called the Dirichlet problem; in the second case, we have to solve a Neumann problem.

1.4.2 *Modeling of heat transfer: The heat equation*

Let $\Omega \subset \mathbb{R}^N$ be the domain occupied by a conducting material of density ρ and let $u = u(x,t)$ be its temperature at any point $x \in \Omega$ and at time t. Denote by $n = n(x)$ the outward unit normal vector to $\partial\Omega$ at the point x. The Fourier law describing the heat propagation in the direction $-n$ and in the interval of time $[t, t+dt]$ states that the quantity dq of heat passing through an elementary surface $d\sigma$ (containing x) of $\partial\Omega$ is given by

$$dq = k(x)\frac{\partial u}{\partial n}(x,t)\, d\sigma\, dt,$$

where k stands for the thermal conductivity coefficient of the material. Consequently, by summing and then integrating by parts (in fact, by applying formula (3.27) in Chapter 3), the total heat in an interval of time T, passing through a surface S delimiting a volume V in Ω, is

$$q = \int_T \int_S k(x) \frac{\partial u}{\partial n}(x, \tau)\, d\sigma\, d\tau = - \int_T \int_V \operatorname{div}\Big(k(x) \frac{\partial u}{\partial x}(x, \tau)\Big)\, dx\, d\tau.$$

Suppose that we are given a heat source $f = f(x,t)$ in Ω. The quantity of heat in V produced by this source in the interval T is obviously

$$\int_T \int_V f(x, \tau)\, dx\, d\tau.$$

The energy conservation law in V in the time interval $[t, t+\delta t]$ implies that

$$\frac{1}{\delta t} \int_V c(x)\rho(x)\big(u(x, t+\delta t) - u(x,t)\big)\, dx = \frac{1}{\delta t} \int_t^{t+\delta t} \int_S f(x,\tau)\, dx\, d\tau$$
$$+ \frac{1}{\delta t} \int_t^{t+\delta t} \int_V \operatorname{div}\Big(k(x) \frac{\partial u}{\partial x}(x,\tau)\Big)\, dx\, d\tau,$$

where $c = c(x)$ is the thermal capacity of the medium. Passing to the limit as $\delta t \to 0$ and as V shrinks to the point x, we obtain (since x was chosen arbitrarily),

$$c(x)\rho(x) \frac{\partial u}{\partial t}(x,t) - \operatorname{div}\Big(k(x) \frac{\partial u}{\partial x}\Big)(x,t) = f(x,t),\ x \in \Omega,\ t \in (0,T). \tag{1.9}$$

If the material is homogeneous, that is, ρ, c, and k do not depend on x, the former equation becomes

$$\frac{\partial u}{\partial t} - a^2 \Delta u = F, \tag{1.10}$$

where $a^2 = k/c\rho$ and $F = f/c\rho$. In the steady state case, that is when u and f do not depend on time, (1.10) is simply the Poisson equation (1.8).

To this PDE, one has to add a boundary condition on $\partial\Omega$ that can be one of those discussed in Example 1.4. It is obvious that at any point x of Ω and at any time t, the temperature u will depend on the initial temperature of the medium $g = g(x)$, a given data of the problem. This takes the form

$$u(x,0) = g(x), \qquad \forall x \in \Omega.$$

Let us mention that equation (1.10) models physical phenomena other than heat transfer. In particular, it also describes diffusion processes in gas or fluids and it appears in electrical problems (in this case u is the potential of the electrical field). Of course, a^2 will have a different meaning in each case, depending on the physical phenomenon under consideration.

1.4.3 *The Black–Scholes equation*

Well-known in financial mathematics is the famous Black–Scholes equation which is in fact a modified heat equation. Let us denote by

- S the price of the underlying asset of a European option,
- T the maturity of the option,
- $V = V(S,t)$ the value at time t of the option $(0 \le t \le T)$,
- σ the volatility of the underlying asset,
- r the (continuously compounded) risk-free interest rate.

Introduce the set

$$D_V = \Big\{(S,t) \mid S \ge 0,\ \ 0 \le t \le T\Big\}.$$

The Black–Scholes equation is the following:

$$\frac{\partial V}{\partial t} + \frac{1}{2}\sigma^2 S^2 \frac{\partial^2 V}{\partial S^2} + rS\frac{\partial V}{\partial S} - rV = 0 \quad \text{in } D_V. \tag{1.11}$$

The derivation (very complicated) of this model is based on stochastic analysis arguments, mainly on stochastic integrals and Itô calculus. The model was published in 1973 by F. Black and M. Scholes in [2]. Further mathematical aspects of the model have been expanded by R. Merton [25]. In 1997, for their works in this field, R. Merton and M. Scholes were awarded the Nobel Prize in Economic Sciences.

The second order partial differential equation (1.11) holds whenever V is sufficiently smooth, that is, if V admits second order derivatives with respect to S and a first order derivative with respect to t, all continuous in the set D_V.

By performing a suitable change of variables $((S,t) \mapsto (x,\tau))$ in (1.11), one ends up with nothing else but the one-dimensional heat equation (see Example 1.4),

$$\frac{\partial u}{\partial \tau} = \frac{\partial^2 u}{\partial x^2} \quad \text{in } D_u,$$

where

$$D_u = \Big\{(x,\tau) \mid -\infty < x < +\infty,\quad 0 \le \tau \le \frac{\sigma^2}{2}T\Big\}.$$

Remark 1.8. *For an American option,* (1.11) *should be replaced by a partial differential inequality, namely,*

$$\frac{\partial V}{\partial t} + \frac{1}{2}\sigma^2 S^2 \frac{\partial^2 V}{\partial S^2} + rS\frac{\partial V}{\partial S} - rV \le 0 \quad \textit{in } D_V.$$

1.4.4 *Modeling of a vibrating string: The wave equation*

Suppose we are given a string of length L, of mass density ρ, and with fixed endpoints. The aim is to determine its vibrations, that is, the deflection $u(x,t)$ at each point x of the string and for any time $t > 0$. To do so, we shall take into consideration the forces acting on a small portion of the string $[x_1, x_2] \subset (0, L)$ (see Figure 1). As mentioned above, to establish the PDE corresponding to a physical problem, one is led to make some simplifications in order not to obtain a very complicated model. Even in this simple case of a vibrating string, we make several assumptions. We suppose the following:

(1) the string is "homogeneous" (uniform mass per unit length),
(2) the cross-section of the string is small compared to L so that at any moment t, the string is represented by the curve $y = u(x,t)$ in the xy-plane,
(3) the string is perfectly elastic,
(4) it does not oppose any resistance to the bending,
(5) the gravitational force is neglected,
(6) the displacements are small and take place only in the vertical plane.

Assumption 4 implies that the tension τ is tangential at any point of the string. Let t be fixed and consider the graph representing the string at this time t (see Figure 1). If P denotes an arbitrary point of the string with coordinates $(x, u(x,t))$ and $\alpha_P \in (0, \pi/2)$ is the angle that the string makes with the horizontal axis, we have

$$\sin \alpha_P(x,t) = \frac{u_x}{\sqrt{1+u_x^2}}(x,t), \quad \cos \alpha_P(x,t) = \frac{1}{\sqrt{1+u_x^2}}(x,t),$$

where, in order to simplify the notation, we set $u_x = \dfrac{du}{dx}$.

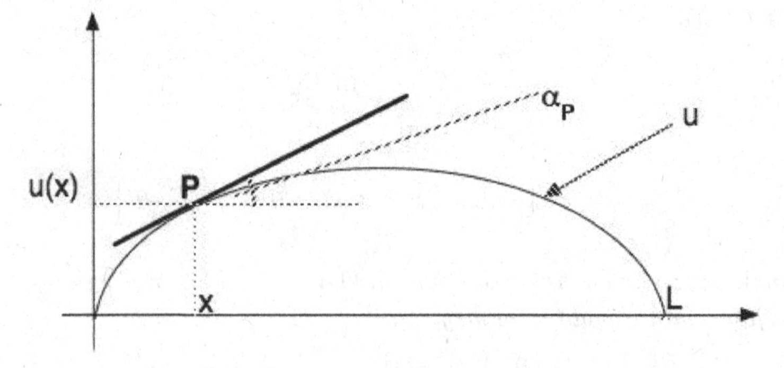

Figure 1. The vibrating string

It follows that the vertical and horizontal components of the tension τ_P at the point P are given, respectively, by

$$\begin{aligned}(\tau_P \sin \alpha_P)(x,t) &= \Big(\tau_P \frac{u_x}{\sqrt{1+u_x^2}}\Big)(x,t),\\ (\tau_P \cos \alpha_P)(x,t) &= \Big(\tau_P \frac{1}{\sqrt{1+u_x^2}}\Big)(x,t).\end{aligned} \tag{1.12}$$

Let us now consider two points $P_1(x_1, u(x_1,t))$ and $P_2(x_2, u(x_2,t))$ on the string and let τ_{P_1} and τ_{P_2} be the tensions at these points, respectively. The condition that the portion (P_1, P_2) of the string must be in equilibrium means that the sum of the forces acting on it is zero. Using assumption 6 and if no exterior force is applied, the only vertical component of the force acting on (P_1, P_2) is the difference between the vertical components of the tensions τ_{P_1} and τ_{P_2}. By Newton's second law of motion (the resulting force at any point between x_1 and x_2 is equal to the mass times the acceleration),

$$(\tau_{P_1} \sin \alpha_{P_1})(x_1,t) - (\tau_{P_2} \sin \alpha_{P_2})(x_2,t) = \int_{x_1}^{x_2} \rho \frac{\partial^2 u}{\partial t^2}(x,t)\, dx,$$

where ρ is the density of the string. Using (1.12) and assumption 1, this identity now becomes

$$-\int_{x_1}^{x_2} \frac{\partial}{\partial x}\Big(\tau_P \frac{u_x}{\sqrt{1+u_x^2}}\Big)(x,t)\, dx = \rho \int_{x_1}^{x_2} \frac{\partial^2 u}{\partial t^2}(x,t)\, dx.$$

Dividing by $(x_2 - x_1)$ and passing to the limit as $(x_2 - x_1) \to 0$ yield

$$\rho \frac{\partial^2 u}{\partial t^2}(x,t) = \frac{\partial}{\partial x}\Big(\tau_P \frac{u_x}{\sqrt{1+u_x^2}}\Big)(x,t). \tag{1.13}$$

Now, due to assumption 5, the horizontal component of forces has to be zero and this means that

$$(\tau_{P_1} \cos \alpha_{P_1})(x_1,t) - (\tau_{P_2} \cos \alpha_{P_2})(x_2,t) = 0,$$

and by (1.12), we obtain

$$\Big(\tau_{P_1} \frac{1}{\sqrt{1+u_x^2}}\Big)(x_1,t) = \Big(\tau_{P_2} \frac{1}{\sqrt{1+u_x^2}}\Big)(x_2,t).$$

Consequently,

$$\int_{x_1}^{x_2} \frac{\partial}{\partial x}\Big(\tau_P \frac{1}{\sqrt{1+u_x^2}}\Big)(x,t\, dx = 0.$$

Thus

$$\tau_P \frac{1}{\sqrt{1+u_x^2}} = \tau_P \cos \alpha_P = \beta,$$

where β is constant with respect to x. By assumption 2, β is positive. This fact used in (1.13) gives

$$\frac{\partial^2 u}{\partial t^2} = c^2 \frac{\partial^2 u}{\partial x^2}, \tag{1.14}$$

which is precisely the one-dimensional wave equation. The notation

$$c^2 = \frac{\beta}{\rho},$$

indicates simply that β/ρ is positive. The fact that the string is fixed at the endpoints implies that

$$u(0,t) = u(L,t) = 0, \quad \forall t > 0,$$

and these are the boundary conditions (of Dirichlet type).

On the other hand, it is clear that the displacement of the string will depend on its initial deflection and initial speed. This makes it necessary to require, for some given functions f and g,

$$u(x,0) = f(x), \quad \forall x \in (0,L),$$
$$\frac{\partial u}{\partial t}(x,0) = g(x), \quad \forall x \in (0,L).$$

To resume, the final problem to be solved in order to obtain the vibrations $u = u(x,t)$ of our string, is the following one:

$$\begin{cases} \dfrac{\partial^2 u}{\partial t^2} = c^2 \dfrac{\partial^2 u}{\partial x^2} & \text{for } (0,L) \times (0,\infty), \\ u(0,t) = u(L,t) = 0 & \text{for } t > 0, \\ u(x,0) = f(x) & \text{for } x \in (0,L), \\ \dfrac{\partial u}{\partial t}(x,0) = g(x) & \text{for } x \in (0,L), \end{cases} \tag{1.15}$$

where the data f and g belong to suitable function spaces.

Remark 1.9.

(1) In the two-dimensional case, equation (1.14) *describes the vibrations of a membrane.*

(2) (Nonhomogeneous wave equation) If the string is nonhomogeneous, its vibrations are described by the equation

$$\frac{\partial}{\partial x}\left(\mu(x)\frac{\partial u}{\partial x}\right) = \rho(x)\frac{\partial^2 u}{\partial t^2}, \tag{1.16}$$

where ρ is the density of the string (mass per unit length) and μ is a coefficient depending on the tension and on the elasticity modulus of the material from which the string is made.

(3) (Waves in periodic media) A more general case of equation (1.14) *is*

$$\frac{\partial^2 u}{\partial t^2} = c^2(x)\frac{\partial^2 u}{\partial x^2},$$

where c^2 is not necessarily a positive constant but a positive function depending on x. If c is periodic, this equation models in particular, the propagation of electromagnetic waves in semiconductors.

1.4.5 *The Helmholtz equation*

The Helmholtz equation is related to the study of the solutions of the wave equation of the form $u(x,t) = e^{i\omega t}w(x)$ where $\omega = kc$ (ω, k, and c are all constants). If such a function u satisfies the wave equation (3) from Example 1.4, it is immediate that w is a solution of the following equation, called the Helmholtz equation:

$$\Delta w + k^2 w = 0. \tag{1.17}$$

This equation appears also in the study of different spectral problems such as that of eigenvalues and eigenfunctions for the Laplace operator. The simplest problem of this type is set in a bounded set $\Omega \subset \mathbb{R}^N$ with Dirichlet condition on the boundary $\partial\Omega$. One looks for u (called eigenfunction) and λ (called eigenvalue) satisfying

$$\begin{cases} -\Delta u = \lambda u & \text{in } \Omega, \\ u = 0 & \text{on } \partial\Omega. \end{cases} \tag{1.18}$$

Multiplying by u and integrating by parts (see the second Green formula (3.27)), one has

$$\int_\Omega |\nabla u|^2\,dx = \lambda \int_\Omega u^2\,dx.$$

This shows that (1.18) has nonzero solutions only if $\lambda > 0$, and in this case, u satisfies a Helmholtz type equation.

Let us also mention that eigenvalue problems of type (1.18) play an essential role in the mathematical formulation of quantum mechanics.

1.4.6 *The Maxwell equations*

The Maxwell equations are fundamental in electromagnetism; they possess a very wide field of applications which include radio, television, phone, and information technologies.

The Maxwell equations consist of a system of four first order linear partial differential equations coupling the (vector) electrical field

$$\vec{E} = \vec{E}(x,t) = \big(E_1(x,t), E_2(x,t), E_3(x,t)\big)$$

and the (vector) magnetic field

$$\vec{B} = \vec{B}(x,t) = \big(B_1(x,t), B_2(x,t), B_3(x,t)\big)$$

and their interactions. The first two equations, expressed in terms of the "curl" operator, describe how currents produce magnetic fields (the Ampère law), and how changes in magnetic fields induce electric fields (the Faraday law). The third and fourth equations, the Gauss laws of electric field and of magnetic field, are expressed in terms of the "divergence" operator. In vacuum, for $x \in \mathbb{R}^3$ and for $t \in \mathbb{R}$, they take the form

$$\begin{cases} \text{(i)} \quad \operatorname{curl} \vec{E} = -\dfrac{\partial \vec{B}}{\partial t}, \\ \text{(ii)} \quad \operatorname{curl} \vec{B} = \mu_0\, \varepsilon_0\, \dfrac{\partial \vec{E}}{\partial t}, \\ \text{(iii)} \quad \operatorname{div} \vec{E} = 0, \\ \text{(iv)} \quad \operatorname{div} \vec{B} = 0, \end{cases} \tag{1.19}$$

where μ_0 is the magnetic permeability while ε_0 is the electric permittivity in vacuum. In (1.19), $\operatorname{curl} \vec{E}$ denotes the vector function

$$\operatorname{curl} \vec{E} = \Big(\frac{\partial E_3}{\partial x_2} - \frac{\partial E_2}{\partial x_3}, \frac{\partial E_1}{\partial x_3} - \frac{\partial E_3}{\partial x_1}, \frac{\partial E_2}{\partial x_1} - \frac{\partial E_1}{\partial x_2}\Big).$$

For the meaning of $\operatorname{div} \vec{E}$, we refer the reader to Section 1.2.

Now observe that one can eliminate either $\vec{B}$ or $\vec{E}$ from system (1.19) to end up with the equation giving the propagation of electromagnetic waves. To do so, recall the following classical formula which holds for any three-dimensional vector function $\vec{J}$:

$$\operatorname{curl}(\operatorname{curl} \vec{J}) = \nabla \operatorname{div} \vec{J} - \Delta \vec{J}.$$

Applying it to $\vec{E}$ and using (1.19)(i) and (iii) yield

$$\operatorname{curl}\Big(- \frac{\partial \vec{B}}{\partial t}\Big) = -\Delta \vec{E}.$$

One can commute the derivatives in time and spatial derivatives (in curl) in the left-hand side term, and applying (1.19)(ii), we get

$$\frac{\partial^2 \vec{E}}{\partial t^2} - c_0^2 \Delta \vec{E} = 0, \tag{1.20}$$

where c_0, known as the speed of light in vacuum, is defined by

$$c_0 = \frac{1}{\sqrt{\mu_0 \varepsilon_0}}.$$

Applying the same computation to $\vec{B}$, one obtains an equation similar to (1.20), namely

$$\frac{\partial^2 \vec{B}}{\partial t^2} - c_0^2 \Delta \vec{B} = 0.$$

As a result, we have two wave equations. They show that $\vec{E}$ and $\vec{B}$ propagate as plane waves with the speed of light c_0 (see Chapter 5 for more details).

1.4.7 *The telegraph equations*

The telegraph equations, which are deduced from the Maxwell equations, are of interest in electricity. They give the relationship between the current strength u and the potential v and their evolution in time in a conductor, namely,

$$\begin{cases} \dfrac{\partial u}{\partial x} + C\dfrac{\partial v}{\partial t} + Gv = 0, \\[2ex] \dfrac{\partial v}{\partial x} + L\dfrac{\partial u}{\partial t} + Ru = 0, \end{cases} \tag{1.21}$$

where x is the variable along the conductor, and C, G, L, and R are, respectively, the capacitance, conductance, inductance, and resistance of the conductor all per unit length.

Similar to the solution of the Maxwell equations, by eliminating successively one of the unknowns from (1.21) and then using a suitable change of functions, the transformed functions u and v both satisfy an equation of the type

$$\frac{\partial^2 w}{\partial t^2} = c^2 \frac{\partial^2 w}{\partial x^2} - \beta w,$$

where β is a positive constant. This is again a wave-type equation.

1.4.8 *The Navier–Stokes equations*

The examples presented above are all second order linear partial differential equations, the scope of this book. For the sake of completeness and in order also to give a flavor of what nonlinearities mean and entail, we shall give also an example of a nonlinear PDE. We chose to end this section with the well-known Navier–Stokes equations, which describe the motion of Newtonian fluids, for which there is linear relationship between the stress tensor and the rate of the deformations tensor (for more details see for instance [9]).

For incompressible fluids, the Navier–Stokes equations consist of a system of nonlinear equations with unknowns the velocity function $\vec{u} = (u_1, \ldots, u_N)$, $N = 2$ or 3, and the pressure in the fluid P, a scalar function, satisfying

$$\begin{cases} \dfrac{\partial \vec{u}}{\partial t} - \mu\Delta\vec{u} + (\vec{u}\cdot\nabla)\vec{u} + \nabla P = \vec{F} \quad \text{in } \Omega\times(0,T), \\ \text{div}\ \vec{u} = 0 \quad \text{in } \Omega\times(0,T), \end{cases} \tag{1.22}$$

where

$$\Delta\vec{u} = (\Delta u_1, \ldots, \Delta u_N),$$

and

$$(\vec{u}\cdot\nabla)\vec{u} = u_1\frac{\partial \vec{u}}{\partial x_1} + \ldots + u_N\frac{\partial \vec{u}}{\partial x_N}.$$

In (1.22), the coefficient μ is the viscosity of the fluid, $\vec{F} = (F_1, \ldots, F_N)$ is the field of exterior forces, and Ω is an open set in $\mathbb{R}^N$. Obviously, one has to add an initial condition for $\vec{u}$ as well as a boundary condition on $\partial\Omega$.

Writing the first equation in (1.22) component by component for the vector functions $\vec{u}$, ∇P, $(\vec{u}\cdot\nabla)\vec{u}$ and $\vec{F}$, leads to the following N equations:

$$\frac{\partial u_i}{\partial t} - \mu\Delta u_i + \sum_{j=1}^{N} u_j\frac{\partial u_i}{\partial x_j} + \frac{\partial P}{\partial x_i} = F_i \quad \text{for } i = 1, \ldots, N.$$

The second equation in (1.22) translates the fact that the fluid is incompressible (its volume is preserved). The presence of the nonlinear term $(\vec{u}\cdot\nabla)\vec{u}$ in system (1.22) renders its study extremely hard. In the two-dimensional case, one has existence and uniqueness of solutions of (1.22) in appropriate function spaces. In the three-dimensional case, one can prove either existence (but not uniqueness) of solutions in some suitable function spaces, or uniqueness (but not existence) in some other function spaces. To prove existence and uniqueness for (1.22) in the same function space is currently one of the most important and challenging open problems in mathematics.

Let us mention that if the nonlinear term in (1.22) is not present, one has the so-called Stokes equations (characterizing in general, fluids with slow velocities). Stokes equations are linear and consequently, their study is easier than the full Navier–Stokes equations.

1.5 The concept of well-posed equations

The concept of well-posedness was introduced by J. Hadamard. From the examples in the previous section, it is clear that the initial and boundary conditions to be prescribed depend on the physical situation and essentially on the order of the equation. If the number of conditions is insufficient, an equation can be satisfied by functions which may not have any relation with the physical phenomenon under study. On the contrary, too many conditions could imply that the equation has no solution (this situation is called overdetermined). In general, a mathematical model is considered satisfactory if for a given data, the boundary value problem (BVP) admits one and only one solution. Nevertheless, even that is not sufficient. For a BVP describing a physical phenomenon, the data originate from measurements which are never perfect. A problem with a unique solution will be considered well-posed only in the case when a small change in the data results in a small change in the solution. This is referred to as stability of a solution with respect to the data.

In a general framework, the notion of well-posedness is defined in the following way:

Definition 1.10. *Let $\mathcal{U}$, $\mathcal{V}$, and $\mathcal{F}$ be topological spaces such that $\mathcal{U} \subset \mathcal{V}$, u a function satisfying a BVP and f a vector function representing the data of the problem, that is, the vector which takes into consideration the initial and boundary conditions, as well as the given exterior forces (the right-hand side term) of the equation. The problem is called well-posed if*

(1) (Existence) For any element $f \in \mathcal{F}$ there exists a solution $u \in \mathcal{U}$ of the problem;

(2) (Uniqueness) The solution is unique;

(3) (Stability) The solution u, considered as an element of $\mathcal{V}$, depends continuously on $f \in \mathcal{F}$.

If one of the above conditions is not satisfied, the problem is said to be ill-posed.

Example 1.11. Hadamard's example. Consider the Laplace equation in $\mathbb{R}^2$ with coordinates (x,t),

$$\begin{cases} \dfrac{\partial^2 u}{\partial t^2} + \dfrac{\partial^2 u}{\partial x^2} = 0, & x \in \mathbb{R},\ 0 < t < +\infty, \\ u(x,0) = 0, & x \in \mathbb{R}, \\ \dfrac{\partial u}{\partial t}(x,0) = \varphi(x), & x \in \mathbb{R}, \end{cases} \tag{1.23}$$

where φ is a given function. One can prove the uniqueness of a solution of this problem, for instance in $C^2(\mathbb{R} \times (0,+\infty))$, if φ is a smooth function. However, this problem is not well-posed, since the stability property is violated. To show it, Hadamard constructed a sequence of initial conditions φ_n close to the zero function, but with the corresponding solutions u_n very "far" from the function $u \equiv 0$, the unique solution of the above equation for $\varphi \equiv 0$.

To do so, Hadamard considered the equation (a variant of (1.23))

$$\begin{cases} \dfrac{\partial^2 u}{\partial t^2} + \dfrac{\partial^2 u}{\partial x^2} = 0, & -\dfrac{\pi}{2} \le x \le \dfrac{\pi}{2},\quad 0 < t < +\infty, \\ u\Big(-\dfrac{\pi}{2}, t\Big) = u\Big(\dfrac{\pi}{2}, t\Big) = 0, & 0 < t < +\infty, \\ u(x,0) = 0, & -\dfrac{\pi}{2} \le x \le \dfrac{\pi}{2}, \\ \dfrac{\partial u}{\partial t}(x,0) = \varphi_n(x), & -\dfrac{\pi}{2} \le x \le \dfrac{\pi}{2}, \end{cases} \tag{1.24}$$

where $\varphi_n(x) = e^{-\sqrt{n}} \cos(nx)$ and n is an odd integer. This is a Cauchy-type problem, that is to say, one which contains both boundary and initial conditions (see Section 2.2 below for more details).

Applying the method of separation of variables as in Chapter 5, Section 5.3.1, one can easily verify that

$$u_n(x,t) = \frac{1}{n}\,\frac{e^{nt} - e^{-nt}}{2}\,e^{-\sqrt{n}}\cos(nx) = \frac{1}{n}\,e^{-\sqrt{n}}\cos(nx)\sinh(nt),$$

is the unique solution of this problem. Observe that φ_n can be considered as a perturbation of the zero function, since it is clear that for any $\varepsilon > 0$, there exists a number N_ε such that

$$\sup_x \left| e^{-\sqrt{n}} \cos(nx) \right| \le \varepsilon \quad \text{for } n \ge N_\varepsilon,$$

the same being true for all the derivatives of $e^{-\sqrt{n}}\cos(nx)$. Nevertheless, for any $t_0 > 0$, no matter how small,

$$\sup_x |u_n(x,t_0)| = \frac{e^{nt_0 - \sqrt{n}} - e^{-nt_0 - \sqrt{n}}}{2n} \to +\infty \quad \text{as } n \to +\infty,$$

due to the presence of the hyperbolic sine function. This is the meaning of u_n being "far" from $u \equiv 0$ in Hadamard's example. ◇

Equation (1.24) is an example of an ill-posed problem, as a small change in the data results in a big change in the solution. As we shall see later on (in Chapter 3 and in a general setting in Chapter 9), the Cauchy problem for elliptic equations is not appropriate with respect to Definition 1.10. They will be solved by prescribing only boundary conditions. Hadamard's example shows the importance of the structure of the equation when a boundary value problem is formulated.

An equally important role is also played by the function spaces where solutions are searched. Obviously, the choice of spaces $\mathcal{U}$, $\mathcal{V}$, and $\mathcal{F}$ depends on the problem to be studied. The following examples are significant in this sense.

Example 1.12. The Cauchy problem for the one-dimensional wave equation. Consider the wave equation in $\mathbb{R}$,

$$\begin{cases} \dfrac{\partial^2 u}{\partial t^2} = c^2 \dfrac{\partial^2 u}{\partial x^2}, & x \in \mathbb{R},\ 0 \leq t \leq T, \\ u(x,0) = \varphi(x), & x \in \mathbb{R}, \\ \dfrac{\partial u}{\partial t}(x,0) = \psi(x), & x \in \mathbb{R}. \end{cases} \tag{1.25}$$

This is called the Cauchy problem for the wave equation in $\mathbb{R}$. Classical theorems (see Chapter 5, Section 5.3) show that this system has a unique solution of class $C^2(\mathbb{R} \times [0,T])$ for any φ in $C^2(\mathbb{R})$ and ψ in $C^1(\mathbb{R})$, and moreover, this solution is explicitly given by the d'Alembert formula

$$u(x,t) = \frac{1}{2}\Big[\varphi(x-ct) + \varphi(x+ct)\Big] + \frac{1}{2c}\int_{x-ct}^{x+ct} \psi(\tau)\, d\tau.$$

It is easy to verify that problem (1.25) is well-posed: the solution u depends continuously on φ and ψ in appropriate norms.

Taking φ in $C_b^k(\mathbb{R})$ and ψ in $C_b^{k-1}(\mathbb{R})$ (see (1.4)) with $k \geq 2$ in (1.25), it follows that u is in $C^k(\mathbb{R} \times [0,T]) \cap C_b^k(\mathbb{R} \times (0,T))$, and moreover,

$$\|u\|_{C_b^k(\mathbb{R}\times(0,T))} \leq C\big(\|\varphi\|_{C_b^k(\mathbb{R})} + \|\psi\|_{C_b^{k-1}(\mathbb{R})}\big),$$

where C is a positive constant independent of the data φ and ψ. This signifies that the Cauchy problem is well-posed in the sense of the definition given above: one can take, for example,

$$\mathcal{U} = \mathcal{V} = C_b^k(\mathbb{R} \times (0,T)), \quad \mathcal{F} = C_b^k(\mathbb{R}) \times C_b^{k-1}(\mathbb{R}),$$

where the role of f is played by the couple $\{\varphi, \psi\}$. ◇

Remark 1.13. *Let us emphasize that by taking different values of k, one can choose other spaces where the Cauchy problem for the one-dimensional wave equation given by (1.25) is still well-posed; for instance, we may take*

$$\mathcal{F} = C^\infty(\mathbb{R}) \times C^\infty(\mathbb{R}), \quad \mathcal{U} = \mathcal{V} = C^\infty(\mathbb{R} \times [0,T]),$$

spaces of infinitely differentiable functions with appropriate topologies.

Remark 1.14. *Example 1.11 above shows that the Cauchy problem (1.23) for the Laplace equation is ill-posed in the C^k-type spaces.*

Example 1.15. The Cauchy problem for the heat equation. Let us now consider the following equation (with α a given constant):

$$\begin{cases} \dfrac{\partial u}{\partial t} = \alpha^2 \Delta u, & x \in \mathbb{R}^N,\ 0 \le t \le T, \\ u(x,0) = \varphi(x), & x \in \mathbb{R}. \end{cases} \tag{1.26}$$

It is called the Cauchy problem for the heat equation. Unlike the preceding example, there is no uniqueness even in the space $C^\infty(\mathbb{R}^N \times [0,T])$. However, if a condition on the behavior of u as $|x| \to +\infty$ is prescribed, then problem (1.26) becomes a well-posed one.

Let $\varphi \in \mathcal{F} = C_b^0(\mathbb{R}^N)$. It follows that there exists a unique solution u of (1.26) such that

$$u \in \mathcal{U} = \mathcal{V} = C^\infty(\mathbb{R}^N \times [0,T]) \cup C_b^0(\mathbb{R}^N \times [0,T]).$$

This solution is explicitly given by Poisson's formula

$$u(x,t) = \frac{1}{(2\alpha\sqrt{\pi t})^N} \int_{\mathbb{R}^N} e^{-|x-y|^2/4\alpha^2 t} \varphi(y)\, dy, \tag{1.27}$$

see for details Chapter 4, Section 4.2. ◇

Remark 1.16. *It is known that this solution satisfies the following inequalities (called the maximum principle)*

$$\inf_{x\in\mathbb{R}^N} \varphi(x) \le u(x,t) \le \sup_{x\in\mathbb{R}^N} \varphi(x).$$

These inequalities have an obvious physical interpretation. The second inequality, for example, says that the maximal temperature attained without any exterior source of heat cannot pass beyond the maximum of the initial temperature. The first inequality has a similar interpretation.

1.6 The Cauchy–Kovalevskaya Theorem

The first proof of the existence and uniqueness of the solution of a Cauchy problem for an ordinary differential equation of the form

$$\begin{cases} \dfrac{du}{dt} = f(t, u), \\ u(t_0) = u_0, \end{cases}$$

was given by Cauchy under the hypothesis that the function f is analytic in a neighborhood of the point (t_0, u_0). This unique solution $u = u(t)$ is analytic in a neighborhood of t_0. Recall that a function is analytic if it can be locally represented by a power series.

This theorem was generalized for partial differential equations by S. V. Kovalevskaya. Her result deals with the so-called Kovalevskaya equations, having the following particular form:

$$\frac{\partial^m u}{\partial t^m} = F\Big(x, t, u, \frac{\partial u}{\partial t}, \dots, \frac{\partial^{m-1} u}{\partial t^{m-1}}, \frac{\partial u}{\partial x}, \dots\Big),$$

where $x = (x_1, \dots, x_N)$, F is a function depending on the partial derivatives of u with respect to x up to order m, but independent of $\partial^m u/\partial t^m$. Moreover, it is supposed that F is analytic in all its arguments. The order of this equation is m, which is also the highest order of the derivative of u with respect to t.

More generally, suppose we are given a system of κ Kovalevskaya-type equations with κ unknowns $u_1, u_2, \dots, u_\kappa$ satisfying

$$\frac{\partial^{n_i} u_j}{\partial t^{n_i}} = F_i\big(x, t, (u)\big), \tag{1.28}$$

where

$$F_i\big(x, t, (u(x,t))\big) = F_i\Big(x, t, u_1, u_2, \dots, u_\kappa, \dots, \partial^k u_j, \dots\Big), \tag{1.29}$$

for $i, j = 1, 2, \dots, \kappa,\ \ k_0 + k_1 + \dots + k_N = k \le n_i,\ \ k_0 < n_i$.

The Cauchy problem for the system of equations (1.28) consists of finding a solution $U = (u_1, \dots, u_\kappa)$ which satisfies the initial conditions

$$\frac{\partial^k u_i}{\partial t^k}(x, 0) = \varphi_i^{(k)}(x), \quad k = 0, 1, \dots, n_i - 1, \tag{1.30}$$

where it is assumed that

H1. The functions $\varphi_i^{(k)}$ are analytic in a neighborhood of the point $x = 0$.

Conditions (1.30) allow us to compute the values of the arguments of F_i $(i = 1, 2, \dots, \kappa)$ in a neighborhood of $x = 0$ for $t = 0$, that is to say, with notation (1.29), to compute $F_i\big(x, 0, (\varphi(x))\big)$ in a neighborhood of $x = 0$.

One requires furthermore the following:

H2. The functions F_i are analytic in a neighborhood of $\big(0, 0, (\varphi(0))\big)$.

We are now able to state the following result:

Theorem 1.17 (Cauchy–Kovalevskaya Theorem). *Under hypotheses* **H1** *and* **H2**, *the Cauchy problem (1.28)–(1.30) has a unique solution* $U = U(x,t)$ *that is analytic in a neighborhood of the point* $(x,t) = (0,0)$.

Consequently, if an analytic function U satisfies system (1.28) and conditions (1.30), then its derivatives of any order are uniquely determined at the point $x = 0$, $t = 0$. Indeed, since in each equation from (1.28) the order of the derivatives with respect to t does not exceed $n_i - 1$, the derivatives with respect to time are all determined by (1.28). The other derivatives can be obtained from the ordinary differential equation (1.30). Therefore all the coefficients of the Taylor series of the solution are well determined at the point $(0,0)$, and this implies the uniqueness of the analytic solution. To establish the existence of such a solution, one proves the convergence of the series built with the former coefficients.

Example 1.18. Example of nonexistence of an analytic solution. Let us consider again the one-dimensional Cauchy problem for the heat equation

$$\begin{cases} \dfrac{\partial u}{\partial t} = \dfrac{\partial^2 u}{\partial x^2}, & x \in (-1,1),\, t > 0, \\ u(x,0) = \dfrac{1}{1-x}, & x \in (-1,1). \end{cases} \tag{1.31}$$

If this problem would have a unique analytic solution in some neighborhood of the origin, its Taylor series would be

$$u(x,t) = \sum_{i,j,} \alpha_{i,j} t^i x^j,$$

where for $i + j = k$ the coefficients $\alpha_{i,j}$ should be equal to $(2k)!$. This is a consequence of the fact that for x in a neighborhood of the origin,

$$\frac{1}{1-x} = 1 + x + x^2 + \dots.$$

The series defining u with the coefficients $(2k)!$ diverges at any point $(0,t)$ for any $t \neq 0$. Therefore the Cauchy problem (1.31) has no analytic solution in a neighborhood of the origin $(0,0)$. The reason is that the right-hand side term of the equation contains a derivative with respect to x of order 2, and this order is greater than the order of the derivative with respect to t in the left-hand side of (1.31). As a matter of fact, the heat equation is not a Kovalevskaya-type equation. ◇

Chapter 2

Classification of Partial Differential Equations

We have seen in Chapter 1 that the Laplace, heat, and wave equations are among the most important partial differential equations. It turns out that they are the representative equations for the three major types of PDEs: elliptic, parabolic, and hyperbolic equations, respectively. As we shall see in the next chapters, there are many results (existence, uniqueness, qualitative properties) concerning these three types of equations. These results are specific for each class of equations. This is why when given a PDE, it is important to know which type it is. The aim of this chapter is to give the rules to answer this question.

We study here primarily second order partial differential equations with unknown u, a real function of two variables defined on an open set Ω in $\mathbb{R}^2$, with emphasis on linear equations. The case of more than two variables will be discussed briefly in the last section.

2.1 Characteristic curves

Following the definitions from the first chapter, a PDE of second order is semilinear if and only if it has the form

$$A\frac{\partial^2 u}{\partial x_1^2} + B\frac{\partial^2 u}{\partial x_1 \partial x_2} + C\frac{\partial^2 u}{\partial x_2^2} + \Phi\Big(x_1, x_2, u, \frac{\partial u}{\partial x_1}, \frac{\partial u}{\partial x_2}\Big) = 0,$$

where A, B, C are functions depending on $(x_1, x_2) \in \Omega$ and $A^2+B^2+C^2 \not\equiv 0$ on the set Ω.

Example 2.1. Suppose that $(x_1, x_2) \in \Omega$ if and only if $x_1^2 + x_2^2 < 1$. The following PDE is semilinear on Ω:

$$\frac{\partial^2 u}{\partial x_1^2} - \sqrt{1 - x_1^2 - x_2^2}\,\frac{\partial^2 u}{\partial x_2^2} + u^2\frac{\partial u}{\partial x_1} = 0. \qquad \diamond$$

Taking into account Definition (1.3) we now give the general form for a linear second order PDE.

Definition 2.2. *A PDE of second order is linear in Ω in $\mathbb{R}^2$ if and only if it has the form*

$$A\frac{\partial^2 u}{\partial x_1^2}+B\frac{\partial^2 u}{\partial x_1 \partial x_2}+C\frac{\partial^2 u}{\partial x_2^2}+D\frac{\partial u}{\partial x_1}+E\frac{\partial u}{\partial x_2}+Fu+G=0, \tag{2.1}$$

where A, B, C, D, E, F, and G are functions depending on $(x_1, x_2) \in \Omega$ and such that $A^2+B^2+C^2 \neq 0$ on Ω.

Example 2.3. The following equations are linear in $\mathbb{R}^2$:

- $\dfrac{\partial^2 u}{\partial x_1^2}-\dfrac{\partial^2 u}{\partial x_2^2}+u=0,$
- $\dfrac{\partial^2 u}{\partial x_1^2}-\dfrac{\partial^2 u}{\partial x_2^2}+x^2\dfrac{\partial u}{\partial x_1}-x_1x_2\dfrac{\partial u}{\partial x_2}=1.$ ◇

As mentioned above, an equation can have several solutions, but all ambiguity is removed if appropriate boundary conditions are given. The notion of characteristic curve plays an essential role in the study of second order PDEs. Let us introduce its definition.

Consider the equation

$$A\frac{\partial^2 u}{\partial x_1^2}+B\frac{\partial^2 u}{\partial x_1 \partial x_2}+C\frac{\partial^2 u}{\partial x_2^2}=F\Big(x_1, x_2, u, \frac{\partial u}{\partial x_1}, \frac{\partial u}{\partial x_2}\Big), \tag{2.2}$$

where A, B, C are functions of x_1 and x_2.

Let γ be a curve in $\mathbb{R}^2$ with parametric representation

$$x_1=\phi(t), \quad x_2=\psi(t). \tag{2.3}$$

Suppose now that all the points of γ are regular, which means that $(\phi')^2+(\psi')^2>0$ at each point of the curve. We are looking for a solution u of (2.2) assuming we are given the values of u and its first order derivatives on γ. Observe that this is equivalent to assuming that the values of u and its normal derivative $\dfrac{\partial u}{\partial n}$ along γ are given (see Section 1.2).

Introduce the notation

$$H(t)=u(\phi(t),\psi(t)), \quad U(t)=\frac{\partial u}{\partial x_1}(\phi(t),\psi(t)), \quad V(t)=\frac{\partial u}{\partial x_2}(\phi(t),\psi(t)).$$

We are in the situation where H, U, and V are given on γ. One would like to determine the second derivatives of u on γ in order to have equation (2.2) satisfied. To find them, one has to add to equation (2.2) the equations obtained by taking the derivatives of U and V, to obtain the following system:

$$\begin{cases} A(\phi(t),\psi(t))\dfrac{\partial^2 u}{\partial x_1^2}(\phi(t),\psi(t)) + B(\phi(t),\psi(t))\dfrac{\partial^2 u}{\partial x_1 \partial x_2}(\phi(t),\psi(t)) \\ \qquad + C(\phi(t),\psi(t))\dfrac{\partial^2 u}{\partial x_2^2}(\phi(t),\psi(t)) = F(\phi(t),\psi(t),H(t),U(t),V(t)), \\ U'(t) = \dfrac{\partial^2 u}{\partial x_1^2}(\phi(t),\psi(t))\phi'(t) + \dfrac{\partial^2 u}{\partial x_1 \partial x_2}(\phi(t),\psi(t))\psi'(t), \\ V'(t) = \dfrac{\partial^2 u}{\partial x_1 \partial x_2}(\phi(t),\psi(t))\phi'(t) + \dfrac{\partial^2 u}{\partial x_2^2}(\phi(t),\psi(t))\psi'(t). \end{cases}$$

Let $\Delta = \Delta(t)$ be the determinant of this system. Then

$$\Delta(t) = \begin{vmatrix} \phi'(t) & \psi'(t) & 0 \\ 0 & \phi'(t) & \psi'(t) \\ A(\phi(t),\psi(t)) & B(\phi(t),\psi(t)) & C(\phi(t),\psi(t)) \end{vmatrix} \tag{2.4}$$

$$= C[\phi'(t)]^2 - B\phi'(t)\psi'(t) + A[\psi'(t)]^2.$$

It is clear that if $\Delta(t) \neq 0$ for any t, the second order derivatives of u are uniquely determined on γ. If $\Delta(t) = 0$ for any t, the system has either an infinity of solutions, or none. In this case, the curve γ is called characteristic. If for any t, $\Delta(t) \neq 0$, one says that γ is not characteristic at any point.

Going back to the formula for Δ, one can give a more precise description of the curve γ whose parametric representation is (2.3). Since γ is regular in the variables x_1, x_2, this curve can be written in the following two forms:

$$\begin{aligned} &x_2 = y(x_1) \quad \text{with } y \text{ a function depending on a single variable,} \\ &x_1 = x(x_2) \quad \text{with } x \text{ a function depending on a single variable.} \end{aligned} \tag{2.5}$$

This says from (2.3) that $\psi(t) = y(\phi(t))$ or $\phi(t) = x(\psi(t))$ so that, by (2.5),

$$(i)\ \psi' = y'\phi' \quad \text{and} \quad (ii)\ \phi' = x'\psi'. \tag{2.6}$$

Using these considerations in (2.4), it follows that three cases may occur.
(1) If $A \neq 0$, by (2.6)(i) the characteristic curves are the solutions of the differential equation

$$A(x_1,x_2)(y')^2 - B(x_1,x_2)y' + C(x_1,x_2) = 0. \tag{2.7}$$

(2) If $C \neq 0$, by (2.6)(ii) the characteristic curves are the solutions of the differential equation

$$C(x_1, x_2)\,(x')^2 - B(x_1, x_2)x' + A(x_1, x_2) = 0. \tag{2.8}$$

(3) If $A = C = 0$, from (2.7), (2.8), and by (2.5) the characteristic curves are the straight lines

$$x_1 = c_1, \quad x_2 = c_2,$$

where c_1, c_2 are arbitrary constants.

Example 2.4. Consider the equation in $\mathbb{R}^2$ given by

$$\frac{\partial^2 u}{\partial x_1^2} - \frac{\partial^2 u}{\partial x_2^2} = 0,$$

for which $A = 1$, $B = 0$ and $C = -1$. Thus we are in the case (1) above and so, the characteristics $x_2 = y(x_1)$ (see (2.5)) are defined by the differential equation

$$(y')^2 - 1 = 0 \Longleftrightarrow (y' \pm 1) = 0 \Longleftrightarrow y = \pm x_1 + const.$$

Consequently, the characteristic curves are

$$x_2 = x_1 + c_1, \quad x_2 = -x_1 + c_2,$$

where c_1, c_2 are arbitrary constants. ◇

It shall be seen in Section 2.3 that characteristics can be used to simplify the form of a given PDE.

2.2 The Cauchy problem and existence theorems

Let us again consider equation (2.2). We are now looking for u satisfying

$$\begin{cases} A\dfrac{\partial^2 u}{\partial x_1^2} + B\dfrac{\partial^2 u}{\partial x_1 \partial x_2} + C\dfrac{\partial^2 u}{\partial x_2^2} = F\Big(x_1, x_2, u, \dfrac{\partial u}{\partial x_1}, \dfrac{\partial u}{\partial x_2}\Big), \\ u\big|_\gamma = \phi_0, \\ \dfrac{\partial u}{\partial n}\Big|_\gamma = \phi_1, \end{cases} \tag{2.9}$$

where ϕ_0 and ϕ_1 are given. This problem is called the Cauchy problem with respect to γ.

Recalling the above results, we state the following consequence of the Cauchy–Kovalevskaya theorem:

Theorem 2.5 (Existence and Uniqueness Theorem). *Assume that the curve γ is not characteristic at any point and A, B, C, F, ϕ_0, and ϕ_1 are analytic in a neighborhood of a point $x_0 \in \gamma$. Then the Cauchy problem* (2.9) *with respect to γ admits an analytic solution in a neighborhood of x_0, and this solution is unique in the class of analytic functions.*

The existence theorem says that if there are two analytic solutions of (2.9), then they coincide in a neighborhood of x_0. One may ask whether there is uniqueness for nonanalytic solutions. For linear equations, an answer was given by the following result due to Holmgren:

Theorem 2.6 (Holmgren's Theorem). *Assume that the curve γ is not characteristic at any point and A, B, C, F, ϕ_0, and ϕ_1 are analytic in a neighborhood of a point $x_0 \in \gamma$. Suppose further that equation (2.2) is linear (that is, it is of the form (2.1)). Then the solution given by Theorem 2.5 in a neighborhood of a point $x_0 \in \gamma$, is unique in the class C^2.*

2.3 Canonical forms: Case of constant coefficients

When having a look at equation (2.1),

$$A\frac{\partial^2 u}{\partial x_1^2} + B\frac{\partial^2 u}{\partial x_1 \partial x_2} + C\frac{\partial^2 u}{\partial x_2^2} + D\frac{\partial u}{\partial x_1} + E\frac{\partial u}{\partial x_2} + Fu + G = 0,$$

with A, B, C, D, E, F, and G functions depending on (x_1, x_2), a natural question is whether there exists a procedure to simplify it, for instance, via an appropriate change of variables. As we shall see in the next sections, the answer is affirmative and the notion of characteristics is essential in order to make this change.

In the following, we shall suppose that A, B, C, D, E, and F are real constants such that

$$A^2 + B^2 + C^2 \neq 0, \tag{2.10}$$

and $G = G(x_1, x_2)$ is a real function defined on an open set Ω in $\mathbb{R}^2$.

To simplify our study, suppose that neither of the constants A, B, C vanishes. We showed in Section 2.1 (see in particular, (2.7)), that the differential characteristic equation associated with (2.1) is

$$A(y')^2 - By' + C = 0. \tag{2.11}$$

Setting $y' = \lambda$, this is equivalent to the equation

$$A\lambda^2 - B\lambda + C = 0, \tag{2.12}$$

whose solutions are

$$\lambda_1 = \frac{B + \sqrt{B^2 - 4AC}}{2A}, \quad \lambda_2 = \frac{B - \sqrt{B^2 - 4AC}}{2A}.$$

The expression $B^2 - 4AC$ is the discriminant of (2.12) and we have the following three cases:

$$B^2 - 4AC > 0, \qquad B^2 - 4AC < 0, \qquad B^2 - 4AC = 0.$$

By analogy with the nature of the conic

$$Ax_1^2 + Bx_1x_2 + Cx_2^2 = 0, \tag{2.13}$$

we introduce the following definition:

Definition 2.7. *The linear equation (2.1) with constant coefficients satisfying (2.11) is called*

(a) elliptic if and only if $B^2 - 4AC < 0$,
(b) parabolic if and only if $B^2 - 4AC = 0$,
(c) hyperbolic if and only if $B^2 - 4AC > 0$.

Remark 2.8. *We discussed in Chapter 1 various partial differential equations. Let us mention that the Laplace and Poisson equations from Section 1.4.1 are the typical models for the class of elliptic equations. The heat equation presented in Section 1.4.2 is the model example for the class of parabolic equations. And finally, the wave equation described in Section 1.4.4, is the model example for the class of hyperbolic equations. We shall present in detail these equations in the classical "strong" formulation in the next three chapters. In their weak "variational" formulation, they will be the object of Chapters 8 and 9 in Part II of this book.*

Remark 2.9. *Notice that the character or the nature (hyperbolic, elliptic or parabolic) of any equation of type (2.1) with constant coefficients depends only on the coefficients of second order derivatives.*

As said previously, it is in connection with the conic (2.13) that Definition 2.7 introduces the notions of elliptic, parabolic, and hyperbolic equations. To understand better why, and taking into account Remark 2.9, let us consider the two-dimensional case. To the Laplacian in two dimensions,

$$\frac{\partial^2}{\partial x_1^2} + \frac{\partial^2}{\partial x_2^2},$$

we associate the characteristic polynomial

$$P(\lambda) = \lambda_1^2 + \lambda_2^2, \quad \text{for } \lambda = (\lambda_1, \lambda_2) \in \mathbb{R}^2.$$

Observe that $P(0) = 0$ and $P(\lambda) > 0$ for any $\lambda \in \mathbb{R}^2$, $\lambda \neq 0$. In general, $P(\lambda) = k$ represents an ellipsoid for any constant k.

To the heat operator

$$\frac{\partial}{\partial t} - \frac{\partial^2}{\partial x_1^2} - \frac{\partial^2}{\partial x_2^2},$$

we associate the polynomial

$$P(\tau, \lambda) = \tau - \lambda_1^2 - \lambda_2^2$$

and we note that $P(\tau, \lambda) = 0$ represents a paraboloid.

Finally, to the wave operator

$$\frac{\partial^2}{\partial t^2} - \frac{\partial^2}{\partial x_1^2} - \frac{\partial^2}{\partial x_2^2},$$

we associate the polynomial

$$P(\tau, \lambda) = \tau^2 - \lambda_1^2 - \lambda_2^2$$

and observe that $P(\tau, \lambda) = 0$ represents a hyperboloid.

Example 2.10.

- The following equation is hyperbolic in $\mathbb{R}^2$, since $B^2 - 4AC = 4$:

$$\frac{\partial^2 u}{\partial x_1^2} - \frac{\partial^2 u}{\partial x_2^2} = 0.$$

- The following equation is elliptic in $\mathbb{R}^2$, since $B^2 - 4AC = -4$:

$$\frac{\partial^2 u}{\partial x_1^2} + \frac{\partial^2 u}{\partial x_2^2} - \frac{\partial u}{\partial x_1} + u = 0.$$

- The following equation is parabolic in $\mathbb{R}^2$ because $B^2 - 4AC = 0$:

$$\frac{\partial^2 u}{\partial x_1^2} + \frac{\partial u}{\partial x_1} - \frac{\partial u}{\partial x_2} + e^{x_1 x_2} = 0. \qquad \diamond$$

In the following we shall be led to make several changes of coordinates in the equation (2.1) (recalled above) of the form,

$$\begin{cases} \Theta_1 = \Theta_1(x_1, x_2), \\ \Theta_2 = \Theta_2(x_1, x_2), \end{cases} \tag{2.14}$$

whose Jabobian determinant J is given by

$$J(\Theta_1, \Theta_2) = \begin{vmatrix} \dfrac{\partial \Theta_1}{\partial x_1} & \dfrac{\partial \Theta_1}{\partial x_2} \\ \dfrac{\partial \Theta_2}{\partial x_1} & \dfrac{\partial \Theta_2}{\partial x_2} \end{vmatrix} = \frac{\partial \Theta_1}{\partial x_1}\frac{\partial \Theta_2}{\partial x_2} - \frac{\partial \Theta_2}{\partial x_1}\frac{\partial \Theta_1}{\partial x_2} \quad \text{for} \quad (x_1, x_2) \in \Omega.$$

The question is now, does such a change modify the character of the equation? The answer is given in the next result.

Proposition 2.11. *Assume that (2.14) is a one-to-one transformation, where the functions Θ_1 and Θ_2 are twice continuously differentiable on Ω, and such that the Jacobian determinant associated to this change does not vanish in any point of Ω, that is,*

$$J(\Theta_1, \Theta_2) = J(\Theta_1(x_1, x_2), \Theta_2(x_1, x_2)) \neq 0, \quad \forall (x_1, x_2) \in \Omega.$$

If $\hat{A}$, $\hat{B}$ and $\hat{C}$ denote the coefficients in the transformed equation, then

$$\hat{B}^2 - 4\hat{A}\hat{C} = J^2(B^2 - 4AC), \tag{2.15}$$

which means that the classification as elliptic, parabolic or hyperbolic of equation (2.1), is invariant under the change of coordinates (2.14).

Proof. Let us see what happens to (2.1) when doing the change (2.14). To do that, we have to compute the derivatives of first and second order of u, when $u(x_1, x_2)$ is replaced by $u(\Theta_1, \Theta_2)$. We have, by the chain rule, the following formulas for the first order derivatives:

$$\frac{\partial u}{\partial x_1} = \frac{\partial u}{\partial \Theta_1}\frac{\partial \Theta_1}{\partial x_1} + \frac{\partial u}{\partial \Theta_2}\frac{\partial \Theta_2}{\partial x_1}, \qquad \frac{\partial u}{\partial \bar{x}_2} = \frac{\partial u}{\partial \Theta_1}\frac{\partial \Theta_1}{\partial x_2} + \frac{\partial u}{\partial \Theta_2}\frac{\partial \Theta_2}{\partial x_2},$$

and for the second order derivatives,

$$\frac{\partial^2 u}{\partial x_1^2} = \frac{\partial^2 u}{\partial \Theta_1^2}\left(\frac{\partial \Theta_1}{\partial x_1}\right)^2 + 2\frac{\partial^2 u}{\partial \Theta_1 \partial \Theta_2}\frac{\partial \Theta_1}{\partial x_1}\frac{\partial \Theta_2}{\partial x_1} + \frac{\partial^2 u}{\partial \Theta_2^2}\left(\frac{\partial \Theta_2}{\partial x_1}\right)^2$$
$$+ \frac{\partial u}{\partial \Theta_1}\frac{\partial^2 \Theta_1}{\partial x_1^2} + \frac{\partial u}{\partial \Theta_2}\frac{\partial^2 \Theta_2}{\partial x_1^2},$$

$$\frac{\partial^2 u}{\partial x_2^2} = \frac{\partial^2 u}{\partial \Theta_1^2}\left(\frac{\partial \Theta_1}{\partial x_2}\right)^2 + 2\frac{\partial^2 u}{\partial \Theta_1 \partial \Theta_2}\frac{\partial \Theta_1}{\partial x_2}\frac{\partial \Theta_2}{\partial x_2} + \frac{\partial^2 u}{\partial \Theta_2^2}\left(\frac{\partial \Theta_2}{\partial x_2}\right)^2$$
$$+ \frac{\partial u}{\partial \Theta_1}\frac{\partial^2 \Theta_1}{\partial x_2^2} + \frac{\partial u}{\partial \Theta_2}\frac{\partial^2 \Theta_2}{\partial x_2^2},$$

$$\frac{\partial^2 u}{\partial x_1 \partial x_2} = \frac{\partial^2 u}{\partial \Theta_1^2}\frac{\partial \Theta_1}{\partial x_1}\frac{\partial \Theta_1}{\partial x_2} + \frac{\partial^2 u}{\partial \Theta_1 \partial X_2}\left(\frac{\partial \Theta_1}{\partial x_1}\frac{\partial \Theta_2}{\partial x_2} + \frac{\partial \Theta_1}{\partial x_2}\frac{\partial \Theta_2}{\partial x_1}\right)$$
$$+ \frac{\partial^2 u}{\partial \Theta_2^2}\frac{\partial \Theta_2}{\partial x_1}\frac{\partial \Theta_2}{\partial x_2} + \frac{\partial u}{\partial \Theta_1}\frac{\partial^2 \Theta_1}{\partial x_1 \partial x_2} + \frac{\partial u}{\partial \Theta_2}\frac{\partial^2 \Theta_2}{\partial x_1 \partial x_2}.$$

Plugging these formulas in (2.1), we obtain the transformed equation under change (2.14),

$$\hat{A}\frac{\partial^2 u}{\partial \Theta_1^2} + \hat{B}\frac{\partial^2 u}{\partial \Theta_1 \partial \Theta_2} + C\frac{\partial^2 u}{\partial \Theta_2^2} + \hat{D}\frac{\partial u}{\partial \Theta_1} + \hat{E}\frac{\partial u}{\partial \Theta_2} + \hat{F}u + \hat{G} = 0, \tag{2.16}$$

where the new coefficients $\hat{A}, \ldots, \hat{F}$ and $\hat{G}$ are given by

$$
\begin{aligned}
\hat{A} &= \Big(\frac{\partial \Theta_1}{\partial x_1}\Big)^2 A + \frac{\partial \Theta_1}{\partial x_1}\frac{\partial \Theta_1}{\partial x_2} B + \Big(\frac{\partial \Theta_1}{\partial x_2}\Big)^2 C, \\
\hat{B} &= 2\frac{\partial \Theta_1}{\partial x_1}\frac{\partial \Theta_2}{\partial x_1} A + \Big[\frac{\partial \Theta_1}{\partial x_1}\frac{\partial \Theta_2}{\partial x_2} + \frac{\partial \Theta_1}{\partial x_2}\frac{\partial \Theta_2}{\partial x_1}\Big] B + 2\frac{\partial \Theta_1}{\partial x_2}\frac{\partial \Theta_2}{\partial x_2} C, \\
\hat{C} &= \Big(\frac{\partial \Theta_2}{\partial x^1}\Big)^2 A + \frac{\partial \Theta_2}{\partial x_1}\frac{\partial \Theta_2}{\partial x_2} B + \Big(\frac{\partial \Theta_2}{\partial x_2}\Big)^2 C, \\
\hat{D} &= \frac{\partial^2 \Theta_1}{\partial x_1^2} A + \frac{\partial^2 \Theta_1}{\partial x_1 \partial x_2} B + \frac{\partial^2 \Theta_1}{\partial x_2^2} + \frac{\partial \Theta_1}{\partial x_1} D + \frac{\partial \Theta_1}{\partial x_2} E, \\
\hat{E} &= \frac{\partial^2 \Theta_2}{\partial x_1^2} A + \frac{\partial^2 \Theta_2}{\partial x_1 \partial x_2} B + \frac{\partial^2 \Theta_2}{\partial x_2^2} + \frac{\partial \Theta_2}{\partial x_1} D + \frac{\partial \Theta_2}{\partial x_2} E, \\
\hat{F} &= F, \qquad \hat{G} = G.
\end{aligned}
\tag{2.17}
$$

A simple computation gives formula (2.15) and then, using the fact that by hypothesis J does not vanish in any point of Ω, concludes the proof. □

Let us choose the following linear change of variables

$$
\begin{cases}
\bar{x}_1 = \alpha_1 x_1 + \beta_1 x_2, \\
\bar{x}_2 = \alpha_2 x_1 + \beta_2 x_2,
\end{cases}
\tag{2.18}
$$

with $\alpha_1 \beta_2 - \alpha_2 \beta_1 \neq 0$. This is a particular case of (2.14) with $(\bar{x}_1, \bar{x}_2)$ as (Θ_1, Θ_2) and $J = \alpha_1 \beta_2 - \alpha_2 \beta_1 \neq 0$. We are so allowed to apply Proposition 2.11 as all its hypotheses are satisfied. To exhibit the transformed equation (2.16) by the change (2.18), we apply formulas (2.17) to get the corresponding new coefficients $\bar{A}$, $\bar{B}$, ..., $\bar{E}$, $\bar{F}$ and $\bar{G}$.

To do so, we have to write down explicitly the coefficients in front of A, B, ..., G in the right-hand side of expressions from (2.17), with $\bar{x}_1$ instead of Θ_1, respectively $\bar{x}_2$ instead of Θ_2. There are only four of them that do not vanish, namely

$$
\frac{\partial \bar{x}_1}{\partial x_1} = \alpha_1, \quad \frac{\partial \bar{x}_1}{\partial x_2} = \beta_1, \quad \frac{\partial \bar{x}_2}{\partial x_1} = \alpha_2, \quad \frac{\partial \bar{x}_2}{\partial x_2} = \beta_2.
$$

Consequently, the new coefficients $\bar{A}, \ldots, \bar{F}$ and $\bar{G}$ are given by

$$
\begin{aligned}
\bar{A} &= \alpha_1^2 A + \alpha_1 \beta_1 B + \beta_1^2 C, \\
\bar{B} &= 2\alpha_1 \alpha_2 A + (\alpha_1 \beta_2 + \alpha_2 \beta_1) B + 2\beta_1 \beta_2 C, \\
\bar{C} &= \alpha_2^2 A + \alpha_2 \beta_2 B + \beta_2^2 C, \\
\bar{D} &= \alpha_1 D + \beta_1 E, \\
\bar{E} &= \alpha_2 D + \beta_2 E, \\
\bar{F} &= F, \qquad \bar{G}(\bar{x}_1, \bar{x}_2) = G(x_1, x_2).
\end{aligned}
\tag{2.19}
$$

Another question arises now: is there a linear transformation such that at least one of the coefficients $\bar{A}$, $\bar{B}$, or $\bar{C}$ is zero? The answer is given with the help of the theory of quadratic forms. Under hypothesis (2.10), the next results give the canonical forms of hyperbolic, parabolic, and elliptic linear equations with constant coefficients.

Proposition 2.12 (Canonical form for the hyperbolic case). *If equation (2.1) is hyperbolic on Ω, then there exists a linear transformation such that in the new coordinates $\bar{x}_1$, $\bar{x}_2$, it takes the form*

$$\frac{\partial^2 u}{\partial \bar{x}_1 \partial \bar{x}_2} = D_1 \frac{\partial u}{\partial \bar{x}_1} + E_1 \frac{\partial u}{\partial \bar{x}_2} + F_1 u + G_1(\bar{x}_1, \bar{x}_2), \tag{2.20}$$

where D_1, E_1, and F_1 are constants and G_1 is a real function.

Proof. Suppose that none of A, B, and C is zero. We would like to have a transformation such that $\bar{A} = \bar{C} = 0$. As seen before, a hyperbolic equation has two families of characteristic curves, given by

$$\begin{aligned} x_1 &= \frac{B + \sqrt{B^2 - 4AC}}{2A} x_1 + C_1, \\ x_2 &= \frac{B - \sqrt{B^2 - 4AC}}{2A} x_1 + C_2, \end{aligned} \tag{2.21}$$

where C_1, C_2 are arbitrary constants. So, if we apply the linear change of variables

$$\begin{cases} \bar{x}_1 = \dfrac{-B + \sqrt{B^2 - 4AC}}{2A} x_1 + x_2, \\ \bar{x}_2 = \dfrac{-B - \sqrt{B^2 - 4AC}}{2A} x_1 + x_2, \end{cases} \tag{2.22}$$

the characteristics (2.21) become the straight lines

$$\bar{x}_1 = C_1, \qquad \bar{x}_2 = C_2.$$

Therefore the choice of α_1, α_2, β_1, and β_2 given by (2.22), when substituted in the formulas given by (2.19), gives $\bar{A} = \bar{C} = 0$ as well as all the values of the other coefficients in the transformed equation (2.20).

If $A = 0$, $B \neq 0$, $C \neq 0$, the characteristic equation is

$$-By' + C = 0,$$

so that the characteristics are $x_1 = C_1$ and $x_1 - (B/C)x_2 = C_2$. With

$$\begin{cases} \bar{x}_1 = x_1, \\ \bar{x}_2 = x_1 - \dfrac{B}{C} x_2, \end{cases}$$

we are led to the preceding case.

The situation where two of the coefficients A, B or C vanish at the same time is treated in an analogous way. $\square$

A variant of the proposition above is the following result:

Proposition 2.13. *If equation (2.1) is hyperbolic on Ω, then there exists a linear transformation such that, in the new coordinates $\hat{x}_1$, $\hat{x}_2$, it has the form*

$$\frac{\partial^2 u}{\partial \hat{x}_1^2} - \frac{\partial^2 u}{\partial \hat{x}_2^2} = D_1 \frac{\partial u}{\partial \hat{x}_1} + E_1 \frac{\partial u}{\partial \hat{x}_2} + F_1 u + G_1(\hat{x}_1, \hat{x}_2), \tag{2.23}$$

where D_1, E_1, and F_1 are constants, and G_1 is a real function.

Proof. From the preceding proposition, we know that equation (2.1) can be written in the form

$$\frac{\partial^2 u}{\partial \bar{x}_1 \partial \bar{x}_2} = \bar{D}_1 \frac{\partial u}{\partial \bar{x}_1} + \bar{E}_1 \frac{\partial u}{\partial \bar{x}_2} + \bar{F}_1 u + G_1(\bar{x}_1, \bar{x}_2).$$

It suffices now to apply another change of variable, namely,

$$\begin{cases} \hat{x}_1 = \bar{x}_1 + \bar{x}_2, \\ \hat{x}_2 = \bar{x}_1 - \bar{x}_2, \end{cases}$$

to obtain (2.23). Recalling (2.22), it means that the change of variables to be done in (2.1) to get (2.23) is actually

$$\begin{cases} \hat{x}_1 = \dfrac{-B}{A} x_1 + 2x_2, \\ \hat{x}_2 = \dfrac{\sqrt{B^2 - 4AC}}{A} x_1. \end{cases} \tag{2.24}$$

This ends the proof of the proposition. □

Example 2.14. The equation

$$\frac{\partial^2 u}{\partial x_1^2} - 5 \frac{\partial^2 u}{\partial x_1 \partial x_2} - \frac{\partial u}{\partial x_1} + \frac{\partial u}{\partial x_2} = 0,$$

is hyperbolic in $\mathbb{R}^2$ because $B^2 - 4AC = 25 > 0$. By the change of variables

$$\begin{cases} \bar{x}_1 = 5x_1 + x_2, \\ \bar{x}_2 = x_2, \end{cases}$$

it becomes

$$\frac{\partial^2 u}{\partial \bar{x}_1 \partial \bar{x}_2} = -\frac{4}{25} \frac{\partial u}{\partial \bar{x}_1} + \frac{1}{25} \frac{\partial u}{\partial \bar{x}_2}. \qquad \diamond$$

Example 2.15. Consider again the equation from the preceding example. Making the linear change (see (2.24)),

$$\begin{cases} \hat{x}_1 = 5x_1 + 2x_2, \\ \hat{x}_2 = 5x_1, \end{cases}$$

we get

$$\frac{\partial^2 u}{\partial \hat{x}_1^2} - \frac{\partial^2 u}{\partial \hat{x}_2^2} = -\frac{3}{25}\frac{\partial u}{\partial \hat{x}_1} - \frac{1}{5}\frac{\partial u}{\partial \hat{x}_2}. \qquad \diamond$$

Proposition 2.16 (Canonical Form for the Elliptic Case). *If (2.1) is elliptic in Ω, then there exists a linear transformation such that, in the new coordinates $\bar{x}_1$, $\bar{x}_2$, it has the form*

$$\frac{\partial^2 u}{\partial \bar{x}_1^2} + \frac{\partial^2 u}{\partial \bar{x}_2^2} = D_1 \frac{\partial u}{\partial \bar{x}_1} + E_1 \frac{\partial u}{\partial \bar{x}_2} + F_1 u + G_1(\bar{x}_1, \bar{x}_2), \tag{2.25}$$

where D_1, E_1, and F_1 are constants, and G_1 is a real function.

Proof. Since $B^2 - 4AC < 0$, neither A nor C is zero. As seen above in the analysis of the differential equation (2.11), an elliptic equation has no characteristic curve in the real plane. So, the methods used in the preceding proofs do not apply. Equation (2.12) now has actually two complex conjugate roots λ_1 and λ_2. By setting

$$y' = -\lambda_1, \quad y' = -\lambda_2,$$

and integrating, one gets the characteristic curves

$$x_2 + \lambda_1 x_1 = C_1 \quad \text{and} \quad x_2 + \lambda_2 x_1 = C_2,$$

where C_1, C_2 are two arbitrary complex constants.

Now take $C_1 = c_1 + ic_2$, $C_2 = c_1 - ic_2$ with c_1 and c_2 arbitrary constants and set $\bar{x}_1 = c_1$, $\bar{x}_2 = c_2$. Then doing the change of variables

$$\begin{cases} \bar{x}_1 = x_2 - \dfrac{B}{2A} x_1, \\ \bar{x}_2 = \dfrac{\sqrt{4AC - B^2}}{2A} x_1, \end{cases} \tag{2.26}$$

we can follow the outlines of the proofs above for the hyperbolic case to get the result. $\square$

Example 2.17. Consider the equation

$$\frac{\partial^2 u}{\partial x_1^2} + \frac{\partial^2 u}{\partial x_1 \partial x_2} + \frac{\partial^2 u}{\partial x_2^2} - u = 0,$$

which is elliptic in $\mathbb{R}^2$. By the transformation

$$\begin{cases} \bar{x}_1 = -\frac{1}{2}x_1 + x_2, \\ \bar{x}_2 = \frac{\sqrt{3}}{2}x_2, \end{cases}$$

it becomes

$$\Delta u = \frac{\partial^2 u}{\partial \bar{x}_1^2} + \frac{\partial^2 u}{\partial \bar{x}_2^2} = \frac{4}{3}u. \qquad \diamond$$

Proposition 2.18 (Canonical Form for the Parabolic Case). *Suppose $A \neq 0$. If equation (2.1) is parabolic in Ω, then there exists a linear transformation such that, in the new coordinates $\bar{x}_1$, $\bar{x}_2$, it has the form*

$$\frac{\partial^2 u}{\partial \bar{x}_1^2} = D_1 \frac{\partial u}{\partial \bar{x}_1} + E_1 \frac{\partial u}{\partial \bar{x}_2} + F_1 u + G_1(\bar{x}_1, \bar{x}_2), \tag{2.27}$$

where D_1, E_1, and F_1 are constants, and G_1 is a real function.

Proof. In the parabolic case, we have $\lambda_1 = \lambda_2 = B/2A$ and the equation admits one family of characteristics

$$x_2 - \frac{B}{2A}x_1 = C,$$

where C is an arbitrary constant. So, it is natural to set

$$\bar{x}_2 = -\frac{B}{2A}x_1 + x_2.$$

The only delicate point in the choice of the linear transformation is now to be sure that by setting $\bar{x}_1 = \alpha_1 x_1 + \beta_1 x_2$, one can find α_1 and β_1 such that

$$\alpha_1 + \frac{B}{2A}\beta_1 \neq 0.$$

A simple analysis of this inequality leads to the following linear transformation:

$$\begin{cases} \bar{x}_1 = \frac{B}{2A}x_1 + x_2, \\ \bar{x}_2 = \frac{-B}{2A}x_1 + x_2, \end{cases}$$

that immediately gives the canonical form (2.27). $\square$

Example 2.19. The equation

$$\frac{\partial^2 u}{\partial x_1^2} + 4\frac{\partial^2 u}{\partial x_1 \partial x_2} + 4\frac{\partial^2 u}{\partial x_2^2} + \frac{\partial u}{\partial x_1} = 0,$$

is parabolic in $\mathbb{R}^2$. By the change of variables

$$\begin{cases} \bar{x}_1 = 2x_1 + x_2, \\ \bar{x}_2 = -2x_2 + x_2, \end{cases}$$

it becomes

$$\frac{\partial^2 u}{\partial \bar{x}_1^2} = -\frac{1}{8}\frac{\partial u}{\partial \bar{x}_1} + \frac{1}{8}\frac{\partial u}{\partial \bar{x}_2}. \qquad \diamond$$

Final conclusions. Equations (2.20), (2.23), (2.25), and (2.27) are called *canonical forms* of the linear equation (2.1). By using linear transformations, it is enough to study only the canonical forms, without losing the generality of equations. Actually, since equations (2.20) and (2.23) are equivalent, only three canonical forms are to be taken into consideration. They are the following:

Hyperbolic case
$$\frac{\partial^2 u}{\partial x_1^2} - \frac{\partial^2 u}{\partial x_2^2} = D_1\frac{\partial u}{\partial x_1} + E_1\frac{\partial u}{\partial x_2} + F_1 u + G_1,$$

Elliptic case
$$\frac{\partial^2 u}{\partial x_1^2} + \frac{\partial^2 u}{\partial x_2^2} = D_1\frac{\partial u}{\partial x_1} + E_1\frac{\partial u}{\partial x_2} + F_1 u + G_1,$$

Parabolic case
$$\frac{\partial^2 u}{\partial x_1^2} = D_1\frac{\partial u}{\partial x_1} + E_1\frac{\partial u}{\partial x_2} + F_1 u + G_1.$$

Notice that the wave equation treated as a model case in Chapter 1, is a particular hyperbolic equation with $x_1 = t$. The heat and Black–Scholes equations correspond to the parabolic case with $x_2 = t$.

2.4 Canonical forms: Case of nonconstant coefficients

We now briefly discuss the case of the general second order linear PDE given by:

$$A\frac{\partial^2 u}{\partial x_1^2} + B\frac{\partial^2 u}{\partial x_1 \partial x_2} + C\frac{\partial^2 u}{\partial x_2^2} + D\frac{\partial u}{\partial x_1} + E\frac{\partial u}{\partial x_2} + Fu + G = 0, \qquad (2.28)$$

where A, B, C, D, E, and F are functions depending on x_1 and x_2. Like in the case of constant coefficients, the quantity $B^2 - 4AC$, which is now a function defined on a bounded open set Ω, determines the type of equation (2.28). Let us denote by $\mathcal{F}$ this function, that is,

$$\mathcal{F}(x) = B^2(x) - 4A(x)C(x) \qquad \text{for } x = (x_1, x_2) \in \Omega. \tag{2.29}$$

Accordingly (recall the analysis performed above for the case of constant coefficients), the following definition is introduced:

Definition 2.20. *Equation (2.28) is said to be*
(a) hyperbolic in Ω if $\mathcal{F} > 0$ in Ω,
(b) elliptic in Ω if $\mathcal{F} < 0$ in Ω,
(c) parabolic in Ω if $\mathcal{F} = 0$ in Ω.

As for the case of constant coefficients, suitable linear transformations lead to canonical forms in these three cases.

• **Hyperbolic case.** Let $\phi(x_1, x_2) = C_1$ and $\psi(x_1, x_2) = C_2$ be the two families of characteristic curves of (2.28) and set

$$\bar{x}_1 = \phi(x_1, x_2), \qquad \bar{x}_2 = \psi(x_1, x_2).$$

Then equation (2.28) becomes

$$\frac{\partial^2 u}{\partial \bar{x}_1 \partial \bar{x}_2} = \mathcal{G}\Big(\frac{\partial u}{\partial \bar{x}_1}, \frac{\partial u}{\partial \bar{x}_2}, u, \bar{x}_1, \bar{x}_2\Big).$$

• **Parabolic case.** In this case, equation (2.28) has one family of characteristics, $\phi(x_1, x_2) = C$. By taking $\bar{x}_1 = \phi(x_1, x_2)$ and $\bar{x}_2$ a function independent of $\bar{x}_1$, equation (2.28) is written in the following canonical form:

$$\frac{\partial^2 u}{\partial \bar{x}_2^2} = \mathcal{G}\Big(\frac{\partial u}{\partial \bar{x}_1}, \frac{\partial u}{\partial \bar{x}_2}, u, \bar{x}_1, \bar{x}_2\Big).$$

• **Elliptic case.** Suppose that equation (2.28) is elliptic in Ω. Let $\phi(x_1, x_2) = K_1, \quad \psi(x_1, y) = K_2$ be the complex solutions of

$$y' = \frac{B \pm i\sqrt{4AC - B^2}}{2A},$$

and set $\bar{x}_1 + i\bar{x}_2 = \phi(x_1, x_2)$, $\bar{x}_1 - i\bar{x}_2 = \psi(x_1, x_2)$. We obtain the following canonical form for equation (2.28):

$$\frac{\partial^2 u}{\partial \bar{x}_1^2} + \frac{\partial^2 u}{\partial \bar{x}_2^2} = \mathcal{G}\Big(\frac{\partial u}{\partial \bar{x}_1}, \frac{\partial u}{\partial \bar{x}_2}, u, \bar{x}_1, \bar{x}_2\Big).$$

The nature of equation (2.28) depends on the domain where the function $\mathcal{F}$ is studied. Clearly, equations can change types, depending on the domain being considered. These are called equations of mixed type. Let us give two examples of such a situation.

Example 2.21.
(1) Consider the Tricomi equation in $\mathbb{R}^2$,

$$\frac{\partial^2 u}{\partial x_1^2} + x_1 \frac{\partial^2 u}{\partial x_2^2} = 0,$$

which degenerates on the line $x_1 = 0$. This equation appears in the study of transonic flows in $\mathbb{R}^2$, being a simple mathematical model of the transition from subsonic to supersonic speeds in aerodynamics. Here $u = u(x_1, x_2)$ represents the stream function of the flow.

If $x_1 < 0$, the equation corresponds to the supersonic regime. It is hyperbolic and its characteristics are the following two families of semicubical parabolas:

$$(x_2 - C)^2 = -\frac{4}{9} x_1^3 \iff x_2 \pm \frac{2}{3} x_1^{\frac{3}{2}} = C,$$

where C, an integration constant, is any real number in the hyperplane $x_1 < 0$. These curves have cusp points (see Example 3.17 in Chapter 3) on the line $x_1 = 0$. On the other hand, if $x_1 > 0$, the equation is elliptic and corresponds to the subsonic regime.
(2) Another example of a mixed-type equation is the following one, known as Keldysh equation,

$$\frac{\partial^2 u}{\partial x_1^2} + x_2 \frac{\partial^2 u}{\partial x_2^2} = 0.$$

It is elliptic in the half-plane $x_2 > 0$ and hyperbolic if $x_2 < 0$. It degenerates on the line $x_1 = 0$. Its characteristics are the parabolas

$$x_2 = -\frac{1}{4}(x_1 - C)^2,$$

where as above, C stands for any real constant in the hyperplane $x_1 < 0$. This equation has important applications in gas dynamics.

The corresponding canonical forms are

$$\frac{\partial^2 u}{\partial \bar{x}_1^2} + \frac{\partial^2 u}{\partial \bar{x}_2^2} - \frac{1}{\bar{x}_2} \frac{\partial u}{\partial \bar{x}_2} = 0,$$

and

$$4 \frac{\partial^2 u}{\partial \bar{x}_1 \partial \bar{x}_2} + \frac{2}{\bar{x}_1 - \bar{x}_2} \left(\frac{\partial u}{\partial \bar{x}_1} - \frac{\partial u}{\partial \bar{x}_2} \right) = 0.$$ ◇

2.5 Case of more than two variables

A second order equation in N independent variables is of the form

$$\sum_{i,j=1}^{N} a_{ij} \frac{\partial^2 u}{\partial x_i \partial x_j} + \sum_{i=1}^{N} b_i \frac{\partial u}{\partial x_i} + cu = F, \tag{2.30}$$

where the coefficients a_{ij}, b_i, c, and the right-hand side term F are functions of $x_1, x_2, \ldots, x_N$. From now on, we suppose that

$$\text{the matrix } \mathcal{A} = (a_{ij})_{1 \le i,j \le N} \text{ is symmetric.}$$

When looking for a solution u of class C^2, this is not restrictive. Indeed, in this case, since $\dfrac{\partial^2 u}{\partial x_i \partial x_j} = \dfrac{\partial^2 u}{\partial x_j \partial x_i}$, one can replace a_{ij} and a_{ji} by $\dfrac{a_{ij} + a_{ij}}{2}$.

Like in the case of two variables, one can define the nature of equation (2.30) by taking into account the matrix $\mathcal{A}$.

Going back to equation (2.1) with constant coefficients, one can notice that its "matrix" is actually

$$\begin{pmatrix} A & B/2 \\ B/2 & C \end{pmatrix} \tag{2.31}$$

whose eigenvalues λ satisfy

$$\begin{vmatrix} A - \lambda & B/2 \\ B/2 & C - \lambda \end{vmatrix} = 0, \tag{2.32}$$

or equivalently,

$$\lambda^2 - (A + C)\lambda - \frac{1}{4}(B^2 - 4AC) = 0.$$

This equation has two real roots λ_1, λ_2. Moreover, as

$$\lambda_1 \lambda_2 = -\frac{1}{4}(B^2 - 4AC) = -\frac{1}{4}\Delta,$$

where Δ is the determinant introduced in (2.4), one has

(a) $\Delta > 0 \iff \lambda_1 \neq 0, \lambda_2 \neq 0$ and are of opposite signs,
(b) $\Delta = 0 \iff$ at least one of λ_1 and λ_2 is zero,
(c) $\Delta < 0 \iff \lambda_1 \neq 0, \lambda_2 \neq 0$ and have the same sign.

Hence the sign of the eigenvalues of matrix (2.31), the principal part of equation (2.1), depends on the sign of Δ. As seen above, this sign characterizes the nature of equation (2.1). It is again this scheme which is used to determine the nature of equation (2.30) with n variables.

The eigenvalues of the matrix $(a_{ij})_{1\leq i,j\leq N}$ are the roots of the equation

$$\begin{vmatrix} a_{11}-\lambda & a_{12} & \dots & a_{1N} \\ a_{21} & a_{22}-\lambda & \dots & a_{2N} \\ \dots & \dots & \dots & \dots \\ a_{N1} & a_{N2} & \dots & a_{NN}-\lambda \end{vmatrix} = 0. \tag{2.33}$$

A classical result in linear algebra says that if the matrix is symmetric, its eigenvalues are all real. Moreover, the number of negative, positive, or zero eigenvalues at an arbitrary point $x_0 \in \mathbb{R}^N$ is invariant under linear transformations of coordinates.

Based on these remarks, the following definition is given:

Definition 2.22. *Let $\lambda_1, \lambda_2, \dots, \lambda_N$ be the eigenvalues given by (2.33). Then equation (2.30) is called*

(a) elliptic at the point x_0 if $\lambda_1, \lambda_2, \dots, \lambda_N$ are not zero and have the same sign at x_0,
(b) hyperbolic at the point x_0 if $\lambda_1, \lambda_2, \dots, \lambda_N$ are not zero and are of the same sign except one at x_0,
(c) ultrahyperbolic at the point x_0 if $\lambda_1, \lambda_2, \dots, \lambda_N$ are not zero and at least two of them are negative and two of them are positive,
(d) parabolic at the point x_0 if one of $\lambda_1, \lambda_2, \dots, \lambda_n$ is zero at x_0.

Equation (2.30) is called elliptic, hyperbolic, parabolic, or ultrahyperbolic on Ω if it is elliptic, hyperbolic, parabolic, or ultrahyperbolic at every point of Ω, respectively.

Example 2.23. The Laplace equation in $\mathbb{R}^N$ is elliptic. The wave equation is hyperbolic in $\mathbb{R}^N$, while the heat equation is parabolic in $\mathbb{R}^N$. ◇

In the previous section, we have seen that by making a change of variables, equation (2.1) with two variables can be reduced to a canonical form. In general, this is not possible in the case of an equation with nonconstant coefficients and with more than two variables. Nevertheless, by using a classical theorem in linear algebra, one can show that at each given point $P = (x_1, \dots, x_N)$ of Ω, it is always possible to make a linear transformation of the form

$$\xi_i = \sum_{k=1}^{N} b_{ik} x_k, \quad i = 1, \ldots, N, \tag{2.34}$$

which reduces (2.30) to a canonical form. The canonical form has a diagonal principal matrix at this point, that is, the transformed equation

$$\sum_{i,j=1}^{N} A_{ij} \frac{\partial^2 u}{\partial \xi_i \partial \xi_j} + \sum_{i=1}^{N} B_i \frac{\partial u}{\partial \xi_i} + Cu = D, \tag{2.35}$$

has

$$A_{ij}(P) = \begin{cases} 0 & \text{if } i \neq j, \\ 1, \ -1 \text{ or } 0, & \text{if } i = j = 1, \ldots, N. \end{cases} \tag{2.36}$$

Let us point out that if a_{ij} are not constants, the coefficients A_{ij} are also not constant and so, at points other than P, they may have values different from those given by (2.36).

In the case where the coefficients in equation (2.30) are constants, the transformed coefficients A_{ij} have the values given in (2.36) at every point of Ω. General results give the precise canonical forms for this situation.

Theorem 2.24. *Suppose equation (2.30) has constant coefficients a_{ij} in a set $\Omega \subseteq \mathbb{R}^N$. Then there is always a change of variables (2.34) with a nonsingular matrix $(b_{ij})_{1 \leq i,j \leq N}$, such that equation (2.35) has the following canonical form in the new coordinates:*

$$\sum_{i=1}^{N} A_{ii} \frac{\partial^2 u}{\partial \xi_i^2} + \sum_{i=1}^{N} B_i \frac{\partial u}{\partial \xi_i} + Cu = D,$$

where $A_{11}, \ldots, A_{NN}$ are equal to one of the values $+1$, -1, 0 in Ω. In particular,

(1) if (2.30) is elliptic, its canonical form in Ω is

$$\sum_{i=1}^{N} \frac{\partial^2 u}{\partial \xi_i^2} + \sum_{i=1}^{N} B_i \frac{\partial u}{\partial \xi_i} + Cu = D.$$

(2) if (2.30) is parabolic, its canonical form in Ω is

$$\sum_{i=1}^{N-1} \frac{\partial^2 u}{\partial \xi_i^2} - \frac{\partial u}{\partial \xi_N} + \sum_{i=1}^{N-1} B_i \frac{\partial u}{\partial \xi_i} + Cu = D.$$

(3) if (2.30) is hyperbolic, its canonical form in Ω is

$$\sum_{i=1}^{N-1} \frac{\partial^2 u}{\partial \xi_i^2} - \frac{\partial^2 u}{\partial \xi_N^2} + \sum_{i=1}^{N} B_i \frac{\partial u}{\partial \xi_i} + Cu = D.$$

In view of this result, let us have a look at the standard form of some often-used equations in mathematical physics. To do so, we need the following definition:

Definition 2.25. *Let $\mathcal{B} = (b_{ij})_{1\le i,j\le N}$ be a symmetric matrix. The polynomial*

$$Q_{\mathcal{B}}(X) = \sum_{i,j=1}^{N} b_{ij}X_iX_j, \quad X = (X_1, \ldots, X_N) \in \mathbb{R}^N,$$

is called the quadratic form associated with the matrix $\mathcal{B}$. This form is positive definite if

$$Q_{\mathcal{B}}(X) > 0 \quad \text{for every } X \in \mathbb{R}^N.$$

A classical result in linear algebra states that the eigenvalues of a symmetric matrix M are all positive if and only if Q_M is positive definite. Based on this fact, another definition of the ellipticity of an equation in an open set Ω, equivalent to that given in Theorem 2.24, can be formulated.

To do so, suppose that $a_{11} \neq 0$ (which is always possible by changing the names of the variables) and that $a_{11} > 0$ in equation (2.30) (which is always possible by simply changing signs).

Definition 2.26. *Equation (2.30) is elliptic if the quadratic form $Q_{\mathcal{A}(x)}$ is positive definite for every $x \in \Omega$.*

2.5.1 *Concluding remarks*

In this chapter, we have given the standard form for different types of second order partial differential equations, as used in the literature.

In view of what we did above, let us conclude by writing down our model PDEs, namely, elliptic equations, heat-type equations, and wave-type equations in the general cases. Comparing with the notation in Theorem 2.24, for the heat and wave equations, we shall call t the last variable ξ_N.

In the sequel, $\mathcal{A} = \mathcal{A}(x) = (a_{ij}(x))_{1\le i,j\le N}$ is a symmetric matrix with its quadratic form positive definite, i.e., $Q_{\mathcal{A}(x)} > 0$ (see Definition 2.25 above).

To finish, we now list the main PDEs that we introduced in this chapter.

Elliptic equations appear naturally in the study of steady-state (that is, independent of time) physical problems, as shown by our various examples. In particular, we have seen that if $u(x)$ is the temperature at a point x of a nonhomogeneous isotropic body, u has to satisfy the following second order equation:

$$\sum_{i=1}^{N} \frac{\partial}{\partial x_i}\left[k(x)\frac{\partial u}{\partial x_i}\right] = 0,$$

which is elliptic, since the function $k = k(x)$, being the conductivity coefficient, is positive at every point x. If the body is homogeneous, k is constant and the former equation is actually the Laplace equation. More generally, an elliptic problem is of the form

$$-\sum_{i,j=1}^{N} a_{ij}\frac{\partial^2 u}{\partial x_i \partial x_j} + \sum_{i=1}^{N} b_i \frac{\partial u}{\partial x_i} + cu = f, \tag{2.37}$$

to which one has to add boundary conditions.

The heat equation in $\mathbb{R}^N$ has the following general form:

$$\frac{\partial u}{\partial t} - \sum_{i,j=1}^{N} a_{ij}\frac{\partial^2 u}{\partial x_i \partial x_j} + \sum_{i=1}^{N} b_i \frac{\partial u}{\partial x_i} + cu = f, \tag{2.38}$$

completed with boundary conditions and one initial condition. The Black–Scholes equation (1.11) is precisely of this form.

The wave equation in $\mathbb{R}^N$ is hyperbolic and is of the form

$$\frac{\partial^2 u}{\partial t^2} - \sum_{i,j=1}^{N} a_{ij}\frac{\partial^2 u}{\partial x_i \partial x_j} + \sum_{i=1}^{N} b_i \frac{\partial u}{\partial x_i} + cu = f, \tag{2.39}$$

completed with boundary conditions and two initial conditions.

Chapter 3

Elliptic Equations

3.1 The Laplace equation

The simplest elliptic partial differential equation in $\mathbb{R}^N$ ($N \geq 2$) is the Laplace equation

$$\Delta u = \frac{\partial^2 u}{\partial x_1^2} + \frac{\partial^2 u}{\partial x_2^2} + \ldots + \frac{\partial^2 u}{\partial x_N^2} = 0. \tag{3.1}$$

A solution of the Laplace equation, that is, a function $u \in C^2(\Omega)$ satisfying (3.1) in a bounded open set $\Omega \subset \mathbb{R}^N$, is called a harmonic function (on Ω).

We shall give some examples of harmonic functions. Recall that by the superposition principle, any linear combination of harmonic functions is still a harmonic function.

Denote by r the radial distance in $\mathbb{R}^3$ between any point $x = (x_1, x_2, x_3) \neq 0$ and the origin (recall that $r^2 = x_1^2 + x_2^2 + x_3^2$).

A simple computation shows that the radial function

$$u = \frac{1}{r},$$

is harmonic in $\mathbb{R}^3 \setminus \{0\}$. A natural idea is then to try to find all possible radial harmonic functions in $\mathbb{R}^N$.

As detailed in Section 3.1.1, the Laplace operator in $\mathbb{R}^2$ in polar coordinates

$$x_1 = r\cos\theta, \quad x_2 = r\sin\theta \quad \text{for } r \in (0, +\infty),\ \theta \in [0, 2\pi),$$

has the form

$$\Delta = \frac{1}{r}\frac{\partial}{\partial r}\left(r\frac{\partial}{\partial r}\right) + \frac{1}{r^2}\frac{\partial^2}{\partial \theta^2} = \frac{\partial^2}{\partial r^2} + \frac{1}{r^2}\frac{\partial^2}{\partial \theta^2} + \frac{1}{r}\frac{\partial}{\partial r}. \tag{3.2}$$

In spherical coordinates in $\mathbb{R}^3$,

$$x_1 = r\sin\phi\cos\theta, \quad x_2 = r\sin\phi\sin\theta, \quad x_3 = \cos\phi, \tag{3.3}$$

for $r \in [0, +\infty)$, $\theta \in [0, 2\pi)$ and $\phi \in [0, \pi)$, one has

$$\Delta = \frac{1}{r^2}\frac{\partial}{\partial r}\left(r^2\frac{\partial}{\partial r}\right) + \frac{1}{r^2}\Lambda_3, \tag{3.4}$$

where Λ_3 is a second order operator containing only angular derivatives (i.e., with respect to θ and ϕ), and is given by

$$\Lambda_3 = \frac{1}{\sin\phi}\frac{\partial}{\partial\phi}\left(\sin\phi\frac{\partial}{\partial\phi}\right) + \frac{1}{\sin^2\phi}\frac{\partial^2}{\partial\theta^2}. \tag{3.5}$$

Definition 3.1. *Any solution $u(r)$ of*

$$\Delta u = \frac{1}{r^{N-1}}\frac{\partial}{\partial r}\left(r^{N-1}\frac{\partial u}{\partial r}\right) = 0. \tag{3.6}$$

is called a radial harmonic function u in $\mathbb{R}^N$.

Remark 3.2. *In $\mathbb{R}^2$, observe that*

$$\frac{1}{r}\frac{\partial}{\partial r}\left(r\frac{\partial u}{\partial r}\right) = 0, \tag{3.7}$$

is an ordinary differential equation. The functions

$$1, \quad \log r,$$

are two solutions of this equation. Hence they are radial harmonic functions in $\mathbb{R}^2$. Moreover, 1 and $\log r$ are linearly independent and so, a general solution of (3.7) consists of linear combinations of these functions.

In $\mathbb{R}^N$, $N > 2$, two linearly independent solutions of equation (3.6) are

$$1, \quad \frac{1}{r^{N-2}} \quad (N > 2).$$

Let us mention that $\log r$ and $1/r^{N-2}$ are harmonic in the whole of $\mathbb{R}^N$ except the origin. It is why one says that $\log r$ is a harmonic function in the whole space with a pole at the origin.

3.1.1 *The Laplacian in polar coordinates*

For the reader's convenience, we end this section with the detailed computation leading to formula (3.2) of the Laplacian in the polar coordinates (3.3).

Step 1. Let $f = (x_1, x_2)$ be a function twice differentiable and set

$$F = F(r,\theta) = f(r\cos\theta, r\sin\theta). \tag{3.8}$$

In this first step, we will establish the formulas giving the first order derivatives of f with respect to x, in terms of the first order derivatives of F with respect to r and θ.

To begin with, let us take the first derivatives of F in (3.8). We get the following formulas for the couple (f, F) defined by (3.8):

$$\frac{\partial F}{\partial r}(r,\theta) = \cos\theta \frac{\partial f}{\partial x_1}(r\cos\theta, r\sin\theta) + \sin\theta)\frac{\partial f}{\partial x_2}(r\cos\theta, r\sin\theta),$$
$$\frac{\partial F}{\partial \theta}(r,\theta) = -r\sin\theta \frac{\partial f}{\partial x_1}(r\cos\theta, r\sin\theta) + r\cos\theta)\frac{\partial f}{\partial x_2}(r\cos\theta, r\sin\theta),$$

which can be seen as a system of equations with two unknowns, the first order derivatives of f. We are allowed to solve it since its determinant equals $r^2 \neq 0$, to get

$$\begin{aligned} \frac{\partial f}{\partial x_1}(r\cos\theta, r\sin\theta) &= \cos\theta\,\frac{\partial F}{\partial r}(r,\theta) - \frac{1}{r}\sin\theta\,\frac{\partial F}{\partial \theta}(r,\theta) = F_1(r,\theta), \\ \frac{\partial f}{\partial x_2}(r\cos\theta, r\sin\theta) &= \sin\theta\,\frac{\partial F}{\partial r}(r,\theta) + \frac{1}{r}\cos\theta\,\frac{\partial F}{\partial \theta}(r,\theta) = F_2(r,\theta). \end{aligned} \tag{3.9}$$

Step 2. To write the Laplacian in polar coördinates, we now have to compute the second order derivatives of f in terms of derivatives of F, as we have done for the first order ones in (3.9) above. To do so, we shall proceed by the classical chain rule. Introduce the functions f_1 and f_2 defined as

$$f_1(x_1,x_2) = \frac{\partial f}{\partial x_1}(x_1,x_2), \qquad f_2(x_1,x_2) = \frac{\partial f}{\partial x_2}(x_1,x_2).$$

With notation (3.8) and using (3.9), it follows that

$$\begin{aligned} f_1(r\cos\theta, r\sin\theta) &= \frac{\partial f}{\partial x_1}(r\cos\theta, r\sin\theta) = F_1(r,\theta), \\ f_2(r\cos\theta, r\sin\theta) &= \frac{\partial f}{\partial x_2}(r\cos\theta, r\sin\theta) = F_2(r,\theta). \end{aligned} \tag{3.10}$$

Recalling (3.8), we are now in position to apply the first relation of (3.9) to the couple (f_1, F_1), and the second one to (f_2, F_2), to obtain

$$\begin{aligned} \frac{\partial}{\partial x_1}\left(\frac{\partial f}{\partial x_1}\right)(r\cos\theta, r\sin\theta) &= \cos\theta\,\frac{\partial F_1}{\partial r}(r,\theta) - \frac{1}{r}\sin\theta\,\frac{\partial F_1}{\partial \theta}(r,\theta), \\ \frac{\partial}{\partial x_2}\left(\frac{\partial f}{\partial x_2}\right)(r\cos\theta, r\sin\theta) &= \sin\theta\,\frac{\partial F_2}{\partial r}(r,\theta) + \frac{1}{r}\cos\theta\,\frac{\partial F_2}{\partial \theta}(r,\theta). \end{aligned} \tag{3.11}$$

To go further, we need to write down the first order partial derivatives of F_1 and F_2. Using their expressions from (3.9) we have,

$$\frac{\partial F_1}{\partial r} = \cos\theta \frac{\partial^2 F}{\partial r^2} + \frac{1}{r^2}\sin\theta\,\frac{\partial F}{\partial \theta} - \frac{1}{r}\sin\theta\,\frac{\partial^2 F}{\partial r \partial \theta},$$
$$\frac{\partial F_1}{\partial \theta} = -\sin\theta\,\frac{\partial F}{\partial r} + \cos\theta\,\frac{\partial^2 F}{\partial r \partial \theta} - \frac{1}{r}\cos\theta\,\frac{\partial F}{\partial \theta} - \frac{1}{r}\sin\theta\,\frac{\partial^2 F}{\partial \theta^2},$$

and

$$\frac{\partial F_2}{\partial r} = \sin\theta \frac{\partial^2 F}{\partial r^2} - \frac{1}{r^2}\cos\theta \frac{\partial F}{\partial \theta} + \frac{1}{r}\cos\theta \frac{\partial^2 F}{\partial r \partial \theta},$$

$$\frac{\partial F_2}{\partial \theta} = \cos\theta \frac{\partial F}{\partial r} + \sin\theta \frac{\partial^2 F}{\partial r \partial \theta} - \frac{1}{r}\sin\theta \frac{\partial F}{\partial \theta} + \frac{1}{r}\cos\theta \frac{\partial^2 F}{\partial \theta^2}.$$

Step 3. We substitute the last formulas above in (3.11), this yields

$$\frac{\partial^2 f}{\partial x_1^2}(r\cos\theta, r\sin\theta) = \cos^2\theta \frac{\partial^2 F}{\partial r^2}(r,\theta) - \frac{2\sin\theta\cos\theta}{r}\frac{\partial^2 F}{\partial r \partial \theta}(r,\theta)$$

$$+ \frac{\sin^2\theta}{r^2}\frac{\partial^2 F}{\partial \theta^2}(r,\theta) + \frac{\sin^2\theta}{r}\frac{\partial F}{\partial r}(r,\theta) + \frac{2\sin^2\theta\cos\theta}{r^2}\frac{\partial F}{\partial \theta}(r,\theta),$$

as well as

$$\frac{\partial^2 f}{\partial x_2^2}(r\cos\theta, r\sin\theta) = \sin^2\theta \frac{\partial^2 F}{\partial r^2}(r,\theta) + \frac{2\sin\theta\cos\theta}{r}\frac{\partial^2 F}{\partial r \partial \theta}(r,\theta)$$

$$+ \frac{\cos^2\theta}{r^2}\frac{\partial^2 F}{\partial \theta^2}(r,\theta) + \frac{\cos^2\theta}{r}\frac{\partial F}{\partial r}(r,\theta) - \frac{2\sin^2\theta\cos\theta}{r^2}\frac{\partial F}{\partial \theta}(r,\theta).$$

Summing up these two equalities gives

$$\Delta f(r\cos\theta, r\sin\theta) = \frac{\partial^2 F}{\partial r^2}(r,\theta) + \frac{1}{r^2}\frac{\partial^2 F}{\partial \theta^2}(r,\theta) + \frac{1}{r}\frac{\partial F}{\partial r}(r,\theta),$$

which is precisely (3.2).

Formula (3.4) for the Laplacian in spherical coordinates is obtained in the same way, the steps being in the spirit of those from the computation above, with obviously more technicalities (but not difficulties) because of the presence of one more coordinate.

3.2 Harmonic functions by the method of separation of variables

In this section, we shall make use of the method of separation of variables in order to get (other than radial) elementary harmonic functions.

3.2.1 *The two-dimensional case*

Recalling (3.2), consider in $\mathbb{R}^2$ the Laplace equation in polar coordinates

$$\frac{1}{r}\frac{\partial}{\partial r}\left(r\frac{\partial u}{\partial r}\right) + \frac{1}{r^2}\frac{\partial^2 u}{\partial \theta^2} = 0. \tag{3.12}$$

The method of separation of variables applied to equation (3.12) consists of searching for solutions $u = u(r,\theta)$ of the form

$$u(r,\theta) = \mathcal{F}(r)\mathcal{G}(\theta),$$

where $\mathcal{F}$ and $\mathcal{G}$ are real functions. By substituting u in (3.12), we obtain

$$\mathcal{F}''\,\mathcal{G} + \frac{1}{r}\mathcal{F}'\,\mathcal{G} + \frac{1}{r^2}\mathcal{F}\,\mathcal{G}'' = 0,$$

where $'$ denotes an ordinary derivative for a function depending on only one variable. This equation can be written in the following equivalent form:

$$\frac{r^2\,\mathcal{F}'' + r\,\mathcal{F}'}{\mathcal{F}} = -\frac{\mathcal{G}''}{\mathcal{G}}.$$

Observe that the expression on the left-hand side of this equation is a function of the variable r only, while the right-hand side depends on θ only. This is not possible unless these two expressions are equal to the same constant. This leads to solving two ordinary differential equations, namely,

$$\begin{aligned} &r^2\,\mathcal{F}'' + r\,\mathcal{F}' - C\,\mathcal{F} = 0, \\ &\mathcal{G}'' + C\,\mathcal{G} = 0, \end{aligned} \tag{3.13}$$

where C is a real constant.

The first equation in (3.13) has two families of linearly independent solutions given by

$$\mathcal{F}_C(r) = \begin{cases} 1 \text{ and } \log r & \text{if } C = 0, \\ r^{\sqrt{C}} \text{ and } r^{-\sqrt{C}} & \text{if } C \neq 0. \end{cases}$$

The second equation in (3.13) has also two families of linearly independent solutions

$$\mathcal{G}_C(\theta) = \begin{cases} 1 \text{ and } \theta & \text{if } C = 0, \\ \cos(\sqrt{C}\theta) \text{ and } \sin(\sqrt{C}\theta) & \text{if } C \neq 0. \end{cases}$$

If $C < 0$, the functions in these formulas are complex which implies that the real and the imaginary parts of these functions will be pairs of real linearly independent solutions.

On the other hand, the product $\mathcal{F}_C(r)\mathcal{G}_C(\theta)$ is a harmonic function in an arbitrary set Ω of $\mathbb{R}^2$ only if this product is a "well-defined" function of class C^2 in Ω. If Ω contains closed curves around the origin (for instance, when Ω is the whole of $\mathbb{R}^2$, or if it contains discs or annuli around the origin), it is clear that the functions $\mathcal{G}_C(\theta)$ have to be periodic with period 2π in order to have the same value once a turn is done around the origin,

$$\mathcal{G}_C(\theta + 2\pi) = \mathcal{G}_C(\theta).$$

If $C = 0$, it is only the constant function 1 that satisfies the periodicity condition. If $C \neq 0$, the definition of $\mathcal{G}_C(\theta)$ implies now that the periodicity is possible only if

$$\sqrt{C} = n, \quad n = 1, 2, \ldots.$$

Consequently, the harmonic functions $\mathcal{G}_C(\theta)$ have the form

$$\mathcal{G}_C(\theta) = \cos(n\theta) \quad \text{or} \quad \mathcal{G}_C(\theta) = \sin(n\theta), \qquad n = 0, 1, 2, \ldots.$$

Thus we have

$$u_n(r, \theta) = \begin{cases} 1, \quad r^n \cos(n\theta), \quad r^n \sin(n\theta), \qquad n = 1, 2, \ldots, \\ \log r, \quad r^{-n} \cos(n\theta), \quad r^{-n} \sin(n\theta), \quad n = 1, 2, \ldots. \end{cases}$$

If Ω does not contain the origin, all these functions are harmonic. If Ω contains the origin, only the functions from the first line are harmonic.

3.2.2 *The three-dimensional case*

Let us now consider the case of $\mathbb{R}^3$ with the Laplacian in spherical coordinates from (3.4). We have to solve

$$\frac{1}{r^2}\frac{\partial}{\partial r}\left(r^2 \frac{\partial u}{\partial r}\right) + \frac{1}{r^2}\Lambda_3(u) = 0, \tag{3.14}$$

where Λ_3 is given by (3.5). We are looking for solutions u of the form

$$u(r, \theta, \phi) = \mathcal{F}(r)\mathcal{G}(\theta, \phi).$$

By the same argument as above, we are led to solve the following ordinary differential equations:

$$\begin{cases} (r^2 \mathcal{F}')' - C\mathcal{F} = 0, \\ \Lambda_3(\mathcal{G}) + C\mathcal{G} = 0, \end{cases} \tag{3.15}$$

where C is an arbitrary real constant.

The first equation has two linearly independent solutions

$$r^{\alpha_1} \quad \text{and} \quad r^{\alpha_2},$$

where α_1 and α_2 are the solutions of the equation

$$\alpha(\alpha + 1) - C = 0.$$

The study of the second equation in (3.15) is more complicated. If one would like to find solutions $\mathcal{G}(\theta, \phi)$ of class C^2 on the unit sphere $S(0, 1)$

centered at the origin, they have to be periodic in θ with period 2π. It can be shown that under certain conditions, the only possible constants C are

$$C_n = n(n+1), \quad n = 0, 1, 2, \ldots.$$

For each C_n, there exist $2n+1$ linearly independent solutions, called Laplace spherical harmonics, $\mathcal{G}_n^k(\theta, \phi)$, $k = 1, 2, \ldots, 2n+1$. The corresponding radial functions are

$$u_{n,k}(r,\theta,\phi) = \begin{cases} r^n \mathcal{G}_n^k(\theta,\phi), & k = 1,2,\ldots,2n+1;\ n = 0,1,2,\ldots, \\ r^{-n-1}\mathcal{G}_n^k(\theta,\phi), & k = 1,2,\ldots,2n+1;\ n = 0,1,2,\ldots. \end{cases}$$

Like in the two-dimensional case, if Ω does not contain the origin, all the above functions are harmonic. Otherwise, one should keep only the functions from the first line.

3.3 The Dirichlet problem in special geometries

3.3.1 *The solution for the unit disc*

Let $\Omega = \{x \in \mathbb{R}^2 |\ \mathrm{dist}(0,x) < 1\}$ be the unit disc whose points in polar coordinates are of the form

$$x_1 = r\cos\theta, \quad x_2 = r\sin\theta,$$

where $0 \le r < 1$ and $0 \le \theta < 2\pi$. One looks for a harmonic function u such that $u = g$ on $\partial\Omega$, where g is a given function, periodic of period 2π and of class C^2. In polar coordinates, this means that $u = u(r,\theta)$ and satisfies the following nonhomogeneous Dirichlet problem for the Laplace equation:

$$\begin{cases} \Delta u = \dfrac{1}{r}\dfrac{\partial}{\partial r}\left(r\dfrac{\partial u}{\partial r}\right) + \dfrac{1}{r^2}\dfrac{\partial^2 u}{\partial \theta^2} = 0 & \text{for } 0 \le r < 1, \quad 0 \le \theta < 2\pi, \\ u(1,\theta) = g(\theta) & \text{for } 0 \le \theta < 2\pi. \end{cases}$$

One would like to build this solution by superposing the harmonic functions

$$r^n\cos(n\theta), \quad r^n\sin(n\theta), \quad n = 0,1,2,\ldots,$$

that we have obtained by the method of separation of variables. To do so, we apply the Fourier method, that is to say, we look for u in the form

$$u(r,\theta) = \frac{A_0}{2} + \sum_{n=1}^{\infty} r^n \big(A_n\cos(n\theta) + B_n\sin(n\theta)\big),$$

where the coefficients A_n, B_n have to be determined. Let

$$g(\theta) = \frac{\alpha_0}{2} + \sum_{n=1}^{\infty} \Big(\alpha_n \cos(n\theta) + \beta_n \sin(n\theta) \Big),$$

be the Fourier series of g. The Fourier coefficients of g are given by

$$\alpha_n = \frac{1}{\pi} \int_0^{2\pi} g(\phi) \cos(n\phi)\, d\phi, \quad n = 0, 1, \ldots,$$

$$\beta_n = \frac{1}{\pi} \int_0^{2\pi} g(\phi) \sin(n\phi)\, d\phi, \quad n = 1, 2, \ldots.$$

Since $u(1, \theta) = g(\theta)$, it follows that

$$\frac{A_0}{2} = \frac{\alpha_0}{2} = \frac{1}{2\pi} \int_0^{2\pi} g(\phi)\, d\phi, \quad A_n = \alpha_n, \quad B_n = \beta_n,\ n = 1, 2, \ldots.$$

Consequently,

$$\begin{aligned} u(r, \theta) = {} & \frac{1}{\pi} \sum_{n=1}^{\infty} r^n \int_0^{2\pi} g(\phi) \big[\cos(n\theta) \cos(n\phi) + \sin(n\theta) \sin(n\phi) \big]\, d\phi \\ & + \frac{1}{2\pi} \int_0^{2\pi} g(\phi)\, d\phi. \end{aligned} \tag{3.16}$$

It is the required solution, provided that the series above converges and u is of class C^2. This is true, due to a classical result in the theory of Fourier series, if the function g is twice continuously differentiable on $(0, 2\pi)$ and its third order derivative is piecewise continuous.

One can give another form of the solution u. Indeed, by a simple computation, from (3.16) we get

$$\begin{aligned} u(r, \theta) &= \frac{1}{2\pi} \int_0^{2\pi} g(\phi)\, d\phi + \frac{1}{\pi} \sum_{n=1}^{\infty} r^n \int_0^{2\pi} g(\phi) \cos(n(\theta - \phi))\, d\phi \\ &= \int_0^{2\pi} g(\phi) \mathcal{P}(r, \phi)\, d\phi, \end{aligned}$$

where for any fixed θ, the function $\mathcal{P}$ is given by

$$\mathcal{P}(r, \phi) = \frac{1}{2\pi} \Big(1 + 2 \sum_{n=1}^{\infty} r^n \cos(n(\theta - \phi)) \Big).$$

The function $\mathcal{P}$ is called the Poisson kernel and is harmonic for any ϕ. It is easy to see that

$$\mathcal{P}(r, \phi) = \frac{1}{2\pi} \frac{1 - r^2}{1 + r^2 - 2r \cos(\theta - \phi)},$$

so u can be written in the following form, called Poisson's integral,

$$u(r, \theta) = \frac{1}{2\pi} \int_0^{2\pi} g(\phi) \frac{1 - r^2}{1 + r^2 - 2r \cos(\theta - \phi)}\, d\phi.$$

3.3.2 *The solution of the exterior Dirichlet problem in $\mathbb{R}^3$*

In this section we solve the Dirichlet problem in a so-called exterior domain $\mathcal{O}$ which is introduced in the next definition.

Definition 3.3. *A set $\mathcal{O}$ in $\mathbb{R}^N$ is called an exterior domain if it is the complement of the closure of a simply connected bounded open set Ω. One has $\partial\mathcal{O} = \partial\Omega$.*

We are now looking for a function u such that

$$\begin{cases} \Delta u = 0 & \text{in } \mathcal{O}, \\ u(x) = g(x) & \text{for } x \in \partial\mathcal{O}, \\ \lim\limits_{|x| \to +\infty} u(x) = 0. \end{cases} \tag{3.17}$$

If $N = 2$, the last condition (the condition at infinity) can be replaced by

$$u(x) \text{ is bounded as } |x| \to +\infty.$$

In both cases, the problem is well-posed. It follows that there exists a unique solution of (3.17).

Let us give a physical problem leading to an exterior Dirichlet problem. For simplicity, we take the ball $B(0,R)$ in $\mathbb{R}^3$ as Ω, and suppose that its boundary is maintained at a given fixed electrical potential g. If u denotes the potential at any point of the domain, it means that

$$u(R,\theta,\phi) = g(\phi) \quad \text{on } \partial B(0,R), \tag{3.18}$$

where r, θ, and ϕ are the spherical coordinates. The aim now is to find the potential u without any additional charge. The potential on $B(0,R)$ is independent of θ and so is u, and moreover it will satisfy (see (3.4) and (3.5))

$$\Delta u = \frac{\partial}{\partial r}\left(r^2 \frac{\partial u}{\partial r}\right) + \frac{1}{\sin\phi}\frac{\partial}{\partial\phi}\left(\sin\phi\frac{\partial u}{\partial\phi}\right) = 0. \tag{3.19}$$

At infinity, u has to vanish, so

$$\lim_{r \to +\infty} u(r,\phi) = 0. \tag{3.20}$$

Consequently, we have to solve (3.19) with the boundary condition given by (3.18) and with (3.20) as condition at infinity.

We shall now solve the problem from this example by the method of separation of variables. We are looking for u satisfying (3.18)–(3.20) and of the form

$$u(r,\phi) = \mathcal{F}(r)\mathcal{G}(\phi).$$

As we have previously seen, we obtain two equations, one involving $\mathcal{F}$ and the other $\mathcal{G}$,

$$\begin{cases} (r^2\,\mathcal{F}')' - C\mathcal{F} = 0, \\ \dfrac{1}{\sin\phi}\dfrac{\partial}{\partial\phi}\left(\sin\phi\dfrac{\partial\mathcal{G}}{\partial\phi}\right) + C\mathcal{G} = 0, \end{cases} \tag{3.21}$$

where $C = C_n = n(n+1)$, $n = 0, 1, 2, \ldots$.

The first equation has the following solutions:

$$\mathcal{F}_n(r) = r^n, \quad \mathcal{F}^\star_n(r) = r^{-(n+1)}.$$

In the second equation, we introduce a change of variable $\cos\phi = w$, and since $d/d\phi = (-\sin\phi)d/dw$, one has to solve

$$(1-w^2)\frac{d^2\mathcal{G}}{dw^2} - 2w\frac{d\mathcal{G}}{dw} + n(n+1)\mathcal{G} = 0,$$

which is the classical Legendre equation. It is well known that its solutions are the Legendre polynomials $\mathcal{G}(w) = P_n(w)$ defined by

$$P_n(x) = \sum_{m=0}^{[n/2]}(-1)^m\frac{(2n-2m)!}{2^n m!(n-m)!(n-2m)!}x^{n-2m},$$

where $[n/2]$ denotes the integer part of $n/2$.

In conclusion, there are two families of solutions of the Laplace equation (3.19) for $n = 0, 1, \ldots$,

$$\begin{aligned} u_n(r,\phi) &= A_n r^n P_n(\cos\phi), \\ u^\star_n(r,\phi) &= B_n r^{-(n+1)} P_n(\cos\phi), \end{aligned} \tag{3.22}$$

where A_n and B_n are some constants.

Consider the first family in (3.22). Then

$$u(r,\phi) = \sum_{n=0}^{\infty} u_n(r,\phi),$$

is still a solution of the Laplace equation. If we require that u satisfies the boundary condition given by (3.18), then

$$u(R,\phi) = g(\phi) = \sum_{n=0}^{\infty} A_n R^n P_n(\cos\phi),$$

which is the Fourier–Legendre series of $g(\phi)$. Classical results imply that

$$A_n R^n = \frac{2n+1}{2}\int_{-1}^{1}\tilde{g}(w)\,P_n(w)\,dw,$$

where $\tilde{g}(w)$ is the transformed g via the change $w = \cos\phi$. Going back to the variable ϕ, one has

$$A_n = \frac{2n+1}{2R^n} \int_0^\pi g(s)\, P_n(\cos s)\, \sin s\, ds,$$

so that

$$u(r,\phi) = \sum_{n=0}^{\infty} \frac{(2n+1)r^n}{2R^n} P_n(\cos\phi) \int_0^\pi f(s)\, P_n(\cos s)\, \sin s\, ds.$$

This function is a solution of the Laplace equation either in the interior or at the exterior of the ball $B(0,R)$, and satisfies boundary condition (3.18). However, it does not satisfy the condition at infinity given by (3.20). Consequently, with the first family of functions in (3.22) one can only construct a solution for the interior Dirichlet problem (it is actually what we did by using Fourier series in the preceding section).

To solve the exterior problem, we shall make use of the second set of functions in (3.22). By the same argument as above, we obtain

$$u(r,\phi) = \sum_{n=0}^{\infty} B_n r^{-(n+1)} P_n(\cos\phi), \tag{3.23}$$

with

$$B_n = \frac{2n+1}{2} R^{n+1} \int_0^\pi g(s)\, P_n(\cos s)\, \sin s\, ds.$$

An easy computation shows that (3.23) is the solution of exterior problem (3.18)–(3.20).

3.4 Green formulas and related properties of harmonic functions

3.4.1 *Green formulas*

The three standard Green formulas are all consequences of the Ostrogradski formula (also known as the flux or divergence-flux formula) that we recall below for the three-dimensional case.

Let Ω be a bounded open set in $\mathbb{R}^N$. In the following, we require that the outward unit normal vector $\vec{n}$ be defined at each point of the boundary $\partial\Omega$. To do so, let us introduce the following definition:

Definition 3.4. *The bounded open set $\Omega \subset \mathbb{R}^N$ is of class C^1 if for any point $x_0 \in \partial\Omega$, there exists $r > 0$ and a function $\varphi \in C^1(B(r, x_0))$ such* *that*

$$\begin{cases} \nabla\varphi(x) \neq 0, \quad \forall x \in B(r, x_0), \\ \Omega \cap B(r, x_0) = \{x \mid x \in B(r, x_0),\ \varphi(x) < 0\}, \\ (\mathbb{R}^N \setminus \overline{\Omega}) \cap B(r, x_0) = \{x \mid x \in B(r, x_0),\ \varphi(x) > 0\}. \end{cases}$$

This definition implies that the normal exists at each point of $\partial\Omega$. In particular, c can be a union of a finite number of (overlapping) regular surfaces (that is, having a tangent plane at each of their points).

Example 3.5. Let Ω be the open subset of $\mathbb{R}^2$ defined by

$$\Omega = \left\{(x, y) \in \mathbb{R}^2 \;\middle|\; \frac{x^2}{a^2} + \frac{y^2}{b^2} < 1\right\}.$$

Its boundary $\partial\Omega$ is an ellipse centered at the origin which can be described by the parametric equations

$$\begin{cases} x = \varphi(t) = a\cos t, \\ y = \psi(t) = b\sin t, \end{cases}$$

for $t \in [0, 2\pi]$.

A classical result in planar curves, says that the tangent to $\partial\Omega$ at a point $(x_0, y_0) = (\varphi(t_0), \psi(t_0))$, is given by

$$\varphi'(t_0)(x - x_0) + \psi'(t_0)(y - y_0) = 0,$$

which implies that the equation of the normal at the point (x_0, y_0) reads

$$\varphi'(t_0)(x - x_0) + \psi'(t_0)(y - y_0) = 0.$$

Hence, the two unit vectors of the normal $\vec{n}$ at the point (x_0, y_0) are

$$\left(\pm\frac{\psi'(t_0)}{\sqrt{\varphi'(t_0)^2 + \psi'(t_0)^2}},\ \mp\frac{\varphi'(t_0)}{\sqrt{\varphi'(t_0)^2 + \psi'(t_0)^2}}\right). \tag{3.24}$$

For a suitable choice of the signs $(+,-)$ or $(-,+)$, this formula furnishes the outward normal unit vector in any point of the boundary.

Applying (3.24) to our example, it follows that the unit vectors of the normal to $\partial\Omega$ at any point $(x_0, y_0) = (a\cos t_0, b\sin t_0)$ are given by

$$\left(\pm\frac{b\cos t_0}{\sqrt{b^2\cos^2 t_0 + a^2\sin^2 t_0}},\ \pm\frac{a\sin t_0}{\sqrt{b^2\cos^2 t_0 + a^2\sin^2 t_0}}\right).$$

If for instance $t_0 = \pi/4$, the outward unit vector $n(x_0, y_0)$, obtained by choosing the signs (+,+), is given by

$$\vec{n}(x_0, y_0) = \left(\frac{b}{\sqrt{b^2 + a^2}}, \frac{a}{\sqrt{b^2 + a^2}}\right). \qquad \diamond$$

Theorem 3.6 (Ostrogradski Theorem). *Let $\vec{V} = (V_1, \dots, V_N)$ be a vector function of class C^1 in a bounded open $\Omega \subset \mathbb{R}^N$ and continuous on $\overline{\Omega}$. Then*

$$\int_\Omega div\, \vec{V}\, dx = \int_{\partial\Omega} (\vec{V}, \vec{n})_{\mathbb{R}^N}\, d\sigma, \tag{3.25}$$

where $\vec{n}$ is the outward unit normal vector at every point of $\partial\Omega$ and $d\sigma$ denotes the surface element.

Green formulas are then obtained by applying (3.25) for some particular choices of functions $\vec{V}$.

First Green Formula

Let us introduce $\vec{V} = \text{grad}\, \varphi$. Suppose that φ is of class $C^2(\overline{\Omega})$. From (3.25) and since $\text{div}(\text{grad}\, \varphi) = \Delta\varphi$, we obtain

$$\int_\Omega \Delta\varphi\, dx = \int_{\partial\Omega} \frac{\partial\varphi}{\partial n}\, d\sigma. \tag{3.26}$$

Second Green Formula

Let φ be of class $C^1(\overline{\Omega})$ and ψ be of class $C^2(\overline{\Omega})$. The Ostrogradski formula applied to the vector function $\vec{V} = \varphi\, \text{grad}\, \psi$, leads to

$$\int_\Omega \varphi\, \Delta\psi\, dx + \int_\Omega (\text{grad}\, \varphi, \text{grad}\, \psi)_{\mathbb{R}^N}\, dx = \int_{\partial\Omega} \varphi\, \frac{\partial\psi}{\partial n}\, d\sigma, \tag{3.27}$$

since it is easily seen that $\text{div}\, \vec{V} = (\text{grad}\, \varphi, \text{grad}\, \psi)_{\mathbb{R}^N} + \varphi\, \Delta\psi$.

Third Green Formula

Interchanging the roles of φ and ψ, both of class $C^2(\overline{\Omega})$, in (3.27) and subtracting gives

$$\int_\Omega (\varphi\, \Delta\psi - \psi\, \Delta\varphi)\, dx = \int_{\partial\Omega} \left(\varphi\, \frac{\partial\psi}{\partial n} - \psi\, \frac{\partial\varphi}{\partial n}\right) d\sigma. \tag{3.28}$$

Remark 3.7. *The following properties of harmonic functions are consequences of the Green formulas:*
(1) Let u be a harmonic function in Ω and choose $\varphi = \psi = u$ in (3.27), the second Green formula. Since $\Delta u = 0$, one has

$$\int_\Omega |grad\, u|^2\, dx = \int_{\partial\Omega} u\, \frac{\partial u}{\partial n}\, d\sigma.$$

Therefore any harmonic function satisfies the inequality

$$\int_{\partial\Omega} u\, \frac{\partial u}{\partial n}\, d\sigma \geq 0.$$

(2) If $\varphi = u$ and $\psi = 1$ in (3.28), the third Green formula, then

$$\int_{\partial\Omega} \frac{\partial u}{\partial n}\, d\sigma = 0,$$

for any harmonic function u. This property is known as the Gauss theorem for harmonic functions.

3.4.2 *Green representation theorem and consequences*

A direct consequence of the third Green formula is an integral representation which, for any function v of class $C^2(\overline{\Omega})$, gives its value at each point in terms of the volume and surface integrals of v, its first derivatives, and its Laplacian. This formula, known as Green representation, allows us to show that each harmonic function has the average property: the value of this function at a point x_0 is the average of its values on a sphere centered at x_0. This property has, as a consequence, the maximum principle for harmonic functions.

To state and prove the representation formula, we need to introduce the fundamental solution of the Laplace equation which plays an essential role in the study of harmonic functions. To do so, we need to recall some formulas related to the areas of spheres and the volumes of balls in $\mathbb{R}^N$ for $N \geq 3$. For $N = 2$, the corresponding notions are the circumference and the surface of a circle.

In the following $\mathcal{S}_N$ and $\mathcal{S}_N(R)$, stand for the area of the $(N-1)$-dimensional sphere of radius 1 and R, respectively. They are given by

$$\mathcal{S}_N = \frac{2\pi^{\frac{N}{2}}}{\Gamma(\frac{N}{2})}, \qquad \mathcal{S}_N(R) = \mathcal{S}_N R^{N-1} = \frac{2\pi^{\frac{N}{2}}}{\Gamma(\frac{N}{2})} R^{N-1}, \tag{3.29}$$

where $\Gamma = \Gamma(\zeta)$ is the Euler gamma function

$$\Gamma(\zeta) = \int_0^{+\infty} s^{\zeta-1}\, e^{-s}\, ds.$$

The explicit formula of $\mathcal{S}_N(R)$ for $N \geq 3$ in (3.29) is actually,

$$\mathcal{S}_N(R) = \begin{cases} \dfrac{(2\pi)^{\frac{N}{2}}}{2\cdot 4\cdots(N-2)}\, R^{N-1} & \text{for } N \text{ even}, \\[2ex] \dfrac{2\,(2\pi)^{\frac{N-1}{2}}}{1\cdot 3\cdots(N-2)}\, R^{N-1} & \text{for } N \text{ odd}. \end{cases} \tag{3.30}$$

The volume $\mathcal{V}_N(R)$ of the N-ball of radius R is given by multiplying (3.30) by R/N, that is

$$\mathcal{V}_N(R) = \begin{cases} \dfrac{(2\pi)^{\frac{N}{2}}}{2 \cdot 4 \cdots N} R^N & \text{for } N \text{ even}, \\ \dfrac{2\,(2\pi)^{\frac{N-1}{2}}}{1 \cdot 3 \cdots N} R^N & \text{for } N \text{ odd}. \end{cases} \tag{3.31}$$

Consider the Laplace problem,

$$\Delta u = 0 \quad \text{in } \mathbb{R}^N \quad \text{for } N \geq 2. \tag{3.32}$$

We are now able to give the following definition:

Definition 3.8. *The fundamental (or elementary) solution of* (3.32) *is the function E defined on $\mathbb{R}^N \setminus \{0\}$ by*

$$E(x) = \begin{cases} -\dfrac{1}{2\pi} \log \dfrac{1}{|x|} = \dfrac{1}{2\pi} \log |x| & \text{for } N = 2, \\ \dfrac{1}{(N-2)\mathcal{S}_N |x|^{N-2}} & \text{for } N \geq 3, \end{cases} \tag{3.33}$$

with $\mathcal{S}_N$ given by (3.29).

A simple computation in (3.33) shows that

$$\begin{aligned} \Delta_x E(x-y) &= 0 \qquad \text{for every } y \neq x, \\ \Delta_y E(x-y) &= 0 \qquad \text{for every } x \neq y. \end{aligned}$$

Observe also that by definition, E belongs to $C^\infty(\mathbb{R}^N \setminus \{0\})$.

Theorem 3.9 (Representation Theorem). *Let Ω be a bounded open set in $\mathbb{R}^N$ for $N \geq 2$, with a boundary of class C^1 and let $v \in C^2(\overline{\Omega})$. Then the value of the function v at a point x_0 in Ω is given by*

$$\begin{aligned} v(x_0) = \int_{\partial\Omega} \Big[v(x) \frac{\partial E}{\partial n}(x - x_0) - E(x - x_0) \frac{\partial v(x)}{\partial n} \Big]\, d\sigma \\ + \int_{\Omega} \Delta v(x)\, E(x - x_0)\, dx. \end{aligned} \tag{3.34}$$

Proof. Let $B(x_0, \varepsilon)$ be the ball centered at the point $x_0 \in \Omega$ and of radius ε sufficiently small in order to have $\overline{B}(x_0, \varepsilon) \subset \Omega$. We now apply the third Green formula (3.28) to the functions v (as ϕ) and $E(\cdot - x_0)$ (as ψ), in the open set

$$\Omega_\varepsilon = \Omega \setminus B(x_0, \varepsilon) \quad \text{with boundary} \quad \partial\Omega_\varepsilon = \partial\Omega \cup \partial B(x_0, \varepsilon).$$

With this choice, from (3.28) we get

$$-\int_{\Omega_\varepsilon} E(x-x_0)\,\Delta v(x)\,dx = \int_{\partial\Omega\cup\partial B(x_0,\varepsilon)} \left[v(x)\,\frac{\partial E}{\partial n}(x-x_0) - E(x-x_0)\,\frac{\partial v(x)}{\partial n}\right] d\sigma. \tag{3.35}$$

From (3.29) and (3.30), it is easily seen that for $x \in \partial B(x_0,\varepsilon)$ and $N \geq 3$,

$$E(x-x_0) = \frac{1}{(N-2)\mathcal{S}_N|\varepsilon|^{N-2}},$$
$$\frac{\partial E}{\partial n}(x-x_0) = -\frac{1}{\mathcal{S}_N|\varepsilon|^{N-1}}.$$

Using these values in (3.35) yields

$$\begin{aligned} -\int_{\Omega_\varepsilon} E(x-x_0)\,\Delta v(x)\,dx &= \int_{\partial\Omega} \left[v(x)\,\frac{\partial E}{\partial n}(x-x_0) - E(x-x_0)\,\frac{\partial v(x)}{\partial n}\right] d\sigma \\ &\quad + \frac{1}{(N-2)\mathcal{S}_N|\varepsilon|^{N-2}} \int_{\partial B(x_0,\varepsilon)} \frac{\partial v(x)}{\partial n}\,d\sigma \\ &\quad - \frac{1}{\mathcal{S}_N|\varepsilon|^{N-1}} \int_{\partial B(x_0,\varepsilon)} v\,d\sigma, \end{aligned} \tag{3.36}$$

where we pass to the limit as $\varepsilon \to 0$. This is straightforward for the volume integral in the left-hand side of (3.36) and the result is the last integral in formula (3.34).

As for the surface integrals, observe that by (3.29)

$$|\partial B(x,\varepsilon)| = \mathcal{S}_N|\varepsilon|^{N-1}. \tag{3.37}$$

Therefore, by the averaging (intermediate) properties of the integral, there is some x_ε in $\partial B(x_0,\varepsilon)$ such that

$$I_1 = \frac{1}{\mathcal{S}_N|\varepsilon|^{N-1}} \int_{\partial B(x_0,\varepsilon)} v\,d\sigma = \frac{1}{|\partial B(x,\varepsilon)|} \int_{\partial B(x_0,\varepsilon)} v\,d\sigma = v(x_\varepsilon).$$

Similarly, there is some $\bar{x}_\varepsilon$ in $\partial B(x_0,\varepsilon)$ such that

$$I_2 = \frac{1}{(N-2)\mathcal{S}_N|\varepsilon|^{N-2}} \int_{\partial B(x_0,\varepsilon)} \frac{\partial v(x)}{\partial n}\,d\sigma = \frac{\varepsilon}{N-2}\,\frac{\partial v}{\partial n}(\bar{x}_\varepsilon).$$

Passing to the limit in these expressions, we obtain

$$\lim_{\varepsilon\to 0} I_1 = v(x_0) \quad \text{and} \quad \lim_{\varepsilon\to 0} I_2 = 0$$

which, used in (3.36), give the result for $N \geq 3$. For $N = 2$, we make use of the same arguments but with E defined by (3.33) for this case. ☐

Corollary 3.10. *Let Ω be a bounded open set in $\mathbb{R}^N (N \geq 2)$ with a boundary of class C^1. If $u \in C^2(\overline{\Omega})$ and is harmonic in Ω, then*

$$u(x_0) = \int_{\partial\Omega} \left[u(x) \frac{\partial E}{\partial n}(x - x_0) - E(x - x_0) \frac{\partial v(x)}{\partial n} \right] d\sigma,$$

for any x_0 in Ω.

This result says that a harmonic function in Ω is completely determined by its values and the values of its normal derivatives on the boundary $\partial\Omega$.

Another consequence of the Green representation theorem is the following important result in the theory of harmonic functions:

Theorem 3.11 (Average Theorem for Harmonic Functions). *Let $B(x_0, \delta)$ be the ball of radius δ and centered at a point x_0 in Ω. If $u \in C^2(\Omega)$ is a harmonic function in Ω, then*

$$\begin{aligned} u(x_0) &= \frac{1}{|B(x_0, \delta)|} \int_{B(x_0,\delta)} u(x)\, dx, \\ u(x_0) &= \frac{1}{|\partial B(x_0, \delta)|} \int_{\partial B(x_0,\delta)} u(x)\, d\sigma, \end{aligned} \tag{3.38}$$

for any $\delta > 0$ such that $\overline{B}(x_0, \delta) \subset \Omega$.

Remark 3.12. *The average formula (3.38) shows that a harmonic function in an open set is of class C^∞ (so it has continuous derivatives of any order).*

Remark 3.13. *Using* (3.30) *and* (3.31) *in formulas* (3.38), *yields the following nice representations for $u(x_0)$:*

$$u(x_0) = \frac{1}{\pi\delta^2} \int_{B(x_0,\delta)} u(x)\, dx, \qquad u(x_0) = \frac{1}{2\pi\delta} \int_{\partial B(x_0,\delta)} u(x)\, d\sigma,$$

for $N = 2$ and

$$u(x_0) = \frac{3}{4\pi\delta^3} \int_{B(x_0,\delta)} u(x)\, dx, \qquad u(x_0) = \frac{1}{4\pi\delta^2} \int_{\partial B(x_0,\delta)} u(x)\, d\sigma,$$

for $N = 3$.

One can also prove a converse result to the Average Theorem 3.11. This will imply that the average property (3.38) can be taken as an alternate definition of a harmonic function.

Theorem 3.14. *Let Ω be a bounded open set in $\mathbb{R}^N$. If u is a continuous function satisfying (3.38) in Ω, then u is a harmonic function.*

From Theorem 3.11, one derives one of the most important results for the Laplace equation, the so-called maximum principle. It implies that the Dirichlet problem is well-posed, so its solution if it exists, is unique.

Theorem 3.15 (Strong Maximum Principle). *Let Ω be a connected bounded open set in $\mathbb{R}^N$ and $u \in C^2(\Omega) \cap C^0(\overline{\Omega})$ a harmonic function in Ω. If u attains its maximal or minimal value in Ω, u is constant in Ω.*

Consequently, u attains its maximal and minimal values on the boundary $\partial\Omega$, that is, for any $x \in \Omega$,

$$\min_{y \in \partial\Omega} u(y) \leq u(x) \leq \max_{y \in \partial\Omega} u(y).$$

Proof. Suppose that there exists a point $x_0 \in \Omega$ such that

$$u(x_0) = \max_{\overline{\Omega}} u = M,$$

so that the set $B = \{x \in \Omega \,|\, u(x) = M\}$ is not empty. Since $u - M$ is also harmonic in Ω, from Theorem 3.11 it follows that

$$0 = u(x_0) - M = \frac{1}{|B(x_0, \delta)|} \int_{B(x_0,\delta)} \big(u(x) - M\big)\, dx \leq 0,$$

with δ sufficiently small in order to have $B(x_0, \delta) \subset \Omega$. Consequently,

$$u - M \equiv 0 \quad \text{in } B(x_0, \delta).$$

Thus B is open in Ω. On the other hand, as u is a continuous function, B is also relatively closed in Ω. But Ω is connected, therefore $B = \Omega$. This implies that u is constant in Ω. The last assertion is obvious, since $u \in C^0(\overline{\Omega})$. □

3.5 Green functions and the Laplace equation

Let Ω be a bounded open set in $\mathbb{R}^N$ with a sufficiently smooth boundary $\partial\Omega$. Suppose we are given a continuous function g defined on $\partial\Omega$. We are searching for a continuous function u defined on $\overline{\Omega}$, harmonic in Ω and such that u is equal to g on $\partial\Omega$, that is, satisfying

$$\begin{cases} \Delta u = 0 & \text{in } \Omega, \\ u = g & \text{on } \partial\Omega, \end{cases} \tag{3.39}$$

which is a nonhomogeneous Dirichlet problem for the Laplace equation.

Example 3.16. For a given constant c, consider the following nonhomogeneous Dirichlet problem:

$$\begin{cases} \Delta u = 0 & \text{in } \Omega, \\ u = c & \text{on } \partial\Omega. \end{cases}$$

It is clear that $u = c$ is a solution of this problem. As we shall see later, this solution is unique. ◇

Consider the equation $\Delta u = 0$ in (3.39). If instead of zero, we are given a continuous function f defined on Ω, we have to solve what is called the nonhomogeneous Dirichlet problem for the Poisson equation

$$\begin{cases} \Delta u = f & \text{in } \Omega, \\ u = g & \text{on } \partial\Omega. \end{cases} \tag{3.40}$$

We are also interested in the nonhomogeneous Neumann problem for the Laplace equation which consists of searching for a continuous function u defined on Ω and satisfying

$$\begin{cases} \Delta u = 0 & \text{in } \Omega, \\ \dfrac{\partial u}{\partial n} = g & \text{on } \partial\Omega. \end{cases}$$

Most properties of the solutions of the Laplace and Poisson equations can be generalized for elliptic equations of the general form (2.25). The peculiarity of the Laplace equation is that one can give explicit solutions by using the Green formulas.

From what we have seen above, it is clear that the existence of a solution of the Dirichlet problem will depend on the geometry of the set Ω. The following example, due to Lebesgue, is a counterexample for the existence of continuous solutions for the Laplace equation.

Example 3.17. Suppose we are given a ball in $\mathbb{R}^N$ with a deformable surface. Let P be an arbitrary point on this surface and push it to the interior of the ball (see Figure 2).

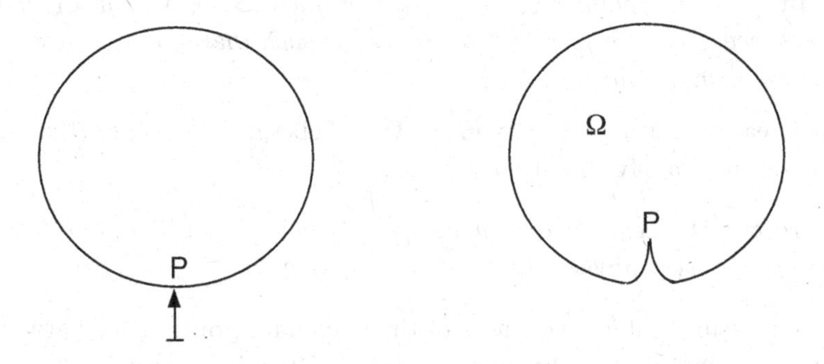

Figure 2. A cusp point

In a neighborhood of this point, we have a conical surface obtained by rotating the curve,

$$x_2 = \begin{cases} e^{-1/x_1} & \text{if } x_1 > 0, \\ 0 & \text{if } x_1 < 0. \end{cases}$$

Lebesgue proved that the Dirichlet problem does not admit a continuous solution in this particular domain Ω in the given figure (Figure 2). Physical reasons can justify this result. Indeed, suppose that we are looking for the temperature u of an isotropic homogeneous body occupying Ω. Assume that on the boundary $\partial\Omega$ we are given a continuous temperature g, which is zero in a neighborhood $\mathcal{V}$ of P and equal to a positive constant C_0 sufficiently large outside $\mathcal{V}$. From the properties listed above, it follows that u has to be close to C_0 at any point in Ω. However, this would imply that $u(x)$ for $x \in \Omega$ cannot approach 0 as x approaches P. One can say that in some sense, there is no sufficient surface around the cusp P in order to maintain the temperature close to zero in any neighborhood of it. $\diamond$

We are concerned here only with "smooth" bounded open sets (in particular, those not containing cusp points such as P in Example 3.17 above). To define such sets, let us consider all the conic surfaces in $\mathbb{R}^N$ obtained by rotating the curves

$$x_k = x_1^p, \qquad x_1 \geq 0 \quad \text{for } p \geq 1, \quad k = 2, \ldots, N,$$

around the x_1-axis. These surfaces delimit "conic" volumes $\mathcal{C}_k$, more and more pointed as p increases (if $N = 2$, one has to replace the curves by segments of straight lines).

Assume now that the Ω satisfies the following condition:

Condition $\mathcal{C}$. *If a point x of Ω is a peak of some conical volume $\mathcal{C}_k$ and if the distance $d(x,y)$ from x to y for $y \in \mathcal{C}_k$ is such that $d(x,y) \leq \rho$ with ρ a positive number, then $y \notin \Omega$.*

It is easily seen that the open set Ω in Lebesgue's example (Example 3.17) does not satisfy condition $\mathcal{C}$.

Theorem 3.18. *Let Ω be a bounded open set in $\mathbb{R}^N$ satisfying condition $\mathcal{C}$. Then the Dirichlet problem* (3.39) *has a solution.*

The next results, all consequences of the maximum principle for harmonic functions, show that the Dirichlet problem (3.39) is well-posed in the sense of Definition 1.10.

Corollary 3.19 (Uniqueness). *The Dirichlet problem (3.39) has a unique solution.*

Proof. If u_1 and u_2 are two solutions of (3.39), their difference w is a harmonic function and $w = 0$ on $\partial\Omega$. The maximum principle (Theorem 3.15) implies that $w \equiv 0$ in $\overline{\Omega}$ and so, $u_1 \equiv u_2$. □

Corollary 3.20 (Continuous Dependence on Data). *Let g_1 and g_2 be two continuous functions defined on $\partial\Omega$. Let u_1 be the solution of (3.39) with $g = g_1$, and u_2 the solution corresponding to $g = g_2$. If*

$$|g_1(x) - g_2(x)| < \varepsilon, \quad \forall x \in \partial\Omega,$$

then

$$|u_1(x) - u_2(x)| \leq \varepsilon, \quad \forall x \in \Omega. \tag{3.41}$$

Proof. The difference $w = u_1 - u_2$ is a harmonic function and $w = g_1 - g_2$ on $\partial\Omega$. Again, the maximum principle gives the result. □

We would like to use the representation formula (3.34) in order to solve some problems, as for instance, problem (3.39) or problem (3.40). However, this formula has a "defect": it contains the values of the normal derivatives of v on the boundary $\partial\Omega$, but in Dirichlet-type problems, we are given only the values of u on the boundary. To get rid of this difficulty, let us introduce, for any fixed x in Ω, the harmonic function $h = h(y)$ which is a solution of the following Dirichlet problem:

$$\begin{cases} \dfrac{\partial^2 h}{\partial y_1^2} + \dfrac{\partial^2 h}{\partial y_2^2} = \Delta_y h = 0 & \text{in } \Omega, \\ h(y) = E(x-y) & \text{for } y \in \partial\Omega. \end{cases} \tag{3.42}$$

Since h also depends on x, it will be denoted $h(x, y)$.

Applying the third Green formula (3.28) to v and h gives

$$0 = \int_{\partial\Omega} \left(E(x-y)\frac{\partial u}{\partial n}(y) - u(y)\frac{\partial h}{\partial n}(x,y) \right) d\sigma_y,$$

which, added to (3.34) leads to the formula

$$v(x) = \int_{\partial\Omega} v(y)\frac{\partial}{\partial n}\left[E(x-y) - h(x,y)\right] d\sigma_y.$$

The function

$$G(x,y) = -E(x-y) + h(x,y), \quad \text{for } x, y \in \overline{\Omega},\ x \neq y,$$

is called the Green function of Ω. With this definition, one has another integral representation for the function v in $C^2(\overline{\Omega})$, namely

$$v(x) = -\int_{\partial\Omega} v(y)\,\frac{\partial G}{\partial n}(x,y)\, d\sigma_y.$$

3.5.1 *Properties of Green functions and Poisson integral*

In this section we give some properties of the Green functions. First of all, let us mention that under some conditions on the geometry of Ω, the solution h of (3.42) exists, so that the Green function is well-defined. Moreover, it can be given explicitly for the three-dimensional case, as can be seen below.

Notice that by definition, $G(x, y) = 0$ if $y \in \partial\Omega$ and, as a function of y, G is harmonic in Ω except for $x = y$. If $y \to x$, then $G(y, x) \to +\infty$ as the logarithm of the inverse of the distance between x and y. One can also easily check that $G(x, y) = G(y, x)$.

What is of great interest about the Green function is actually the fact that the following function:

$$u(x) = -\int_{\partial\Omega} g(y) \frac{\partial G(x, y)}{\partial n}\, d\sigma_y, \tag{3.43}$$

is a solution of the Dirichlet problem

$$\begin{cases} \Delta u = 0 & \text{if } x \in \Omega, \\ u = g & \text{if } x \in \partial\Omega. \end{cases} \tag{3.44}$$

By the maximum principle, this solution is unique.

For some particular sets $\Omega \subset \mathbb{R}^3$, for instance balls, one can furnish explicit formulas for the Green function. Let $B(0, R)$ be the ball in $\mathbb{R}^3$ with center 0 and radius R. The function $1/|x - y|$ is harmonic in $B(0, R)$ with a pole at $x = y$. By taking the inverse with respect to $B(0, R)$, one gets the function

$$F(y, x) = \frac{1}{\left|\frac{R}{|y|} y - \frac{|y|}{R} x\right|},$$

which is harmonic and moreover,

$$F(y, x) = \frac{1}{|x - y|} \quad \text{if } y \in B(0, R).$$

(Recall that two points X and Y are inverse with respect to $B(0, R)$ if $|X||Y| = R^2$.)

Observe now that

$$g(y, x) = -\frac{1}{4\pi} F(y, x),$$

is the solution of (3.42).

Let α be the angle formed by the vectors $\vec{0x}$ and $\vec{0y}$, and set

$$\rho = |y - x| = \Big(|y|^2 + |x|^2 - 2|y||x|\cos\alpha\Big)^{1/2},$$

$$\rho' = \Big|R|y|y - |y|Rx\Big| = \Big(R^2 + |y|^2R^2|x|^2 - 2|y||x|\cos\alpha\Big)^{1/2}.$$

The Green function for the ball $B(0,R)$ is simply

$$G(x,y) = \frac{1}{4\pi}\left(\frac{1}{\rho} - \frac{1}{\rho'}\right).$$

Substitute this value in (3.43). Since the normal direction to $B(0,R)$ is the radial direction, a simple computation leads to the following expression of u for every $x \in B(0,R)$:

$$u(x) = \frac{1}{4\pi R}\int_{\partial B(0,R)} \frac{(R^2 - |x|^2)}{\big(R^2 + |x|^2 - 2R|x|\cos\alpha\big)^{3/2}} f(y)\, d\sigma_y. \tag{3.45}$$

Define $\mathcal{P}$, called Poisson kernel, as

$$\mathcal{P}(y,x) = \frac{1}{4\pi R}\frac{(R^2 - |x|^2)}{\big(R^2 + |x|^2 - 2R|x|\cos\alpha\big)^{3/2}}.$$

Then (3.45) can be expressed as follows:

$$u(x) = \int_{\partial B(0,R)} \mathcal{P}(y,x)\, f(y)\, d\sigma_y \quad \text{for } x \in B(0,R),$$

formula that is known as the Poisson integral.

3.5.2 *The Neumann problem in the unit disc in* $\mathbb{R}^2$

We end this section with a brief discussion on the nonhomogeneous Neumann problem for the Laplace equation already mentioned at the beginning of Section 3.5, namely

$$\begin{cases} \Delta u = 0 & \text{in } \Omega, \\ \dfrac{\partial u}{\partial n} = g & \text{on } \partial\Omega. \end{cases} \tag{3.46}$$

Notice first that if u is a solution of this equation, $u + C$ is still a solution for an arbitrary constant C. Moreover, one can immediately see that if u is a solution, then

$$\int_{\partial\Omega} g\, d\sigma = 0. \tag{3.47}$$

This is called *the compatibility condition* for equation (3.46). One important result is that condition (3.47) is not only necessary, but it is also sufficient for the existence of a solution of the Neumann interior (or even exterior) problem for the Laplace equation.

To solve (3.46) in some special domains, one can use some of the techniques that were described in the preceding sections. For the particular case when Ω is the unit disc, one may apply, as for the Dirichlet problem, the method of Fourier series. It can be shown that

$$u(r,\theta) = \frac{A_0}{2} + \sum_{n=1}^{\infty} r^n \Big(A_n \cos(n\theta) + B_n \sin(n\theta) \Big),$$

where the coefficients A_n, B_n are given by

$$A_n = \frac{1}{n\pi} \int_{-\pi}^{\pi} f(\phi) \cos(n\phi)\, d\phi, \quad n = 1, 2, \ldots,$$

$$B_n = \frac{1}{n\pi} \int_{-\pi}^{\pi} f(\phi) \sin(n\phi)\, d\phi, \quad n = 1, 2, \ldots,$$

is the required solution of the Neumann problem in the unit disc.

3.6 General elliptic equations

Let us now consider the general case of elliptic equations of the form (2.37), that is,

$$Lu = -\sum_{i,j=1}^{N} a_{ij} \frac{\partial^2 u}{\partial x_i \partial x_j} + \sum_{i=1}^{N} b_i \frac{\partial u}{\partial x_i} + cu = f, \tag{3.48}$$

where for all $i, j \in \{1, 2, \ldots, N\}$, the functions a_{ij}, b_i and c are in $C^0(\overline{\Omega})$, the matrix $(a_{ij})_{1 \leq i,j \leq N}$ being positive definite according to Definitions 2.25 and 2.26. Most of the results given for the Laplace equation can be generalized to (3.48). Here, as traditionally done, we chose to put the sign $-$ in the first term of L.

In particular, the maximum principle states that if $c(x) \leq 0$, $f(x) \leq 0$ a.e. for x in Ω, and if u is a solution of class $C^2(\Omega) \cap C^0(\overline{\Omega})$ with $\Omega \subset \mathbb{R}^N$ bounded, then u reaches its negative minimum on $\partial\Omega$. If $c(x) \leq 0$ and $f(x) \geq 0$ a.e. for x in Ω, then u reaches a positive maximum on the boundary as well. Consequently, if $c \leq 0$ one has uniqueness of the solution.

We now turn to the following general Dirichlet problem:

$$\begin{cases} Lu = f & \text{in } \Omega, \\ u = g & \text{on } \partial\Omega. \end{cases} \tag{3.49}$$

The uniqueness of the solution of this problem is straightforward from the maximum principle, but the question of the existence of a solution is a delicate topic (since in particular, it strongly depends on the choice of functional spaces where one hopes to find solutions). The following example (see for details [40], Chapter 3) shows that for the homogeneous Dirichlet problem for the Laplace equation, requiring f to belong to the space $C^0(\overline{\Omega})$ is not sufficient to ensure the existence of solutions.

Example 3.21. Let $B(0,1/2)$ be the disk in $\mathbb{R}^2$ centered at the origin of radius $1/2$ and consider the problem

$$\begin{cases} -\Delta u = f & \text{in } B(0,1/2), \\ \quad u = 0 & \text{on } \partial B(0,1/2), \end{cases}$$

where

$$f(x) = \begin{cases} \dfrac{x_1^2 - x_2^2}{2|x|^2} \left[\dfrac{4}{(-\ln|x|)^{1/2}} + \dfrac{1}{2(-\ln|x|)^{3/2}} \right] & \text{for } x \in B(0,1/2) \setminus \{0\}, \\ 0 & \text{for } x = 0. \end{cases}$$

One can easily check that the solution of this Dirichlet problem is explicitly given by

$$u(x) = \begin{cases} (x_1^2 - x_2^2)\big(-\ln|x|\big)^{1/2} & \text{for } x \in B(0,1/2) \setminus \{0\}, \\ 0 & \text{for } x = 0. \end{cases}$$

Observe now that the function f belongs to $C^0\big(\overline{B}(0,1/2)\big)$. Nevertheless, the solution u does not belong to $C^2\big(\overline{B}(0,1/2)\big)$. Indeed, a simple computation shows that the derivative $\dfrac{\partial^2 u}{\partial x_1 \partial x_2}$ is not even bounded in any neighborhood of the origin. ◇

We now state an existence and uniqueness theorem for problem (3.49) in an appropriate space, under some additional assumptions on the coefficients in (3.48), on the data F and G, and on the open set Ω and its boundary. To formulate this general theorem, some definitions and notations have to be introduced.

Definition 3.22. *Let $\alpha \in (0,1]$ and $\mathcal{O}$ be an open set in $\mathbb{R}^N$. A function f is said to be Hölder continuous with exponent α on $\overline{\mathcal{O}}$ if there is a constant c such that*

$$\big|f(x) - f(y)\big| \leq c\,|x-y|^\alpha, \quad \forall x, y \in \overline{\mathcal{O}}, \tag{3.50}$$

that is, if

$$\mathcal{H}_{\alpha,\overline{\mathcal{O}}}(f) = \sup_{\overline{\mathcal{O}},x\neq y} \frac{|f(x) - f(y)|}{|x-y|^\alpha} < +\infty.$$

Obviously, a Hölder continuous function is continuous.

Remark 3.23. *If f satisfies* (3.50) *with $\alpha = 1$, one says that f is a Lipschitz function.*

Example 3.24. Let us show that for $\alpha \in (0,1]$, the function

$$f(x) = |x|^\alpha,$$

is a Hölder continuous function on $\mathbb{R}$. To do so, it suffices to prove that

$$\big||x|^\alpha - |y|^\alpha\big| \leq |x-y|^\alpha, \quad \forall x \neq 0,\ \forall y \neq 0. \tag{3.51}$$

Let $y \neq 0$ be fixed and consider the function

$$f(t) = \big(t + |y|\big)^\alpha - t^\alpha - |y|^\alpha, \quad \forall t \in \mathbb{R}_0^+.$$

As $\alpha \in (0,1]$, it is easily seen that $f'(t) \leq 0$ in $\mathbb{R}^+$, so that f is decreasing on $\mathbb{R}^+$. Since $f(0) = 0$ and f is continuous for $t = 0$, we deduce that $f(t) \leq 0$ on $\mathbb{R}_0^+$. Thus for $t = |x-y| > 0$, we have

$$|x|^\alpha \leq \big(|x-y| + |y|\big)^\alpha \leq |x-y|^\alpha + |y|^\alpha.$$

This proves (3.51), since we can exchange x and y in the computation. ◇

Recalling Definition 1.1 and notation (1.1), if Ω is a bounded open set in $\mathbb{R}^N$, we introduce the following space:

$$\mathcal{C}^{k,\alpha}(\overline{\Omega}) = \Big\{ u \in C^k(\overline{\Omega}) \ \Big|\ \forall \beta \in \mathbb{N}_0^N \text{ with } |\beta| = k,\ \mathcal{H}_{\alpha,\overline{\Omega}}(D^\beta(u)) < +\infty \Big\},$$

for $k \in \mathbb{N}_0$ and $\alpha \in (0,1]$.

It is easily seen that $\mathcal{C}^{k,\alpha}(\overline{\Omega})$ is a Banach space for the norm

$$\|u\|_{\mathcal{C}^{k,\alpha}(\overline{\Omega})} = \|u\|_{C^k(\overline{\Omega})} + \sum_{|\beta|=k} \mathcal{H}_{\alpha,\overline{\Omega}}\big(D^\beta(u)\big),$$

where $\|u\|_{C^k(\overline{\Omega})}$ is given by (1.5). We also introduce the space

$$\mathcal{C}^{k,\alpha}(\Omega) = \Big\{ u \in C^k(\Omega) \ \Big|\ u \in \mathcal{C}^{k,\alpha}(K), \text{ for every compact } K \subset \Omega \Big\}.$$

Consider now the operator L from (3.48).

Definition 3.25. *The operator L is called uniformly elliptic in $\overline{\Omega}$ if there exists a constant $a > 0$ such that*

$$\sum_{i,j=1}^N a_{ij}(x)\xi_i\xi_j \geq a|\xi|^2 \quad \textit{for } x \in \overline{\Omega},\ \forall \xi \in \mathbb{R}^N.$$

The following two general theorems are the main results for problem (3.49) when the operator L is uniformly elliptic. Their proofs are complex and technical, they can be found for example in [22], [19], and [40].

Theorem 3.26. *Let Ω be a bounded open set in $\mathbb{R}^N$ with $\partial\Omega$ of class C^1. Suppose that the operator L is uniformly elliptic for some $a > 0$ and that the coefficients of L are all in $C^{0,\alpha}(\overline{\Omega})$ for $\alpha \in (0,1]$ and that $c \geq 0$. If f is in $C^{0,\alpha}(\overline{\Omega})$ and g is in $C^0(\partial\Omega)$, then problem (3.49) has a unique solution u such that*

$$u \in C^{2,\alpha}(\Omega) \cap C^0(\overline{\Omega}).$$

A natural question now is: do the solutions furnished by this theorem depend continuously on the data f and g? In other words, is problem (3.49) well-posed in the functional framework of Theorem 3.26? There is no answer to this question. What was shown is that under additional smoothness hypotheses on the boundary $\partial\Omega$ and on g, the solution u given by Theorem 3.26 is more regular, and this ensures the continuous dependence on the data.

Theorem 3.27 (Schauder's Estimates). *Let Ω be a bounded open set in $\mathbb{R}^N$ with $\partial\Omega$ of class $C^{2,\alpha}$ for $\alpha \in (0,1]$, and suppose that the operator L and its coefficients satisfy the hypotheses of Theorem 3.26. If f is in $C^{0,\alpha}(\overline{\Omega})$ and g is in $C^{2,\alpha}(\partial\Omega)$, then the unique solution u of (3.49) belongs to the space $C^{2,\alpha}(\overline{\Omega})$.*

Furthermore, there exists a positive constant C, independent of the data f and g such that

$$\|u\|_{C^{2,\alpha}(\overline{\Omega})} \leq C\big(\|f\|_{C^{0,\alpha}(\overline{\Omega})} + \|g\|_{C^{2,\alpha}(\partial\Omega)}\big).$$

Chapter 4

Parabolic Equations

The most well-known parabolic equation is the heat equation

$$\frac{\partial u}{\partial t} - c^2 \Delta u = f \quad \text{in } \Omega,$$

where $u = u(x,t)$ is the temperature in a homogeneous body, c^2 is the thermal diffusion coefficient, and Ω is a set in $\mathbb{R}^N$. Some of the difficulties when studying this equation come from the fact that obviously, the temporal and spatial variables do not play the same role. So, the question is, in what kind of function spaces do we search for a classical solution? What is natural to expect is that for a fixed t, say in the interval of time $(0,T)$, the solution u must be of class C^2 (as in the case of elliptic equations). This leads us to introduce the more sophisticated vector-valued space $C^2(\Omega;\, C^0([0,T]))$ which precisely has the functional requirements for the solution u. We shall not treat this framework in this chapter. In a more general setting, this will be the subject of Section 10.2 with applications to parabolic equations in Sections 10.3.1 and 10.4.1.

We shall confine ourselves to the case of the heat equation in some particular domains. By applying the method of separation of variables and Fourier transforms, explicit solutions are constructed. We treat in detail the one-dimensional case and just give the outline for the N-dimensional case. Finally, we show some results for the general form of parabolic equations (2.38), that is,

$$\frac{\partial u}{\partial t} - \sum_{i,j=1}^{N} a_{ij} \frac{\partial^2 u}{\partial x_i \partial x_j} + \sum_{i=1}^{N} b_i \frac{\partial u}{\partial x_i} + cu = F.$$

For more details concerning parabolic equations in the context of classical solutions, we refer the reader to [15], [18], [30], and [39].

4.1 The one-dimensional heat equation

As a first application, let us consider the heat equation in a cylindrical homogeneous beam of length L. It is assumed that its surface is perfectly isolated so the temperature is transmitted only in the direction of the beam, hence u depends on t and x. Suppose that at the two extremities $x = 0$ and $x = L$ of the beam, the temperature is known, and that the initial temperature is a given function $\varphi = \varphi(x)$. One is led to solve the boundary value problem

$$\begin{cases} \dfrac{\partial u}{\partial t} - c^2 \dfrac{\partial^2 u}{\partial x^2} = 0 & \text{for } 0 < x < L, \ \ 0 < t < T, \\ u(0,t) = f_1(t) & \text{for } 0 \le t < T, \\ u(L,t) = f_2(t) & \text{for } 0 \le t < T, \\ u(x,0) = \varphi(x) & \text{for } 0 \le x \le L, \end{cases} \tag{4.1}$$

where T is given (it can also take the value $+\infty$).

The following result holds:

Theorem 4.1 (Maximum Principle). *Let u be a solution of problem (4.1) in the closed rectangle $\overline{R} = [0, L] \times [0, T]$. Assume that u is in $C^2(\overline{R})$. Then u attains its minimal and maximal values at $t = 0$ or on the vertical faces $x = 0$, $x = L$ of $\overline{R}$.*

For the proof of this result, we need to recall some basic notions in the theory of functions $f = f(x, y)$ depending on two variables, as is the case of the functions we deal with in this section. For the sake of simplicity, we confine ourselves to the case of the rectangle R.

Definition 4.2. *The point $(x_0, y_0) \in R$ is called a critical point for a function f if*

$$\frac{\partial f}{\partial x}(x_0, y_0) = \frac{\partial f}{\partial y}(x_0, y_0) = 0. \tag{4.2}$$

A classical result makes a connection between the critical points of f and its minimum and maximum values.

Proposition 4.3. *If the function $f : \overline{R} \to \mathbb{R}$ has a local maximum or local minimum at a point $(x_0, y_0) \in R$, then this point is a critical one for f.*

Remark 4.4. *We shall not give here the proof of this result, this being beyond the scope of the present book. Let us just mention that in the proof a key point is the fact that (x_0, y_0) is interior to R. The situation is different* *if f admits its extremum on the boundary ∂R. For example, if the point (x_0, y_0) is on one side of $\overline{R}$, the derivatives in (4.2) will not both vanish. What will vanish is only the derivative in the direction parallel to the side under consideration. To be more precise, if for example, the extremum point is on a right vertical side of $\overline{R}$ (see Figure 3 below), it is the derivative of f with respect to y in this point which vanishes, but not the derivative with respect to x.*

Let us point out that the converse of Proposition 4.3 is not true. As a matter of fact there are several "quasi"-converse results but under additional hypotheses on the second order partial derivatives of f. For the reader's convenience, let us give one of them.

Suppose that f is of class $C^2(\overline{R})$. Its Hessian matrix is given by

$$\begin{pmatrix} \dfrac{\partial^2 f}{\partial x^2} & \dfrac{\partial^2 f}{\partial x \partial y} \\ \\ \dfrac{\partial^2 f}{\partial x \partial y} & \dfrac{\partial^2 f}{\partial y^2} \end{pmatrix}$$

whose determinant D is

$$D = \frac{\partial^2 f}{\partial x^2}\frac{\partial^2 f}{\partial y^2} - \left(\frac{\partial^2 f}{\partial x \partial y}\right)^2.$$

One has the following result:

Proposition 4.5. *Let $(x_1, y_1) \in R$ be a critical point for f.*
(i) If $D < 0$, then (x_1, y_1) is not an extremum for f.
(ii) If $D > 0$ and

$$\frac{\partial^2 f}{\partial x^2}(x_1, y_1) < 0,$$

then f admits a local maximum in (x_1, y_1). If $D > 0$ and

$$\frac{\partial^2 f}{\partial x^2}(x_1, y_1) > 0,$$

then f admits a local minimum in (x_1, y_1).
(iii) If $D = 0$, no conclusion can be made.

As a consequence (in order to conclude this short overview), the following result hods true:

Proposition 4.6. *Let f be of class $C^2(\overline{R})$. If $(x_0, y_0) \in R$ is a local maximum point for f (hence a critical point), then*

$$\frac{\partial^2 f}{\partial x^2}(x_0, y_0) \leq 0, \qquad \frac{\partial^2 f}{\partial y^2}(x_0, y_0) \leq 0.$$

Proof of Theorem 4.1. We shall proceed by contradiction. Introduce for $\varepsilon > 0$, the function u_ε, defined as

$$u_\varepsilon(x,t) = u(x,t) - \varepsilon x^2,$$

where u is the solution from the statement of the theorem.

By hypothesis u is of class $C^2(\overline{R})$, obviously, the same is true for u_ε and so, it attains its local maximal and minimal values in $\overline{R}$. Suppose that u_ε admits a local maximum in an interior point $(x_0, t_0) \in R$, that is in $(0, L) \times (0, T)$. By Propositions 4.3 and 4.6, one has in particular,

$$\begin{aligned} (i) \quad & \frac{\partial u_\varepsilon}{\partial t}(x_0, t_0) = 0, \\ (ii) \quad & \frac{\partial^2 u_\varepsilon}{\partial x^2}(x_0, t_0) \leq 0. \end{aligned} \tag{4.3}$$

By the definition of u_ε, from (4.1) we derive

$$\frac{\partial u_\varepsilon}{\partial t} - c^2 \frac{\partial^2 u_\varepsilon}{\partial x^2} = \frac{\partial u}{\partial t} - c^2 \frac{\partial^2 u}{\partial x^2} - c^2 \varepsilon = -c^2 \varepsilon.$$

Compute this expression in (x_0, t_0) and use (4.3)(i), to get

$$-c^2 \frac{\partial^2 u_\varepsilon}{\partial x^2}(x_0, t_0) = -c^2 \varepsilon,$$

that is

$$\frac{\partial^2 u_\varepsilon}{\partial x^2}(x_0, t_0) > 0,$$

which contradicts (4.3)(ii).

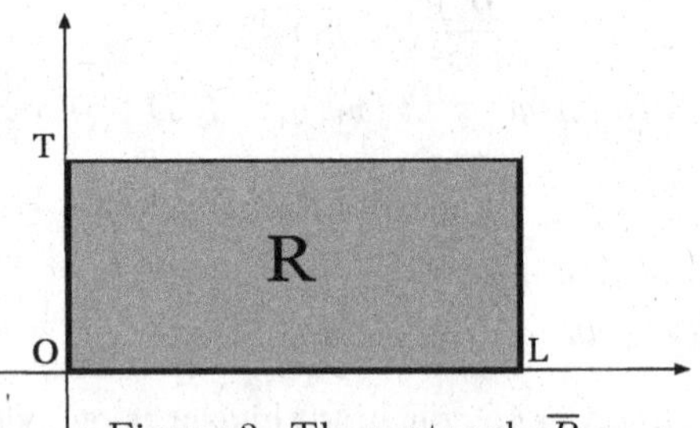

Figure 3. The rectangle $\overline{R}$

Now suppose that u_ε admits a local maximum at a point (x_0, T), belonging to the horizontal side $\{x_0 \in [0, L],, t = T\}$ of $\overline{R}$ (see Figure 3). Therefore, one must have

$$(i) \quad \frac{\partial u_\varepsilon}{\partial x}(x_0, T) = 0 \quad \text{and} \qquad (ii) \quad \frac{\partial^2 u_\varepsilon}{\partial x^2}(x_0, T) \leq 0. \tag{4.4}$$

Observe that because (x_0, T) is on the boundary of R, one does not have the equivalent of (4.3)(i), as noticed in Remark 4.4. Nevertheless, since (x_0, T) is an extremum point of u_ε, one still can obtain information on the derivative $\dfrac{\partial u_\varepsilon}{\partial t}$ at the point (x_0, T). Indeed,

$$\frac{\partial u_\varepsilon}{\partial t}(x_0, T) = \lim_{h \to 0^+} \frac{u_\varepsilon(x_0, T) - u_\varepsilon(x_0, T-h)}{h} \geq 0.$$

This, used in the equation from (4.1) gives, as before

$$\frac{\partial^2 u_\varepsilon}{\partial x^2}(x_0, T) > 0,$$

which contradicts (4.4)(ii).

It follows that the only place where u_ε can attain its maximum, is the union $\mathcal{U}$ of the three remaining sides (the thick lines in Figure 3) of $\overline{R}$,

$$\mathcal{U} = \{x_0 \in [0, L],\, t = 0\} \cup \{x = 0,\, t \in [0, T]\} \cup \{x = L \in [0, T]\}.$$

Let $M = \max\limits_{\mathcal{U}} u$, which is well-defined by hypothesis. Therefore

$$u_\varepsilon(x, t) \leq M + \varepsilon L^2, \quad (x, t) \in \overline{R},$$

so that

$$u(x, t) = u_\varepsilon(x, t) - \varepsilon\, x^2 \leq M + \varepsilon\big(L^2 - x^2\big), \quad (x, t) \in \overline{R}.$$

This inequality being true for any $\varepsilon > 0$, it follows that

$$u(x, t) \leq M, \quad (x, t) \in \overline{R},$$

and this concludes the proof of the theorem. □

As for the Laplace equation, the maximum principle implies the uniqueness of the solution and its continuous dependence on the data, that is, an inequality of the type given by (3.41). If the existence of a solution is proved, this means that problem (4.1) is well-posed in the sense of Hadamard.

Also, like in the elliptic case, the maximum principle has a physical interpretation. Let U be the maximum of the given temperature at one of the two extremities of the bar for $t = 0$ and V the minimum at the other extremity. Then the temperature at any point in the bar cannot exceed U, and cannot be less than V, which agrees with the fact that the temperature diffuses from higher to lower temperatures.

Remark 4.7. *One can show (with a much more complicated proof) that actually, the maximum of u is attained only on one of the three sides composing $\mathcal{U}$, which is a "stronger" maximum principle compared to that stated in Theorem 4.1.*

4.1.1 *The method of separation of variables*

Suppose that in (4.1) $f_1 = f_2 \equiv 0$ and that φ is continuous and vanishes at $x = 0$ and $x = L$. One looks for u of the form

$$u(x,t) = \mathcal{F}(x)\mathcal{G}(t),$$

which substituted in (4.1) gives

$$\frac{\mathcal{G}'}{c^2\mathcal{G}} = \frac{\mathcal{F}''}{\mathcal{F}} = K,$$

where K is a constant. It is easily seen that for $K > 0$, the unique possible solution is $u \equiv 0$. With $K = -\rho^2$, one has to solve two ordinary differential equations

$$\mathcal{F}'' + \rho^2\mathcal{F} = 0, \tag{4.5}$$

and

$$\mathcal{G}' + c^2\rho^2\mathcal{G} = 0. \tag{4.6}$$

The boundary conditions on u imply that $\mathcal{F}$ has to satisfy

$$\mathcal{F}(0) = 0, \quad \mathcal{F}(L) = 0.$$

The general solution of equation (4.5) being $F(x) = \alpha\cos\rho x + \beta\sin\rho x$ (where α and β are constants), it follows that

$$\mathcal{F}(0) = \alpha = 0,$$
$$\mathcal{F}(L) = \beta\sin\rho L, \quad \beta \neq 0.$$

Hence $\sin\rho L = 0$, which implies that

$$\rho = \frac{n\pi}{L} \quad \text{for} \quad n = 1, 2, \ldots,$$

and setting $\beta = 1$, one has the family of solutions

$$\mathcal{F}_n(x) = \sin\frac{n\pi x}{L}, \qquad n = 1, 2, \ldots$$

The general solution of equation (4.6), written with $\rho = n\pi/L$, is

$$\mathcal{G}_n(t) = \beta_n e^{-\lambda_n^2 t}, \quad \lambda_n = \frac{cn\pi}{L},$$

where β_n is a constant. Therefore

$$u_n(x,t) = \beta_n \sin\frac{n\pi x}{L} e^{-\lambda_n^2 t}, \quad n = 1, 2, \ldots$$

By superposition, we obtain the series

$$u(x,t) = \sum_{n=1}^{\infty} u_n(x,t). \tag{4.7}$$

The initial condition for (4.1) implies necessarily that

$$\varphi = \sum_{n=1}^{\infty} \beta_n \sin \frac{n\pi x}{L}, \tag{4.8}$$

hence β_n are the Fourier coefficients of the function φ on the interval $[0, L]$. They are given by the formulas

$$\beta_n = \frac{2}{L} \int_0^L \varphi(\tau) \sin \frac{n\pi\tau}{L}\, d\tau, \quad n = 1, 2, \ldots \tag{4.9}$$

Suppose that the function φ is twice continuously differentiable on $(0, L)$ and that its third order derivative is piecewise continuous. A known result insures the convergence of series (4.8) with coefficients (4.9). As a consequence, the series u given by (4.7) is convergent, of class C^2 with respect to $x \in [0, L]$, and continuous for $t \geq 0$.

Remark 4.8. *A direct consequence of (4.7) is the following regularity effect, a basic property of the solutions of the heat equation: the solution u is of class C^∞ with respect to the variable x for $t > 0$ (without any further regularity assumptions on the data).*

We now consider two cases, that of a finite rod with Neumann boundary conditions first, and then the case of an infinite rod.

4.1.2 *A finite rod with Neumann conditions*

For the one-dimensional case, imposing Neumann conditions at the endpoints of the bar (which physically means that they are insulated) amounts to replacing the boundary condition in (4.1) by

$$\frac{\partial u}{\partial x}(0, t) = \frac{\partial u}{\partial x}(L, t) = 0, \quad 0 \leq t < T.$$

Proceeding by separation of variables as above, one finds that the solution u is of the form

$$u(x, t) = \sum_{n=1}^{\infty} C_n \sin \frac{(2n-1)\pi}{2L} \exp\left(-\frac{(2n-1)^2\pi^2}{4L^2} t\right),$$

where

$$C_n = \frac{2}{L} \int_0^L \varphi(\tau) \sin \frac{(2n-1)\pi\tau}{2L}\, d\tau, \quad n = 1, 2, \ldots$$

As an example, one can also consider a Dirichlet–Neumann problem for a finite rod of the following form:

$$\begin{cases} \dfrac{\partial u}{\partial t} - c^2 \dfrac{\partial^2 u}{\partial x^2} = 0 & \text{for } 0 < x < L,\ \ 0 < t < +\infty, \\ u(0,t) = 0 & \text{for } 0 \leq t < T, \\ \dfrac{\partial u}{\partial x}(L,t) = 0 & \text{for } 0 \leq t < +\infty, \\ u(x,0) = \varphi(x) & \text{for } 0 \leq x \leq L, \end{cases} \tag{4.10}$$

with the initial condition $\varphi = \varphi(x)$ satisfying $\varphi(0) = 0$ (which is compatible with the Dirichlet condition required for u).

For this case, the maximum principle (for which we give a direct proof below) is formulated as follows:

Theorem 4.9 (Maximum Principle). *Assume that the initial temperature in* (4.10) *is such that* $\varphi \in C^0([0, L])$. *If* u *satisfies equation* (4.10), *then*

$$\varphi_{min} \leq u(x,t) \leq \varphi_{max} \quad \textit{for} \ \ 0 < x < L,\ \ 0 < t < +\infty, \tag{4.11}$$

where

$$\varphi_{min} = \min_{x \in [0,L]} \varphi(x), \qquad \varphi_{max} = \max_{x \in [0,L]} \varphi(x). \tag{4.12}$$

Proof. Observe first by hypothesis, φ_{min} and φ_{max} are well-defined.

We shall use in the sequel the notions of the positive part f^+ and the negative part f^- of a real function f, defined respectively, as follows:

$$\begin{aligned} f_+ = \ \ \max\big(f(x), 0\big) &= \begin{cases} f(x) & \text{if } f(x) > 0, \\ 0 & \text{if } f(x) \leq 0. \end{cases} \\ f_- = -\min\big(f(x), 0\big) &= \begin{cases} -f(x) & \text{if } f(x) < 0, \\ \ \ 0 & \text{if } f(x) \geq 0, \end{cases} \end{aligned} \tag{4.13}$$

(for more details on these notions, see for example Proposition 7.38 in Chapter 7).

We now introduce the functions U^- and U^+,

$$\begin{aligned} U^-(x,t) &= \min_{x \in [0,L]} \big(u(x,t) - \varphi_{min}, 0\big), \\ U^+(x,t) &= \max_{x \in [0,L]} \big(u(x,t) - \varphi_{max}, 0\big). \end{aligned} \tag{4.14}$$

Using definitions (4.13), we actually have

$$U^-(x,t) = \min_{x\in[0,L]} \big(u(x,t) - \varphi_{min}, 0\big) = -\big(u(x,t) - \varphi_{min}\big)_-,$$
$$U^+(x,t) = \max_{x\in[0,L]} \big(u(x,t) - \varphi_{max}, 0\big) = \big(u(x,t) - \varphi_{max}\big)_+.$$

From (4.12), it follows that $\varphi(x) - \varphi_{min} \geq 0$ and $\varphi(x) - \varphi_{max} \leq 0$ for $x \in [0, L]$. Consequently,

$$\begin{aligned} U^-(0,t) &= \min_{x\in[0,L]} \big(\varphi(x) - \varphi_{min}, 0\big) = 0 \quad \text{for } 0 < t < +\infty, \\ U^+(0,t) &= \max_{x\in[0,L]} \big(\varphi(x) - \varphi_{max}, 0\big) = 0 \quad \text{for } 0 < t < +\infty. \end{aligned} \tag{4.15}$$

Multiply the equation in (4.10) by U^+ and integrate with respect to x from 0 to L. By integration by parts, we get

$$\begin{aligned} &\int_0^L \frac{\partial u}{\partial t} U^+ \, dx + c^2 \int_0^L \frac{\partial u}{\partial x}\frac{\partial U^+}{\partial x}\, dx \\ &\qquad - c^2 \frac{\partial u}{\partial x}(L,t)\, U^+(L,t) + c^2 \frac{\partial u}{\partial x}(0,t)\, U^+(0,t) \\ &\quad = \int_0^L \frac{\partial u}{\partial t}\, U^+ \, dx + c^2 \int_0^L \frac{\partial u}{\partial x}\frac{\partial U^+}{\partial x}\, dx = 0, \end{aligned} \tag{4.16}$$

where we used the Neumann boundary condition from (4.10) as well as relations (4.15).

According again to definitions above, it is easily seen that

$$\frac{\partial U^+}{\partial t} = \begin{cases} \dfrac{\partial u}{\partial t} & \text{if } u(x,t) - \varphi_{max} > 0, \\ 0 & \text{if } u(x,t) - \varphi_{max} \leq 0, \end{cases}$$

which implies

$$\frac{\partial u}{\partial t} U^+ = \frac{1}{2}\frac{\partial |U^+|^2}{\partial t}. \tag{4.17}$$

Similarly, one also has

$$\frac{\partial U^+}{\partial x} = \begin{cases} \dfrac{\partial u}{\partial x} & \text{if } u(x,t) - \varphi_{max} > 0, \\ 0 & \text{if } u(x,t) - \varphi_{max} \leq 0, \end{cases}$$

and this yields

$$\frac{\partial u}{\partial x}\frac{\partial U^+}{\partial x} = \left|\frac{\partial U^+}{\partial x}\right|^2. \tag{4.18}$$

These last two identities (4.17) and (4.18) used in (4.16), give

$$\frac{1}{2}\int_0^L \frac{\partial |U^+|^2}{\partial t}\,dx + c^2\int_0^L \left|\frac{\partial U^+}{\partial x}\right|^2 dx = 0.$$

Integrating with respect to t from 0 to some $\tau \in (0, +\infty)$, leads to

$$\frac{1}{2}\int_0^L |U^+|^2(x,\tau)\,dx + c^2\int_0^\tau\int_0^L \left|\frac{\partial U^+}{\partial x}\right|^2 dxdt = 0,$$

which, in view of the arbitrariness of τ, obviously implies that $U^+(x,t) = 0$ for $0 < x < L,\ \ 0 < t < +\infty$. This means, considering definition (4.14), that

$$u(x,t) \leq \varphi_{max}, \quad \text{for } \ 0 < x < L,\ \ 0 < t < +\infty,$$

which is the right-hand side inequality in (4.11).

The other one is obtained by multiplying (4.10) by U^- and following along the lines of the proof above. □

4.1.3 *An infinite rod*

The problem to be solved is the following:

$$\begin{cases} \dfrac{\partial u}{\partial t} - c^2\dfrac{\partial^2 u}{\partial x^2} = 0, & -\infty < x < +\infty, \quad 0 < t < T, \\ u(x,0) = \varphi(x), & -\infty < x < +\infty, \end{cases}$$

where φ is a continuous function.

Not imposing other conditions, this problem possesses an infinite number of solutions, which possibly can be unbounded. So it is reasonable to look for solutions which are continuous and bounded for $-\infty < x < +\infty$, and $0 \leq t < T$. One can show that if such a solution exists it satisfies inequality (4.11) from Theorem 4.9, that is if

$$M = \sup_{-\infty<x<+\infty} \varphi(x) \quad \text{and} \quad m = \inf_{-\infty<x<+\infty} \varphi(x),$$

then

$$m \leq u(x,t) \leq M \quad \text{for} \quad t \in (0,T).$$

This maximum principle implies that the problem is well-posed if it has a solution. This solution can be obtained by using the method of separation of variables. As in the first case, we end up with a family of solutions of the form

$$u_\rho(x,t) = (A\cos\rho x + B\sin\rho x)e^{-c^2\rho^2 t},$$

where A and B are constants. All the series of functions of this type are periodic functions in x for $t = 0$. However, if one does not make the hypothesis that φ is periodic, this is impossible. Therefore it is natural to make use of what is called the Fourier integral instead of Fourier series.

On the other hand, since A and B are arbitrary, one can suppose that they depend on ρ. The function

$$u(x,t) = \int_0^{+\infty} u_\rho(x,t)\,d\rho = \int_0^{+\infty} (A(\rho)\cos(\rho x) + B(\rho)\sin(\rho x))\,e^{-c^2\rho^2 t}\,dx,$$

is differentiable and verifies the heat equation.

Since $u(x,0) = \varphi(x)$, we have

$$A(\rho) = \frac{1}{\pi}\int_{-\infty}^{+\infty} \varphi(\tau)\cos\rho\tau\,d\tau, \quad B(\rho) = \frac{1}{\pi}\int_{-\infty}^{+\infty} \varphi(\tau)\sin\rho\tau\,d\tau.$$

This leads to the following Fourier integral formula:

$$u(x,0) = \frac{1}{\pi}\int_0^{+\infty}\left[\int_{-\infty}^{+\infty} \varphi(\tau)\cos(\rho x - \rho\tau)\,d\tau\right]d\rho,$$

and hence,

$$\begin{aligned} u(x,t) &= \frac{1}{\pi}\int_0^{+\infty}\left[\int_{-\infty}^{+\infty} \varphi(\tau)\cos(\rho x - \rho\tau)\,e^{-c^2\rho^2 t}\,d\tau\right]d\rho \\ &= \frac{1}{\pi}\int_{-\infty}^{+\infty} \varphi(\tau)\left[\int_0^{+\infty} e^{-c^2\rho^2 t}\cos(\rho x - \rho\tau)\,d\rho\right]d\tau. \end{aligned}$$

The integral with respect to ρ can be computed explicitly in order to get the following representation for u:

$$u(x,t) = \begin{cases} \dfrac{1}{2c\sqrt{\pi t}}\displaystyle\int_{-\infty}^{+\infty} \varphi(\tau)\,\exp\left(-\frac{(x-\tau)^2}{4c^2 t}\right)d\tau & \text{if } t > 0, \\ \varphi(x) & \text{if } t = 0. \end{cases} \tag{4.19}$$

Remark 4.10. *Suppose that the initial temperature φ is zero, except on a small interval (a,b) where it is positive. Then formula (4.19) becomes*

$$u(x,t) = \frac{1}{2\sqrt{\pi t}}\int_a^b \varphi(\tau)\exp\left(-\frac{(x-\tau)^2}{4t}\right)d\tau, \quad t > 0.$$

Since φ is positive, the temperature u is positive as soon as t is positive. The physical interpretation is that the heat is transmitted with an infinite speed, which is in contradiction with the experiments. The explanation of this paradox comes from the fact that the hypotheses under which the model was obtained are not verified exactly in nature. Nevertheless, for many cases, the model (4.1) is a good approximation of the process of heat transmission.

4.2 The two and three-dimensional heat equations

We will just list here some results for the heat equation in a bounded open set Ω in $\mathbb{R}^N$. It is worthy to mention that almost all the results obtained in the preceding section, in particular the maximum principle for $N = 1$, are true for the case $N > 1$.

Let Ω be a bounded open set in $\mathbb{R}^2$ and suppose that on the boundary $\partial\Omega$ the temperature is given. We consider the Dirichlet problem,

$$\begin{cases} \dfrac{\partial u}{\partial t} - \Delta u = 0 & \text{for } x \in \Omega,\ t > 0, \\ u(x,0) = \varphi(x) & \text{for } x \in \Omega, \\ u(x,t) = 0 & \text{for } x \in \partial\Omega,\ t \geq 0. \end{cases}$$

The maximum principle is still valid and the problem has a unique solution continuously depending on the data. To solve it, one may apply the separation of variables for special geometries of the set Ω.

As an example, we consider the case where $\Omega \subset \mathbb{R}^2$ is the rectangle with sides $[0,a]$ and $[0,b]$. By the method of separation of variables, we obtain the solution

$$u(x,t) = \sum_{m,n=1}^{\infty} A_{mn} \sin\frac{m\pi x_1}{a} \sin\frac{n\pi x_2}{b} \exp\left(-\pi^2\left(\frac{m^2}{a^2} + \frac{n^2}{b^2}\right)t\right),$$

where

$$A_{mn} = \frac{4}{ab}\int_0^a \int_0^b \varphi(\tau) \sin\frac{m\pi\tau_1}{a} \sin\frac{n\pi\tau_2}{b}\, d\tau, \quad m,n = 1,2,\ldots$$

If Ω is the whole space $\mathbb{R}^3$, the problem is well-posed. In determining continuous and bounded solutions, one has an equivalent formula to (4.19),

$$u(x,t) = \begin{cases} \dfrac{1}{2\sqrt{\pi t}} \displaystyle\int_{\mathbb{R}^3} \exp\left(-\frac{\sum_{i=1}^3 (x_i - \tau_i)^2}{4t}\right) \varphi(\tau)\, d\tau & \text{for } 0 < t < T, \\ \varphi(x), & \text{for } t = 0. \end{cases}$$

4.3 The general case of parabolic equations

We end this chapter by just mentioning several results concerning the general case of second order parabolic equations (2.38).

Let $\Omega \subset \mathbb{R}^N$ be a bounded open set and consider the equation,

$$Lu = \frac{\partial u}{\partial t} - \sum_{i,j=1}^{N} a_{ij}\frac{\partial^2 u}{\partial x_i \partial x_j} + \sum_{i=1}^{N} b_i \frac{\partial u}{\partial x_i} + c(t)u = f(x,t),$$

where the functions a_{ij}, b_i and c are continuous on $[0,T]\times\Omega$ and such that

$$\sum_{i,j=1}^{N} a_{ij}(x,t)\xi_i\xi_j \geq \alpha|\xi|^2 \quad \text{a.e. in } \Omega\times(0,T),$$

with a constant $\alpha \geq 0$. Some of the results obtained above can be generalized for the equation $Lu = 0$. In particular, we have the maximum principle for this problem.

Theorem 4.11 (The Maximum Principle). *If $u \in C^0(\overline{\Omega})$ is a solution of $Lu = 0$ with $c \equiv 0$, then*

$$\min_Q u \leq u(x,t) \leq \max_Q u,$$

where

$$Q = \Big\{(x,0) \;\Big|\; x \in \partial\Omega\Big\}.$$

This result implies in particular, the uniqueness of the solution of the Cauchy problem if the coefficients of L are bounded. It also implies that the problem is well-posed.

Consider now the parabolic equation with constant coefficients

$$\begin{cases} \dfrac{\partial u}{\partial t} - \displaystyle\sum_{i,j=1}^{N} a_{ij}\frac{\partial^2 u}{\partial x_i \partial x_j} = 0 & \text{for } x \in \mathbb{R}^N,\ t > 0, \\ u(x,0) = \varphi(x) & \text{for } x \in \Omega, \\ u(x,t) = 0 & \text{for } x \in \partial\Omega,\ t \geq 0, \end{cases} \tag{4.20}$$

where the matrix $(a_{ij})_{1\leq i,j\leq n}$ is positive definite.

A Poisson formula was given for this situation to obtain the solution of the Cauchy problem for (4.20) in the form

$$u(x,t) = \frac{1}{(4\pi t)^{N/2}\sqrt{A}} \int \varphi(y)\exp\left(-\sum_{i,j=1}^{N}\frac{a^{ij}(x_i-y_i)(x_j-y_j)}{4t}\right)dy,$$

where $A = \det(a_{ij})$, and $(a^{ij})_{1\leq i,j\leq n}$ is the inverse matrix of $(a_{ij})_{1\leq i,j\leq n}$.

Under additional hypotheses on the coefficients, the above results have been generalized to the case of nonconstant coefficients in (4.20) (see for instance [15] and [39]).

Chapter 5

Hyperbolic Equations

The wave equation

$$\frac{\partial^2 u}{\partial t^2} - c^2 \Delta u = 0, \tag{5.1}$$

is the most important example of hyperbolic partial differential equations.

We have seen that each second order hyperbolic equation is reduced to a canonical form which, at least at one point, has the wave equation as principal part. This is why they inherit most properties of the wave equation.

From the physical point of view, the wave equation is of major interest in applications since it describes not only the propagation of waves, but also vibration phenomena. The small vibrations of a string are given by the one-dimensional wave equation, as we have shown in Section 1.4. The two-dimensional wave equation describes the vibrations of a membrane. The study of the sound propagation in the atmosphere leads to the wave equation in three dimensions. For further details, we refer the reader to [18], [20], and [30] and other references therein.

The general case of hyperbolic equations is treated in Section 10.4.2. A specific functional framework is introduced in Section 10.2 where the wave equation is well-posed in the sense of Definition 1.10, and where we shall prove existence and uniqueness of a solution.

In this chapter, as in Chapter 4, we confine ourselves to classical solutions. We shall construct explicit solutions for the Cauchy problem for (5.1) in the one-dimensional case and discuss the nice d'Alembert formula. We also do the same for the two-dimensional case on a rectangular membrane. Again, our main tool will be the method of separation of variables combined with Fourier expansions.

5.1 Wave propagation

As we have mentioned, equation (5.1) describes a large variety of propagation of acoustic waves in homogeneous and isotropic media. In this case, c represents the speed of the propagation. The same equation is also satisfied by the components of the electrical and magnetic fields in a vacuum, and here c is nothing else but the speed of light.

5.1.1 *Plane waves*

One can identify immediately some solutions of the wave equation. Let $\vec{\xi} = (\xi_1, \ldots, \xi_N)$ be a unit vector in $\mathbb{R}^N$, that is, $\xi_1^2 + \ldots + \xi_N^2 = 1$. For any fixed t, the equation

$$\xi_1 x_1 + \ldots + \xi_N x_N - ct = \kappa, \tag{5.2}$$

with κ an arbitrary constant, is a hyperplane in the space $\mathbb{R}^N$. The vector $\vec{\xi}$ is orthogonal to this plane. If t increases, the hyperplane moves in the direction $\vec{\xi}$ with speed c.

Now, let $F = F(y)$ be a function of class C^2 depending on only one variable. It is easily seen that

$$u(x,t) = F(\xi_1 x_1 + \ldots + \xi_N x_N - ct), \tag{5.3}$$

is a solution of (5.1) and equals $F(\kappa)$ on the hyperplanes (5.2). A solution of the form (5.3) is called a plane wave.

Consider the one-dimensional wave equation (5.1). From (5.2) and (5.3), it is clear that the "progressive" planes are $x - ct = \kappa$ and $x + ct = \kappa$, so that the plane waves are

$$u_1(x,t) = F(x - ct) \quad \text{and} \quad u_2(x,t) = G(x + ct),$$

where F and G are arbitrary functions of class C^2.

The functions of type u_1 are progressive (traveling) waves moving in the positive direction of the x-axis while that of type u_2 move in the negative direction of the x-axis, both with speed c. The traveling waves u_1 and u_2 are also called *forward* and *backward* waves, respectively.

To see that, suppose c is positive. The function u_1 is constant in the points $(x(t), t)$ characterized by

$$x(t) - ct = constant,$$

which by differentiation implies that $\dfrac{dx}{dt} = c$. This means that u_1 is a wave with constant shape traveling in the positive direction of the x-axis with speed c. Similarly, u_2 is a wave moving in the negative direction of the x-axis with speed c.

Example 5.1. For the two-dimensional wave equation, one has an infinity of unit vectors $\vec{\xi}$. The progressive plane corresponding, for example, to the vector $\vec{\xi} = (1/\sqrt{2}, 1/\sqrt{2})$, is the straight line in $\mathbb{R}^2$

$$\frac{1}{\sqrt{2}}x_1 + \frac{1}{\sqrt{2}}x_2 - ct = \kappa,$$

moving in the direction of its normal vector $(1/\sqrt{2}, 1/\sqrt{2})$ with the speed c. The corresponding plane waves are of the form

$$u(x,t) = F\Big(\frac{1}{\sqrt{2}}x_1 + \frac{1}{\sqrt{2}}x_2 - ct\Big). \qquad \diamond$$

5.1.2 *Spherical waves*

A function $u = u(r,t)$ is called a spherical wave if it satisfies the equation

$$\frac{1}{r^{N-1}}\frac{\partial}{\partial r}\Big(r^{N-1}\frac{\partial u}{\partial r}\Big) - \frac{\partial^2 u}{\partial t^2} = 0.$$

Notice the analogy with the radial harmonic functions.

The spherical waves in $\mathbb{R}^3$ are of the form

$$u_1(r,t) = \frac{F(r-t)}{r}, \qquad u_2(r,t) = \frac{G(r+t)}{r},$$

where F and G are arbitrary functions of class C^2. The waves u_1 are the backward waves, while the waves u_2 are the forward ones.

5.2 The Cauchy problem for the wave equation

Consider equation (5.1) in $\mathbb{R}^n \times \mathbb{R}$, that is to say,

$$\frac{\partial^2 u}{\partial t^2} - c^2\Delta u = 0 \quad \text{for } x \in \mathbb{R}^N, \;\; t \in \mathbb{R}. \tag{5.4}$$

The Cauchy problem (with data φ and ψ) for the wave equation (5.4) consists of searching for solutions satisfying the initial conditions

$$\begin{cases} u(x,0) = \varphi(x) & \text{for } x \in \mathbb{R}^N, \\ \dfrac{\partial u}{\partial t}(x,0) = \psi(x) & \text{for } x \in \mathbb{R}^N. \end{cases} \tag{5.5}$$

A fundamental property of the wave equation is the fact that the value of a solution at any arbitrary point depends only on its initial values in any bounded set of $\mathbb{R}^N$. It was proved that if $u = u(x,t)$ is of class $C^2(\mathbb{R}^{N+1})$ and is a solution of (5.1) for $x \in \mathbb{R}^N$ and $t \in \mathbb{R}$, then for any $x_0 \in \mathbb{R}^N$, $t_0 \in \mathbb{R}$, and $T > 0$,

$$\int_{\overline{B}(x_0,t_0-cT)} \left[\left(\frac{\partial u}{\partial x_1}\right)^2 + \ldots + \left(\frac{\partial u}{\partial x_N}\right)^2 + \left(\frac{\partial u}{\partial t}\right)^2\right]\Big|_{t=T} dx$$
$$\leq \int_{\overline{B}(x_0,t_0)} \left[\left(\frac{\partial u}{\partial x_1}\right)^2 + \ldots + \left(\frac{\partial u}{\partial x_N}\right)^2 + \left(\frac{\partial u}{\partial t}\right)^2\right]\Big|_{t=0} dx,$$

where $\overline{B}(\hat{x}, \beta)$ is the closed ball in $\mathbb{R}^N$ with center $\hat{x}$ and of radius β.

As an immediate consequence, if

$$u(x,0) = \frac{\partial u}{\partial t}(x,0) \equiv 0,$$

then $u \equiv 0$, and this implies the uniqueness of the solutions of class C^2 of the Cauchy problem (5.4)–(5.5).

5.2.1 *Conservation of the energy*

The result above is related to another fundamental property of the wave equation stated in a bounded open set, namely, the conservation of energy. Let us discuss it for the one-dimensional wave equation (for the general case, we refer the reader to [15], [39]).

Theorem 5.2. *Let $L > 0$ and $f \in C^2([0,L])$, $g \in C^1([0,L])$. Consider the following Cauchy problem for the wave equation in the interval $(0,L)$ with data f and g, that is,*

$$\begin{cases} \dfrac{\partial^2 u}{\partial t^2} = c^2 \dfrac{\partial^2 u}{\partial x^2} & \text{for } \; 0 < x < L, \; t > 0, \\ u(0,t) = u(L,t) = 0 & \text{for } \; t \geq 0, \\ u(x,0) = f(x) & \text{for } \; 0 < x < L, \\ \dfrac{\partial u}{\partial t}(x,0) = g(x) & \text{for } \; 0 < x < L. \end{cases} \tag{5.6}$$

Then the energy function $E_u = E_u(t)$ defined as

$$E_u(t) = \frac{1}{2}\int_0^L \left[\left(\frac{\partial u}{\partial t}\right)^2 + c^2\left(\frac{\partial u}{\partial x}\right)^2\right] dx \qquad \text{for } \; t \geq 0, \tag{5.7}$$

is constant. In particular,

$$E_u(t) = E_u(0) = \frac{1}{2}\int_0^L \Big((g(x))^2 + c^2(f'(x))^2\Big) dx \qquad \text{for } \; t > 0. \tag{5.8}$$

Proof. Using the wave equation from (5.6), we have

$$\begin{aligned}\frac{\partial E_u}{\partial t} &= \int_0^L \left(\frac{\partial u}{\partial t}\frac{\partial^2 u}{\partial t^2} + c^2 \frac{\partial u}{\partial x}\frac{\partial^2 u}{\partial x \partial t}\right) dx \\ &= c^2 \int_0^L \left(\frac{\partial u}{\partial t}\frac{\partial^2 u}{\partial x^2} + \frac{\partial u}{\partial x}\frac{\partial^2 u}{\partial x \partial t}\right) dx,\end{aligned}$$

whence,

$$\frac{\partial E_u}{\partial t} = c^2 \int_0^L \frac{\partial}{\partial x}\left(\frac{\partial u}{\partial t}\frac{\partial u}{\partial x}\right) dx.$$

Integrating by parts yields

$$\frac{\partial E_u}{\partial t} = c^2 \left(\frac{\partial u}{\partial t}\frac{\partial u}{\partial x}\right)\Big|_{x=0}^{x=L} = 0, \tag{5.9}$$

since the boundary conditions in (5.6) imply

$$\frac{\partial u}{\partial t}(0,t) = \frac{\partial u}{\partial t}(L,t) = 0.$$

Because of (5.9) the conservation of energy holds true, E is constant.

To end the proof, it remains to compute $E_u(0)$. By definition,

$$E_u(0) = \frac{1}{2}\int_0^L \left[\left(\frac{\partial u}{\partial t}(x,0)\right)^2 + c^2\left(\frac{\partial u}{\partial x}(x,0)\right)^2\right] dx,$$

from which (5.8) is immediate by using the initial conditions in (5.6). □

Corollary 5.3. *Under the hypotheses of Theorem 5.2, if problem* (5.6) *admits a solution, then this solution is unique.*

Proof. Let u_1 and u_2 be two solutions of (5.6). So, $U = u_1 - u_2$ is a solution of (5.6) with data $f = 0$ and $g = 0$. Consequently, the energy of U satisfies, in view of (5.7),

$$E_U(t) = \frac{1}{2}\int_0^L \left[\left(\frac{\partial U}{\partial t}\right)^2 + c^2\left(\frac{\partial U}{\partial x}\right)^2\right] dx.$$

According to (5.8), one has $E_U(t) = E_U(0) = 0$ for $t > 0$ and so

$$\left(\frac{\partial U}{\partial t}\right)^2 + c^2\left(\frac{\partial U}{\partial x}\right)^2 = 0 \quad \text{for } x \in [0,L],\ t \geq 0,$$

which implies that U is constant. Since $U(0,t) = 0$, we get the result. □

A kind of stability of the solutions of problem (5.6) with data satisfying the hypotheses from Theorem 5.2 can be proved by a similar argument.

Corollary 5.4. *Let u_1 and u_2 be two solutions of* (5.6), *corresponding to the data f_1, g_1, respectively to f_2, g_2, satisfying the hypotheses of Theorem 5.2. Assume that for $\varepsilon > 0$ small enough,*

$$\sup_{x\in\mathbb{R}} |g_1(x) - g_2(x)| \le \varepsilon, \qquad \sup_{x\in\mathbb{R}} |(f_1(x) - f_2(x))'| \le \varepsilon.$$

Then

$$E(u_1 - u_2(t)) \le 2T\varepsilon.$$

Proof. As before, $U = u_1 - u_2$ is the solution of (5.6) with data $f_1 - f_2$ and $g_1 - g_2$. In view of (5.8), the energy of U satisfies for $t > 0$,

$$\begin{aligned} E_U(t) &= \frac{1}{2}\int_0^L \left(\frac{\partial U}{\partial t}\right)^2 dx + c^2 \int_0^L \left(\frac{\partial U}{\partial x}\right)^2 dx = E_U(0) \\ &= \frac{1}{2}\int_0^L \left[(g_1(x) - g_2(x))^2 + c^2((f_1'(x) - f_2'(x))^2\right] dx. \end{aligned}$$

Integrating with respect to t from 0 to some T, and taking into account the hypotheses on the data yield

$$\begin{aligned} &\int_0^T \int_0^L \left(\frac{\partial U}{\partial t}\right)^2 dx + c^2 \int_0^T \int_0^L \left(\frac{\partial U}{\partial x}\right)^2 dx \\ &\qquad \le T\big[\sup_{x\in\mathbb{R}} |g_1(x) - g_2(x)| + \sup_{x\in\mathbb{R}} |(f_1(x) - f_2(x))'|\big] \le 2T\varepsilon, \end{aligned}$$

which concludes the proof. □

Observe that the result given in Corollary 5.4 means that small perturbations in the data results in small perturbations of the energies of the solutions. (In the same context, see also Theorem 5.9.)

As mentioned in the introduction of this chapter, a functional framework suitable in particular, for the well-posedness of equation (5.1) in the sense of Definition 1.10, is introduced in Section 10.2.

5.2.2 *Poisson formula*

Consider the wave equation in $\mathbb{R}^3$ with Cauchy data φ and ψ,

$$\begin{cases} \dfrac{\partial^2 u}{\partial t^2} - c^2 \Delta u = 0 & \text{for } x \in \mathbb{R}^3,\ t > 0, \\ u(x, 0) = \varphi(x) & \text{for } x \in \mathbb{R}^3, \\ \dfrac{\partial u}{\partial t}(x, 0) = \psi(x) & \text{for } x \in \mathbb{R}^3. \end{cases} \tag{5.10}$$

Theorem 5.6 below states that a solution of this equation is given by the so-called Poisson formula, namely, (5.19) for $N = 3$ and (5.24) for $N = 2$. The proof of this result is based on the following classical result:

Theorem 5.5. *Let $\Psi \in C^2(\mathbb{R}^3)$ and u_Ψ be the solution of (5.10) with initial data 0 and Ψ, that is,*

$$\begin{cases} \dfrac{\partial^2 u_\Psi}{\partial t^2} - c^2 \Delta u_\Psi = 0 & \text{for } x \in \mathbb{R}^3, \; t > 0, \\ u_\Psi(x,0) = 0 & \text{for } x \in \mathbb{R}^3, \\ \dfrac{\partial u_\Psi}{\partial t}(x,0) = \Psi(x) & \text{for } x \in \mathbb{R}^3. \end{cases} \tag{5.11}$$

Then the function u_Ψ given by the following formula, known as Kirchhoff formula, is a solution of (5.11):

$$u_\Psi(x,t) = \frac{1}{4\pi c^2 t} \int_{\partial B(x,ct)} \Psi(s)\, d\sigma. \tag{5.12}$$

Proof. Let $g = g(s)$ be a function in $C^0(\mathbb{R}^3)$. Define for $x \in \mathbb{R}^3$ and $h > 0$ an arbitrary constant, the function $\mathcal{T}_h(g(x))$, as

$$\mathcal{T}_h(g(x)) = \frac{1}{4\pi h^2} \int_{\partial B(x,h)} g(s)\, d\sigma. \tag{5.13}$$

We first show that $\mathcal{T}_h(g)$ satisfies the equation

$$\Delta \mathcal{T}_h(g(x)) - \frac{1}{h} \frac{\partial^2}{\partial h^2} (h \mathcal{T}_h(g(x))) = 0. \tag{5.14}$$

Since by the product rule

$$\frac{1}{h} \frac{\partial^2}{\partial h^2} (h \mathcal{T}_h(g(x))) = \frac{2}{h} \frac{\partial}{\partial h} (\mathcal{T}_h(g(x))) + \frac{\partial^2}{\partial h^2} (\mathcal{T}_h(g(x))), \tag{5.15}$$

to prove (5.14), one has to compute the first and second derivatives of $\mathcal{T}_h(g)$ with respect to h.

Observe that for any x fixed, the change of variable $s = x + hy$ leads to

$$\mathcal{T}_h(g(x)) = \frac{1}{4\pi} \int_{\partial B(0,1)} g(x + hy)\, d\sigma_y, \tag{5.16}$$

where $d\sigma_y$ is the surface element on $\partial B(0,1)$, the index y indicating that the integration variable is y.

A simple computation shows that for $k \in \{1, 2, 3\}$,

$$\begin{aligned} \frac{\partial g(x + hy)}{\partial h} &= y_k \Big(\frac{\partial g}{\partial s_k} \Big)(x + hy), \\ \frac{\partial g(x + hy)}{\partial y_k} &= h \Big(\frac{\partial g}{\partial s_k} \Big)(x + hy). \end{aligned} \tag{5.17}$$

Using this formula, as well as (1.7) from Section 1.2, and then setting $G(y) = g(x + hy)$ in (5.16), yield

$$\frac{\partial}{\partial h}(\mathcal{T}_h(g(x))) = \frac{1}{4\pi}\int_{\partial B(0,1)} \sum_{k=1}^{3} y_k \frac{\partial g}{\partial s_k}(x+hy)\, d\sigma_y$$
$$= \frac{1}{4\pi h}\int_{\partial B(0,1)} \frac{\partial G}{\partial n}(x+hy)\, d\sigma_y,$$

where we can apply the first Green formula (3.26) to get

$$\frac{\partial}{\partial h}(\mathcal{T}_h(g(x))) = \frac{1}{4\pi h}\int_{B(0,1)} \Delta_y g(x+hy)\, dy.$$

Using (5.17) again, gives

$$\frac{\partial}{\partial h}(\mathcal{T}_h(g(x))) = \frac{h}{4\pi}\int_{B(0,1)} (\Delta g)(x+hy)\, dy. \tag{5.18}$$

As a consequence

$$\frac{\partial^2}{\partial h^2}(\mathcal{T}_h(g(x))) = \frac{1}{4\pi}\int_{B(0,1)} (\Delta g)(x+hy)\, dy$$
$$+ \frac{1}{4\pi}\int_{B(0,1)} \sum_{k=1}^{3} y_k \Big(\frac{\partial(\Delta g)}{\partial s_k}\Big)(x+hy)\, dy.$$

Integrating by parts in the last integral, one has

$$\frac{\partial^2}{\partial h^2}(\mathcal{T}_h(g(x))) = \frac{1}{4\pi}\int_{\partial B(0,1)} (\Delta g)(x+hy)\, d\sigma_y - \frac{1}{2\pi}\int_{B(0,1)} (\Delta g)(x+hy)\, dy.$$

Replacing this expression and (5.18) into (5.15) yields

$$\frac{1}{h}\frac{\partial^2}{\partial h^2}(h\mathcal{T}_h(g(x))) = \frac{1}{4\pi}\int_{\partial B(0,1)} (\Delta g)(x+hy)\, d\sigma_y,$$

which, recalling definition (5.13) and relation (5.18), gives (5.14).

Now choose in Definition 5.13, ct in place of h and Ψ in place of g, and introduce the notation

$$u_\Psi(x,t) = t\, \mathcal{T}_{ct}(\Psi(x)),$$

which, as a matter of fact, is nothing else than formula (5.12). Moreover, writing (5.14) for $\mathcal{T}_{ct}(\Psi(x))$, it is easily seen that u_Ψ satisfies the wave equation from (5.11). By definition, $u_\Psi(x,0) = 0$. Moreover,

$$u'_\Psi(x,0) = \mathcal{T}_0(\Psi(x)) = \Psi(x),$$

and this ends the proof of Theorem 5.5. □

We can now state the existence result for (5.10).

Theorem 5.6. *Assume that* φ *is in* $C^3(\mathbb{R}^3)$ *and* ψ *is in* $C^2(\mathbb{R}^3)$. *Then the solution of* (5.10) *is given by the following formula, known as Poisson formula:*

$$u(x,t) = \frac{1}{4\pi c^2 t}\frac{\partial}{\partial t}\left[\int_{\partial B(x,ct)} \varphi(s)\, d\sigma_t\right] + \frac{1}{4\pi c^2 t}\int_{\partial B(x,ct)} \psi(s)\, d\sigma_t. \quad (5.19)$$

Proof. Let u_ψ be the solution of (5.10) with initial data 0 and ψ. The Kirchhoff formula (5.12) for u_ψ reads

$$u_\psi(x,t) = \frac{1}{4\pi c^2 t}\int_{\partial B(x,ct)} \psi(s)\, d\sigma_t. \quad (5.20)$$

Now, let u_φ be the solution of (5.10) corresponding to the initial data 0 and φ, that is,

$$\begin{cases} \dfrac{\partial^2 u_\varphi}{\partial t^2} - c^2\Delta u_\varphi = 0 & \text{for } x \in \mathbb{R}^3,\ t > 0, \\ u_\varphi(x,0) = 0 & \text{for } x \in \mathbb{R}^3, \\ \dfrac{\partial u_\varphi}{\partial t}(x,0) = \varphi(x) & \text{for } x \in \mathbb{R}^3, \end{cases} \quad (5.21)$$

whose solution is

$$u_\varphi(x,t) = \frac{1}{4\pi c^2 t}\int_{\partial B(x,ct)} \varphi(s)\, d\sigma_t. \quad (5.22)$$

It is easily seen from (5.21) that the derivative $U_\varphi = \dfrac{\partial u_\varphi}{\partial t}$ verifies

$$\frac{\partial U_\varphi}{\partial t}(x,0) = \frac{\partial^2 u_\varphi}{\partial t^2}(x,0) = c^2\Delta u_\varphi(x,0) = 0,$$

and so, for φ in $C^3(\mathbb{R}^3)$, U_φ is a solution of

$$\begin{cases} \dfrac{\partial^2 U_\varphi}{\partial t^2} - c^2\Delta U_\varphi = 0 & \text{for } x \in \mathbb{R}^3,\ t > 0, \\ U_\varphi(x,0) = \varphi(x) & \text{for } x \in \mathbb{R}^3, \\ \dfrac{\partial U_\varphi}{\partial t}(x,0) = 0 & \text{for } x \in \mathbb{R}^3. \end{cases} \quad (5.23)$$

By the superposition principle, the function

$$u = u_\psi + U_\varphi = u_\psi + \frac{\partial u_\varphi}{\partial t},$$

is a solution of (5.10) satisfying the initial conditions (5.5).

As a consequence, using (5.12) and (5.22), the solution of the Cauchy problem for (5.10) in $\mathbb{R}^3$, with $\varphi \in C^3(\mathbb{R}^3)$ and $\psi \in C^2(\mathbb{R}^3)$, is given by the Poisson formula (5.19). □

Using the same argument, one can prove that the solution in $\mathbb{R}^2$ is given by the following formula:

$$
\begin{aligned}
u(x,t) = \frac{1}{2\pi c}\frac{\partial}{\partial t}\left[\int_{\overline{B}(x,ct)} \frac{\psi(x')}{\sqrt{t^2 - d^2(x',x)}}\,dx'\right] \\
+ \frac{1}{2\pi c}\int_{\overline{B}(x,ct)} \frac{\varphi(x')}{\sqrt{t^2 - d^2(x',x)}}\,dx',
\end{aligned} \tag{5.24}
$$

where $d(x',x)$ is the (Euclidean) distance between x and x'.

Notice the difference between formula (5.24) and the Poisson formula (5.19), in the first one we have the sum of two volume integrals in $\overline{B}(x,ct)$, while the Poisson formula in $\mathbb{R}^3$ involves two surface integrals on $\partial B(x,ct)$.

5.2.3 *The Huygens principle and wave diffusion*

An interesting phenomenon appeared above: the solution in $\mathbb{R}^2$ defined by (5.24) depends on the values of the data in the whole ball $B(x,t)$, while the solution in $\mathbb{R}^3$ given by (5.19) depends only on the values of the data on the surface of the ball $B(x,t)$. One says that the solution in $\mathbb{R}^3$ verifies the Huygens principle.

To understand its physical significance, suppose that φ and ψ vanish everywhere except in a small ball $B(x_0,\varepsilon)$. Let x be an arbitrary point of $\mathbb{R}^3$ and let us have a look at the values of $u(x,t)$ for $t \geq 0$. By Poisson formula (5.19), one has $u(x,t) = 0$ for every t for which $\partial B(x,t)$ does not intersect $B(x_0,\varepsilon)$. If d is the distance between x and x_0, $\partial B(x,t)$ intersects $B(x_0,\varepsilon)$ only if $t \in [d-\varepsilon, d+\varepsilon]$. Thus $u(x,t) = 0$ for $t < d-\varepsilon$ and for $t > d+\varepsilon$, and one has $u(x,t) \neq 0$ only for $t \in [d-\varepsilon, d+\varepsilon]$.

This is not the case in $\mathbb{R}^2$: since $B(x_0,\varepsilon)$ intersects $\overline{B}(x,t)$ for all $t \geq d-\varepsilon$, one has $u(x,t) = 0$ for $t < d-\varepsilon$, but $u(x,t) \neq 0$ for all $t \geq d-\varepsilon$. Formula (5.24) shows then that u goes to zero as t increases.

Therefore a small perturbation confined in a ball $B(x_0,\varepsilon)$ of $\mathbb{R}^3$ gives rise to a wave which lies between the surfaces $\partial B(x_0,t+\varepsilon)$ and $\partial B(x_0,t-\varepsilon)$. In $\mathbb{R}^2$, the wave is present everywhere in the interior of $\partial B(x_0,t+\varepsilon)$ and we have what is called the phenomenon of diffusion of waves (this phenomenon does not happen in $\mathbb{R}^3$). One understands better this physical interpretation when comparing a "sound" (the solution in $\mathbb{R}^3$) with a "wave" (the solution in $\mathbb{R}^2$). In the atmosphere, one hears a sound during a finite time; a stone thrown in a pool, gives rise to a plane wave leaving behind it a drag, slowly vanishing in time.

5.3 The one-dimensional wave equation

Let us now again consider the wave equation in $\mathbb{R}$,

$$Lu = \frac{\partial^2 u}{\partial t^2} - c^2 \frac{\partial^2 u}{\partial x^2} = 0 \quad \text{for } x \in \mathbb{R}, \ \ t > 0. \tag{5.25}$$

Observe first that the characteristics of this equation are the straight lines $x + ct = k$ and $x - ct = k$, where k is a constant (see Section 2.3).

A second observation is that any function satisfying $Lu = 0$ is of the form

$$u(x,t) = g(x+ct) + f(x-ct), \tag{5.26}$$

where g and f are functions of class C^2.

This is due to the particular feature of the differential operator L in (5.25) which can be written as the following product:

$$L = \frac{\partial^2}{\partial t^2} - c^2 \frac{\partial^2}{\partial x^2} = \Big(\frac{\partial}{\partial t} + c\frac{\partial}{\partial x}\Big)\Big(\frac{\partial}{\partial t} - c\frac{\partial}{\partial x}\Big). \tag{5.27}$$

Now, if u is a solution of (5.25) and we set

$$\varphi_1 = \frac{\partial u}{\partial t} - c\frac{\partial u}{\partial x}, \tag{5.28}$$

then by (5.27),

$$\frac{\partial \varphi_1}{\partial t} + c\frac{\partial \varphi_1}{\partial x} = 0.$$

It is known (see, for instance, [10]) that for this first order PDE, the solution has the form $\varphi_1(x,t) = G(x-ct)$ for some differentiable function G. It follows (see (5.28)) that u is a solution of the linear first order equation

$$\frac{\partial u}{\partial t} - c\frac{\partial u}{\partial x} = G(x-ct). \tag{5.29}$$

Its solution is obtained by adding to the general solution of the homogeneous equation

$$\frac{\partial v}{\partial t} - c\frac{\partial v}{\partial x} = 0, \tag{5.30}$$

a particular solution of (5.29). To do so, let $\widehat{G}$ be a primitive of G and set

$$\psi(x,t) = -\frac{1}{2c}\widehat{G}(x-ct).$$

A simple computation shows that ψ is a particular solution of (5.29). On the other hand, the general solution of (5.30) is $v(x,t) = g(x+ct)$ for some differentiable function g. Consequently,

$$u(x,t) = g(x+ct) - \frac{1}{2c}\widehat{G}(x-ct),$$

which shows that u is of the form (5.26).

Remark 5.7. *The method we used to solve* (5.25) *can be generalized to all linear PDEs of the form*

$$L_1 L_2 \dots L_p u = \mathcal{F}, \tag{5.31}$$

where

$$L_k = \left(\alpha_k \frac{\partial}{\partial x_1} + \beta_k \frac{\partial}{\partial x_2} + \gamma_k\right)^{m_k},$$

with α_k, β_k, γ_k *constants, and* $m_k \in \mathbb{N}$ *for* $k \in \{1, \dots, p\}$. *Every solution of equation* (5.31) *is of the form* $u = v_1 + \dots + v_p + \mathcal{G}$ *where* $\mathcal{G}$ *is a particular solution of* (5.31), *and the functions* $v_1, \dots, v_p$ *are as follows:*

$$v_k(x_1, x_2) = \exp\left(-\frac{\gamma_k}{\alpha_k} x\right) f(\beta_k x_1 - \alpha_k x_2) \quad \text{if } m_k = 1,$$

$$v_k(x_1, x_2) = \exp\left(-\frac{\gamma_k}{\alpha_k} x\right) \sum_{\ell=1}^{m_k} x^{\ell-1}\, \psi_\ell(\beta_k x_1 - \alpha_k x_2) \quad \text{if } m_k > 1.$$

Formula (5.26) has some nice properties (see for instance, [30]). Let us just recall two of them. The first one is the so-called parallelogram rule. Observe that the functions g and f are constant along the characteristic $x + ct = constant$, respectively, $x - ct = constant$. Thus, if $ABDC$ is a parallelogram formed by lines which are characteristics (see Figure 4), i.e.,

$$AC \,||\, BD, \qquad AB \,||\, CD,$$

so that $g(A) = g(B)$, $g(C) = g(D)$, and $f(A) = f(C)$, $f(B) = f(D)$, then at the corners of the parallelogram, the solution u satisfies

$$u(A) + u(D) = u(B) + u(C).$$

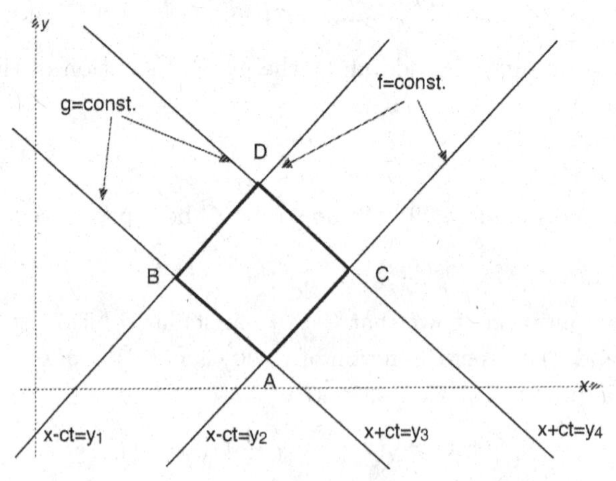

Figure 4. Characteristic parallelogram

The second property concerns the influence of the data on the solution. Formula (5.26) shows that the value of u at a point (x_0, t_0) is only determined by what happens at $t = 0$ in the interval $(x_0 - ct_0, x_0 + ct_0)$, called the domain of dependence of u at the point (x_0, t_0). It is obtained by drawing the characteristic lines $x_0 - ct_0$ and $x_0 + ct_0$ with slopes $1/c$ and $-1/c$, respectively (see Figure 5 below).

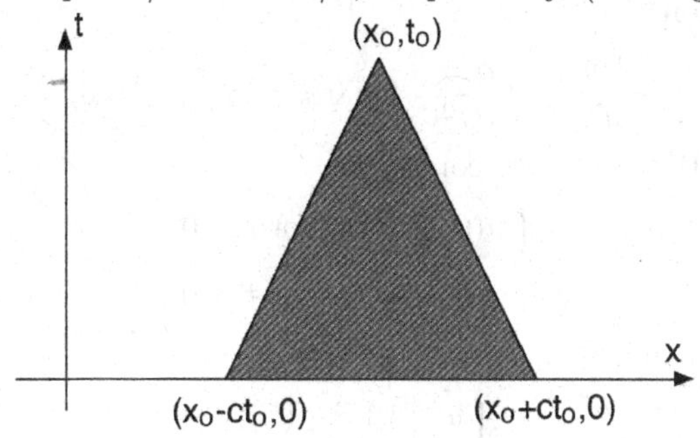

Figure 5. Domain of dependence of u at (x_0, t_0)

The fact that u possesses at each point a domain of dependence is sometimes quoted as: the solutions of the wave equation have a finite signal velocity. This behavior is very different from that of the solutions of the heat equation which depend on the information given along the whole x-axis, from $-\infty$ to $+\infty$. The same property for the wave equation is impossible. Indeed, from the considerations above, this could happen only if $c = +\infty$! This is why one says that the heat equation has an "infinite signal velocity".

On the other hand, the values of u at $t = 0$ in an interval $I = [a, b]$ have an influence on a domain bigger than the corresponding domain of dependence. This domain, called the range of influence of I, is the set of all the points (x, t) whose domain of dependence includes some or all the points of I, as depicted in Figure 6.

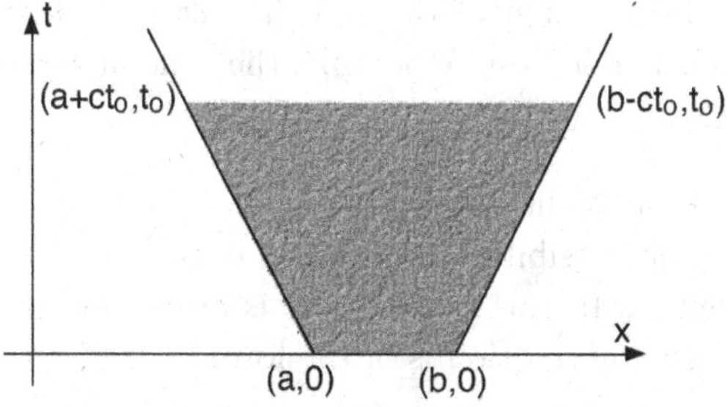

Figure 6. Range of influence of (a, b)

If a point $(\bar{x}, \bar{t})$ is outside the range of influence, then the values of $u(\bar{x}, \bar{t})$ do not depend on I.

5.3.1 *Separation of variables*

The method of separation of variables can also be used to find the solution u of the equation

$$\frac{\partial^2 u}{\partial t^2} = c^2 \frac{\partial^2 u}{\partial x^2} \quad \text{for } 0 < x < L, \ \ t > 0, \tag{5.32}$$

and satisfying the boundary conditions

$$\begin{cases} u(0,t) = 0 & \text{for } t > 0, \\ u(L,t) = 0 & \text{for } t > 0, \end{cases} \tag{5.33}$$

and the initial conditions

$$\begin{cases} u(x,0) = f(x), \\ \dfrac{\partial u}{\partial t}(x,0) = g(x). \end{cases} \tag{5.34}$$

One would like to find a solution u of the form

$$u(x,t) = \mathcal{F}(x)\mathcal{G}(t).$$

From (5.32) one has

$$\mathcal{F}\mathcal{G}'' = c^2 \mathcal{F}''\mathcal{G},$$

which leads to the following two ordinary differential equations:

$$\mathcal{F}'' - k\mathcal{F} = 0, \tag{5.35}$$

and

$$\mathcal{G}'' - c^2 k\mathcal{G} = 0. \tag{5.36}$$

The boundary conditions (5.33) imply $\mathcal{F}(0) = 0$ and $\mathcal{F}(L) = 0$. For $k = 0$, the general solution of (5.35) being $ax + b$, these conditions are satisfied if $\mathcal{F} \equiv 0$. For k positive, $k = \eta^2$, the general solution is

$$\mathcal{F} = A\,e^{\eta x} + B\,e^{-\eta x},$$

and again the initial conditions imply $\mathcal{F} \equiv 0$.

There is also the possibility to have $k < 0$, that is, of the form $k = -p^2$ for some constant p. In this case, (5.35) is an eigenvalue problem of the type (1.18). Its general solution is of the form

$$\mathcal{F}(x) = A\cos(px) + B\sin(px).$$

The boundary conditions need to have $A = 0$ and $B\sin(pL) = 0$, so that

$$p = \frac{n\pi}{L}, \quad n = 1, 2, \ldots.$$

Setting $B = 1$, we have the following family of solutions of (5.35):

$$\mathcal{F}_n(x) = \sin\left(\frac{n\pi}{L}x\right), \quad n = 1, 2, \ldots.$$

For n negative, one has essentially the same solutions but with the negative sign.

We now solve (5.36) with k fixed, $k = -(n\pi/L)^2$. Setting

$$\lambda_n = \frac{cn\pi}{L},$$

the equation becomes

$$\mathcal{G}'' + \lambda_n^2 \mathcal{G} = 0,$$

and the general solution is

$$\mathcal{G}_n(t) = B_n \cos(\lambda_n t) + B_n^\star \sin(\lambda_n t).$$

Consequently, the functions $u_n(x,t) = \mathcal{F}_n(x)\mathcal{G}_n(t)$ given by

$$u_n(x,t) = \Big(B_n \cos(\lambda_n t) + B_n^\star \sin(\lambda_n t)\Big) \sin\left(\frac{n\pi}{L}x\right), \quad n = 1, 2, \ldots. \quad (5.37)$$

satisfy equation (5.32), the boundary conditions (5.33), but not the initial conditions. Since equation (5.32) is linear and homogeneous, every finite sum of u_n is also a solution by the superposition principle.

To obtain a solution of the Cauchy problem, it is worthy to introduce the infinite series

$$u(x,t) = \sum_{n=1}^{\infty} u_n(x,t). \quad (5.38)$$

The initial conditions require necessarily

$$\begin{cases} u(x,0) = \displaystyle\sum_{n=1}^{\infty} B_n \sin\left(\frac{n\pi x}{L}\right) = f(x), \\ \dfrac{\partial u}{\partial t}(x,0) = \displaystyle\sum_{n=1}^{\infty} \frac{cn\pi}{L} B_n^\star \sin\left(\frac{n\pi x}{L}\right) = g(x). \end{cases}$$

It follows that B_n are the Fourier coefficients of f, while $B_n^\star$ are those of g, that is, for $n = 1, 2, \ldots$,

$$\begin{aligned} B_n &= \frac{2}{L} \int_0^L f(\tau) \sin\left(\frac{n\pi\tau}{L}\right) d\tau, \\ B_n^\star &= \frac{2}{cn\pi} \int_0^L g(\tau) \sin\left(\frac{n\pi\tau}{L}\right) d\tau. \end{aligned} \quad (5.39)$$

Suppose now that the functions f and g are twice continuously differentiable on $(0, L)$ and that moreover, their third order derivatives are piecewise continuous on the same interval. A classical result in the theory of Fourier series ensures the convergence of series (5.38) with coefficients (5.39), as well as that of the series obtained by differentiating twice with respect to time and to x. In conclusion, u is the required solution of the Cauchy problem given by (5.32)–(5.34).

Example 5.8. Assume $g \equiv 0$ and f is an odd function, periodic with period $2L$. Applying in (5.38) classical formulas in trigonometry, one gets

$$u(x,t) = \frac{1}{2}\sum_{n=1}^{\infty} B_n \sin\Big(\frac{n\pi}{L}(x-ct)\Big) + \frac{1}{2}\sum_{n=1}^{\infty} B_n \sin\Big(\frac{n\pi}{L}(x+ct)\Big).$$

Recalling the definition of the Fourier series of f, one actually has

$$u(x,t) = \frac{1}{2}\big[f(x-ct) + f(x+ct)\big], \tag{5.40}$$

a formula now familiar to us as seen in the previous discussion, and which is know to have a physical interpretation. The function $f(x-ct)$ is a wave propagating to the right (forward wave) as t increases and $f(x+ct)$ is a wave propagating to the left (backward wave). The solution u is the superposition of these two plane waves. ◇

5.3.2 *Eigenvalues and eigenfunctions*

The functions u_n defined by (5.37) and (5.39), are called eigenfunctions or characteristic functions and $\lambda_n = cn\pi/L$ are the eigenvalues or the characteristic values of the vibrating string. The set $\{\lambda_1, \lambda_2, \ldots\}$ is called the spectrum. Every u_n is harmonic with frequency $\lambda_n/2\pi$ cycles per time unit; this motion is the n *normal mode* of the string.

The first mode is called *fundamental mode*, the other ones are the harmonics or *overtones*; in music they give the octave, the octave plus the fifth, and the other modes. Since

$$\sin\frac{n\pi x}{L} = 0 \quad \text{for} \quad x = \frac{L}{n}, \frac{2L}{n}, \ldots, \frac{(n-1)L}{n},$$

the n-th mode has $n-1$ nodes and these are the points of the string which do not move.

At any time, the string has the form of a sine wave as depicted in the next figure (for $L = 1$):

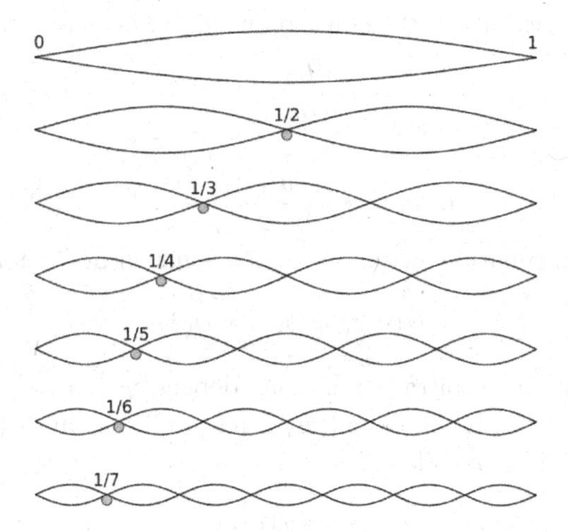

Figure 7. Harmonics of a vibrating string

5.3.3 *The d'Alembert solution*

A solution of the form (5.40) can be obtained by a simpler and more elegant method, the d'Alembert method, by introducing a change of variables.

Consider again the Cauchy problem for the wave equation in $\mathbb{R}$,

$$\begin{cases} \dfrac{\partial^2 u}{\partial t^2} - c^2 \Delta u = 0 & \text{for } x \in \mathbb{R},\ t > 0, \\ u(x,0) = f(x) & \text{for } x \in \mathbb{R}, \\ \dfrac{\partial u}{\partial t}(x,0) = g(x) & \text{for } x \in \mathbb{R}. \end{cases} \tag{5.41}$$

We now give the d'Alembert solution for (5.41).

Theorem 5.9. *Assume $f \in C^2(\mathbb{R})$ and $g \in C^1(\mathbb{R})$. Then the solution of the wave equation (5.41) is given by the d'Alembert formula*

$$u(x,t) = \frac{1}{2}\big[f(x+ct) + f(x-ct)\big] + \frac{1}{2c}\int_{x-ct}^{x+ct} g(s)\,ds. \tag{5.42}$$

This solution is of class C^2 in $\mathbb{R} \times [0, +\infty)$ and is unique.

Proof. Let us introduce the following change of variables:

$$v = x + ct, \quad z = x - ct.$$

With these new variables, the equation in (5.41) becomes simply

$$\frac{\partial^2 U}{\partial v \partial z} = 0,$$

where

$$U(v, z) = u\Big(\frac{v+z}{2}, \frac{v-z}{2c}\Big).$$

Integrating this equation twice, one gets a solution of the form

$$U(v, z) = \phi(v) + \psi(z), \tag{5.43}$$

where ϕ and ψ are arbitrary functions depending on only one variable. They are determined by fulfilling the initial conditions in (5.41). They read in the new variables as follows:

$$\begin{cases} U(v, v) = u(v, 0) = f(v), \\ c\dfrac{\partial U}{\partial v}(v, v) - c\dfrac{\partial U}{\partial z}(v, v) = \dfrac{\partial u}{\partial t}(v, 0) = g(v). \end{cases}$$

Taking into account (5.43), ϕ and ψ do satisfy

$$\begin{cases} \phi(v) + \psi(v) = f(v), \\ c\phi'(v) - c\psi'(v) = g(v). \end{cases} \tag{5.44}$$

Integrating the second equality, we obtain

$$\phi(v) - \psi(v) = c(s_0) + \frac{1}{c}\int_{s_0}^{v} g(s)\,ds,$$

where $c(s_0) = \phi(s_0) - \psi(s_0)$. Solving (5.44), one gets

$$\phi(v) = \frac{1}{2}f(v) + \frac{1}{2c}\int_{s_0}^{v} g(s)\,ds + \frac{1}{2}c(s_0),$$

as well as

$$\psi(v) = \frac{1}{2}f(v) - \frac{1}{2c}\int_{s_0}^{v} g(s)\,ds - \frac{1}{2}c(s_0).$$

Consequently, using again (5.43), one has

$$U(v, z) = \frac{1}{2}f(v) + \frac{1}{2}f(z) + +\frac{1}{2c}\int_{z}^{v} g(s)\,ds.$$

Turning back to the variables x and t,

$$u(x, t) = U(x + ct, x - ct),$$

and the final result is the d'Alembert solution (5.42).

This solution is unique and stable. Indeed, if u_1 and u_2 are the solutions of (5.32) corresponding to the data f_1, g_1 and f_2, g_2, respectively, then from (5.42) one immediately has

$$|u_1(x,t) - u_2(x,t)| \leq \sup_{x\in\mathbb{R}} |f_1 - f_2| + T \sup_{x\in\mathbb{R}} |g_1 - g_2|, \quad \text{for } x \in \mathbb{R},\, t \in [0,T].$$

This means that we have uniform pointwise stability (at least for finite time). In conclusion, the wave equation (5.32) is well-posed in the sense of Definition 1.10 in the space C^2 in $\mathbb{R} \times [0, +\infty)$. □

Remark 5.10. *If $g \equiv 0$, formula* (5.42) *is exactly that given in Example 5.8, where f was supposed to be odd and periodic, hypotheses that were not necessary to obtain the d'Alembert solution. The method used above actually gives the solution for the Cauchy problem for an arbitrary f.*

Remark 5.11.

(1) The d'Alembert formula can be written in the form that we have encountered in the preceding sections (see for example (5.26) *or* (5.40)*),*

$$u(x,t) = F(x+ct) + G(x-ct),$$

with

$$F(x+ct) = \frac{1}{2} f(x+ct) + \frac{1}{2c} \int_0^{x+ct} g(s)\, ds,$$
$$G(x-ct) = \frac{1}{2} f(x-ct) + \frac{1}{2c} \int_{x-ct}^{0} g(s)\, ds.$$

This says again that the solution is the superposition of forward and backward waves.

(2) The d'Alembert formula shows that the value of u at any point (x,t) depends on the values of f at the points $x - ct$ and $x + ct$, and on the values of g on the whole interval $[x - ct,\, x + ct]$ which, as seen above (see Figure 5), is the domain of dependence of (x,t).

In other words, this also means that the values of f and g at a point $\bar{x}$ have a contribution to the values of u at any point (x,t) provided this point is situated in the cone $\bar{x} - ct < x < \bar{x} + ct$ which is the range of influence of $\bar{x}$ (see Figure 6 above).

(3) Since from (5.42)*, one has*

$$f(x_0) = u(x_0, 0) \quad \textit{and} \quad g(x_0) = \frac{\partial u}{\partial t}(x_0, 0),$$

it follows that (5.42) can be rewritten in the form

$$u(x_0, t_0) = \frac{1}{2}\big[u(x_0 + ct_0, 0) + u(x_0 - ct_0, 0)\big] + \frac{1}{2c}\int_{x_0-ct_0}^{x_0+ct_0} \frac{\partial u}{\partial t}(s, 0)\, ds.$$

In particular, if $g = 0$, this formula says that the value of u at the apex (x_0, t_0) of the domain of dependence of the interval

$$I = [x_0 - ct_0, x_0 + ct_0],$$

is simply the average of the values of u at $t = 0$ at the extremities of I.

Consider the figure below,

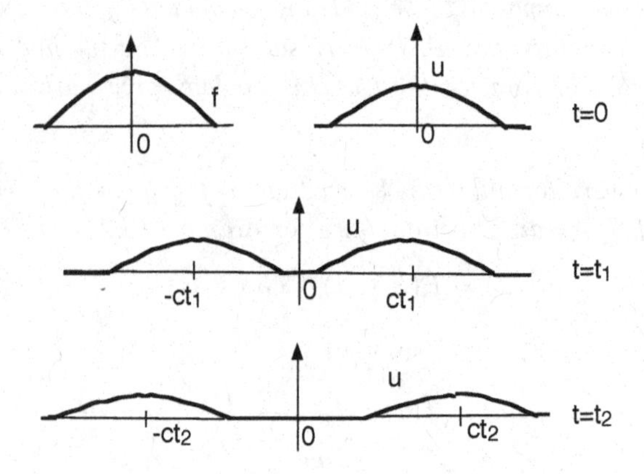

Figure 8. Traveling waves and the solution u

Here, f has compact support, $g \equiv 0$, and u is depicted at the instants $t = 0$, $t_1 > 0$ and $t_2 > t_1$. Observe that at time $t = 0$, u vanishes outside the support of f. Afterwards, as noticed in Example 5.8, the two traveling forward and backward waves $f(x - ct)$ and $f(x + ct)$ are observed, u decreasing slowly to zero as t increases.

Remark 5.12. *Observe that there is no regularity effect for the wave equation as in the case of the heat equation (see Remark 4.8). Under the above assumptions on f and g, the solution u is in general only in $C^2(\mathbb{R} \times [0, +\infty))$.*

Remark 5.13. *Let us point out that the d'Alembert method is specific for the hyperbolic wave equation. It is easily seen that it does not furnish solutions neither for the Laplace equation nor for the heat equation.*

5.4 The two-dimensional wave equation

As mentioned above, the two-dimensional wave equation (5.32) models the vibrations of a membrane. Like in the one-dimensional case, one looks for a solution of the form

$$u(x,t) = \mathcal{F}(x_1, x_2)\mathcal{G}(t).$$

To determine $\mathcal{F}$, we proceed as in the one-dimensional case and obtain the Helmholtz equation

$$\frac{\partial^2 \mathcal{F}}{\partial x_1^2} + \frac{\partial^2 \mathcal{F}}{\partial x_2^2} + \nu^2 \mathcal{F} = 0, \qquad \nu > 0, \tag{5.45}$$

where ν is a constant to be determined.

Remark 5.14. *We introduced this equation in Chapter 1, namely, equations* (1.17) *and* (1.18). *To find $\mathcal{F}$ and ν is again an eigenvalue problem as in the one-dimensional case (see equation (5.35)).*

Equation (5.45) can also be solved by the method of separation of variables. Setting $\mathcal{F}(x_1, x_2) = h(x_1)g(x_2)$, one has to solve two ordinary differential equation in order to determine h and g, namely,

$$h'' + k^2 h = 0,$$
$$g'' + p^2 g = 0,$$

with $k^2 + p^2 = \nu^2$.

In the particular case of a rectangular membrane $[0, a] \times [0, b]$, the solutions of the former two equations are

$$h_m(x_1) = \sin\frac{m\pi x_1}{a} \quad \text{and} \quad g_n(x_2) = \sin\frac{n\pi x_2}{b}, \qquad n, m = 1, 2, \ldots$$

The equation defining $\mathcal{G}$ is

$$\mathcal{G}'' + \lambda^2 \mathcal{G} = 0, \quad \lambda = c\nu.$$

One does have necessarily

$$\lambda_{mn} = c\pi\sqrt{\frac{m^2}{a^2} + \frac{n^2}{b^2}},$$

which are the eigenvalues corresponding to the eigenfunctions of the form

$$u_{mn} = \Big(B_{mn}\cos(\lambda_{mn}t) + \bar{B}_{mn}\sin(\lambda_{mn}t)\Big)\sin\frac{m\pi x_1}{a}\sin\frac{n\pi x_2}{b}.$$

Let us consider the double series

$$u(x,t) = \sum_{m=1}^{\infty}\sum_{n=1}^{\infty} u_{mn}. \tag{5.46}$$

Taking into account the initial values, one gets

$$B_{mn} = \frac{4}{ab} \int_0^a \int_0^b f(x) \sin \frac{m\pi x_1}{a} \sin \frac{n\pi x_2}{b} \, dx,$$

$$\bar{B}_{mn} = \frac{4}{ab\lambda_{mn}} \int_0^a \int_0^b g(x) \sin \frac{m\pi x_1}{a} \sin \frac{n\pi x_2}{b} \, dx,$$

where $n = 1, 2 \ldots$ and $m = 1, 2 \ldots$. Series (5.46) with these coefficients converges and so u is the required solution.

PART II

Variational Partial Differential Equations

Chapter 6

L^p-spaces

In the previous chapters, we have solved well-posed PDEs which admit classical solutions, that is, solutions that are smooth functions possessing continuous derivatives. We have also seen in Section 3.6 that some PDEs do not have solutions. In particular, Example 3.21 shows that the problem $-\Delta u = f$ does not always have a solution when $N \geq 2$ and f is only continuous. Such equations model physical phenomena, hence a solution is expected to exist, but maybe not in the classical sense. In this case, the idea is to consider suitable functional spaces where a solution can exist in a "weaker" sense. Sobolev spaces are the appropriate ones in this setting. We introduce them in the next chapter.

Here, we give some preliminary notions. We start with a short presentation of Banach, Hilbert, and in particular, L^p-spaces. We also give a review of the notions of weak and weak* convergence which will be needed in our discussion. For details on L^p-spaces, we refer the reader to [3], [4], [33], and [34] and for proofs and more details concerning Banach and Hilbert spaces, to [14], [33], [34], [43], and [44].

6.1 Some properties of Banach spaces

Let us begin by recalling the definition of Banach spaces which will be the functional spaces used here.

Definition 6.1. *Let X be a vector space. A mapping $\|\cdot\| : X \mapsto \mathbb{R}_0^+$ is called a norm on X if for every $x, y \in X$, the following are satisfied:*

(1) $\|x\| = 0 \Longleftrightarrow x = 0$,

(2) $\|\lambda x\| = |\lambda|\|x\|$ *for any* $\lambda \in \mathbb{R}$,

(3) $\|x + y\| \leq \|x\| + \|y\|$.

The space X together with its norm denoted by $\|\cdot\|_X$ is called a normed space. If the space X is a complete metric space with respect to the associated metric ρ given by

$$\rho(x,y) = \|x-y\|_X, \quad \forall x,y \in X, \tag{6.1}$$

then X is called a Banach space.

Remark 6.2. *Let X be a normed vector space with norm $\|\cdot\|_X$.*
(1) From (6.1), the convergence of a sequence in a normed space X (also called strong convergence) is given by,

$$x_n \to x \text{ in } X \Longleftrightarrow \|x_n - x\|_X \to 0. \tag{6.2}$$

(2) From (3) of Definition 6.1, it follows that

$$x_n \to x \text{ in } X \Longrightarrow \|x_n\|_X \to \|x\|_X, \tag{6.3}$$

since

$$\Big| \|x_n\|_X - \|x\|_X \Big| \leq \|x_n - x\|_X \to 0.$$

It can be shown that the converse of (6.3) is false.
(3) Let $\|\cdot\|_1$ be another norm on X. The two norms $\|\cdot\|_X$ and $\|\cdot\|_1$ are equivalent if there exist two positive constants α and β such that

$$\alpha\|x\|_X \leq \|x\|_1 \leq \beta\|x\|_X, \qquad \forall x \in X. \tag{6.4}$$

In this case, the convergence of a sequence in $(X, \|\cdot\|_X)$ is equivalent to that in $(X, \|\cdot\|_1)$.
(4) In a finite dimensional vector space (in particular in $\mathbb{R}^N$), all the norms are equivalent.

Example 6.3. The following are Banach spaces:
(1) The space $\mathbb{R}^N$ together with the Euclidean norm

$$\|x\| = \Big(\sum_{i=1}^{N} x_i^2\Big)^{1/2},$$

where $x = (x_1, x_2, \ldots, x_N)$, is a Banach space. This norm induces what is called the Euclidean metric in $\mathbb{R}^N$. As done in Section 1.2, $\|x\|$ is simply denoted $|x|$.

(2) The space of all continuous functions on $[0,1]$, denoted $C^0([0,1])$, is a Banach space with respect to the sup norm given by

$$\|f\| = \sup_{x\in[0,1]} |f(x)|.$$

(3) If $(B_1, \|\cdot\|_1), \ldots, (B_m, \|\cdot\|_m)$ are m Banach spaces, then their Cartesian product $B = B_1 \times \ldots \times B_m$ is also a Banach space with respect to the norm

$$\|x\|_B = \sum_{i=1}^{m} \|x_i\|_i, \qquad \forall\, x = (x_1, \ldots, x_m) \in B. \tag{6.5}$$

(4) The L^p-spaces are an important class of Banach spaces. They will be discussed in the next section. ◇

6.1.1 *Linear operators in normed vector spaces*

We define and give below some basic properties of linear mappings on normed linear spaces.

Definition 6.4. *Let X and Y be normed linear spaces. A mapping $T : X \mapsto Y$ is called a linear map (or a linear operator) if for any $x, y \in X$ and for $\alpha, \beta \in \mathbb{R}$,*

$$T(\alpha x + \beta y) = \alpha T(x) + \beta T(y). \tag{6.6}$$

A linear map T is called a bounded operator if there exists a positive constant c such that

$$\|T(x)\|_Y \leq c\|x\|_X, \quad \forall x \in X.$$

If T is linear and bounded, we set

$$\|T\| = \sup_{x \in X \setminus \{0\}} \frac{\|T(x)\|_Y}{\|x\|_X}, \tag{6.7}$$

which is finite and defines a norm. We denote by $\mathcal{L}(X, Y)$ the set of all linear and bounded maps from X to Y endowed with this norm.

Remark 6.5. *Clearly, $\mathcal{L}(X,Y)$ is a vector space and the norm given by (6.7) is well-defined and implies*

$$\|T(x)\|_Y \leq \|T\|_{\mathcal{L}(X,Y)} \|x\|_X, \quad \forall x \in X. \tag{6.8}$$

The quantity $\|T\|_{\mathcal{L}(X,Y)}$ is the smallest number for which (6.8) holds and is in fact, the norm $\|T\|$ defined by (6.7). Moreover, for $T \in \mathcal{L}(X, Y)$, one can easily verify that this norm is also given by

$$\|T\|_{\mathcal{L}(X,Y)} = \sup_{\substack{x \in X \setminus \{0\} \\ \|x\|_X \leq 1}} \frac{\|T(x)\|_Y}{\|x\|_X} = \sup_{\substack{x \in X \\ \|x\|_X = 1}} \|T(x)\|_Y. \tag{6.9}$$

Example 6.6.
(1) Let $X = Y = C^0([0,1])$ with the sup norm and $x \in X$. The mapping $T : X \mapsto Y$ given by $T(f) = x^2 f$ is a bounded linear operator and

$$\sup_{f \in X \setminus \{0\}} \frac{\|T(f)\|_Y}{\|f\|_X} = 1.$$

(2) Let $X = C^1([0,1])$ be the space of functions on $[0,1]$ with continuous first derivatives and $Y = C^0([0,1])$ together with the sup norm. The differential operator $T : X \mapsto Y$ defined by $T(f) = f'$ for all $f \in X$ is linear. However, it is unbounded. ◇

We also have the following characterization of bounded linear operators:

Theorem 6.7. *Let $T : X \mapsto Y$ be a linear map between two normed spaces X and Y. Then the following statements are equivalent:*
(1) T is bounded.
(2) T is continuous.
(3) T is continuous at a point $x_0 \in X$.

Proof. (1) $\Rightarrow$ (2). Since T is bounded, (6.2) and the linearity of T imply that for all $u, v \in X$

$$\|T(u) - T(v)\|_Y \leq \|T\|_{\mathcal{L}(X,Y)} \|u - v\|_X.$$

Hence T is continuous.

(2) $\Rightarrow$ (3) is obvious.

(3) $\Rightarrow$ (1). We can assume T is continuous at $x_0 = 0$. If T is not bounded, then there exists no M such that $\|T(x)\|_Y \leq M\|x\|_X$ for all $x \in X$. So, for every $n \in \mathbb{N}$, there exists $x_n \in X$ such that

$$\|x_n\|_X = 1 \quad \text{and} \quad \|T(x_n)\|_Y > n\|x_n\|_X.$$

Let $y_n = \dfrac{x_n}{n}$. It follows that $y_n \to 0$. Since T is linear and continuous at 0, $T(y_n) \to T(0) = 0$. This implies that $\|T(y_n)\|_Y \to 0$ and contradicts the fact that

$$\|T(y_n)\|_Y = \left\|T\left(\frac{x_n}{n}\right)\right\|_Y > 1.$$

Hence T must be bounded. □

Proposition 6.8. *Let X, Y be normed spaces. If Y is a Banach space, then $\mathcal{L}(X, Y)$ is a Banach space for the norm given by (6.7).*

Remark 6.9. *If $T : X \mapsto \mathbb{R}$ is a linear operator from a normed space X into $\mathbb{R}$, T is called a linear functional on X.*

6.1.2 *Dual of a normed space*

We now give the important notion of dual of a normed space.

Definition 6.10. *Let X be a normed space. The dual space of X, denoted X',is given by $X' = \mathcal{L}(X, \mathbb{R})$, that is, the set of all linear and continuous maps from X into $\mathbb{R}$. We denote by*

$$\langle x', x \rangle_{X', X}$$

the image $x'(x)$ of $x \in X$ under $x' \in X'$. The duality pairing between X' and X is given by $\langle \cdot, \cdot \rangle_{X', X}$.

The bidual of X, denoted X'', is the dual of the dual space of X. That is, $X'' = (X')'$.

In this book, we shall be considering the dual of Banach spaces and in the following, E denotes a Banach space. Definition 6.10 and Proposition 6.8 imply that the dual of a Banach space is also a Banach space. We state this as a proposition.

Proposition 6.11. *Let E be a Banach space. Then the dual space E' of E is a Banach space for the norm*

$$\|x'\|_{E'} = \sup_{x \in E \setminus \{0\}} \frac{|\langle x', x \rangle_{E', E}|}{\|x\|_E} \quad \text{for every } x' \in E'.$$

Moreover, if $x' \in E'$,

$$|\langle x', x \rangle_{E', E}| \leq \|x'\|_{E'} \|x\|_E \quad \text{for every } x \in E.$$

Remark 6.12. *This proposition shows that E'' is also a Banach space. Although $E \neq E''$ in general, E may be identified with a subspace of E'' through a canonical isometry and then we write $E \subset E''$. We obtain the isometry as follows: for a fixed $x \in E$, introduce the map $f_x : E' \mapsto \mathbb{R}$ defined as*

$$f_x(x') = \langle x', x \rangle_{E', E}.$$

This implies $f_x \in E''$ and if we define the map

$$\mathcal{F} : x \in E \mapsto f_x \in E'', \tag{6.10}$$

we can identify $x \in E$ with $f_x \in E''$, hence E with its image $\mathcal{F}(E) \subset E''$. The map $\mathcal{F}$ is an isometry, that is,

$$\|x\|_E = \|\mathcal{F}(x)\|_{E''} = \|f_x\|_{E''}, \quad \forall\, x \in E.$$

We can now give the following definition:

Definition 6.13. *Let $\mathcal{F}$ be the map defined by (6.10). The space E is called reflexive if $\mathcal{F}(E) = E''$.*

If E is reflexive, then we can identify E with E''.

6.2 Hilbert spaces

Hilbert spaces are a particular class of Banach spaces which play an essential role in partial differential equations.

Definition 6.14. *Let H be a vector space. A mapping*

$$(\cdot\,,\,\cdot)_H : H \times H \longmapsto \mathbb{R},$$

is called a (real) scalar product if for $x,\ y,\ z \in H$,

(1) $(x,x)_H \geq 0$ for any $x \in H$,

(2) $(x,x)_H = 0 \implies x = 0$,

(3) $(x,y)_H = (y,x)_H$,

(4) $(\lambda x + \mu y, z)_H = \lambda(x,z)_H + \mu(y,z)_H$ for any $\lambda,\ \mu \in \mathbb{R}$.

The space H together with this scalar product is called an inner product space. The norm associated with this scalar product is defined as

$$\|x\|_H = \sqrt{(x,x)_H}. \tag{6.11}$$

If H is a Banach space with respect to this norm, then H is called a Hilbert space.

The norm given by (6.11) is also referred to as the norm induced by the scalar product. It can be shown that it is indeed a norm on H. This is a consequence of the following inequality:

Proposition 6.15 (Cauchy–Schwarz Inequality). *Let H be an inner product space. Then for every $x,\ y \in H$,*

$$|(x,y)_H| \leq \sqrt{(x,x)_H}\,\sqrt{(y,y)_H}.$$

Proof. By Definition 6.14, for every $x,\ y \in H$ and $t \in \mathbb{R}$, we have

$$0 \leq (x+ty, x+ty)_H = (x,x)_H + t^2(y,y)_H + 2t(x,y)_H.$$

This defines a second order polynomial function of t which is nonnegative for any value of the variable. Hence its discriminant is either negative or zero. That is,

$$(x,y)_H^2 - (x,x)_H\,(y,y)_H \leq 0$$

which implies the claimed inequality. □

Example 6.16. The following are examples of Hilbert spaces:
(1) The space $\mathbb{R}^N$ together with the inner product

$$(x,y)_{\mathbb{R}^N} = x \cdot y = \sum_{i=1}^{N} x_i y_i,$$

is a Hilbert space. This product induces the Euclidean norm in $\mathbb{R}^N$.
(2) The real $L^2(\mathcal{O})$ space, where $\mathcal{O}$ is an open set in $\mathbb{R}^N$, is a Hilbert space with respect to the following inner product:

$$(f,g)_{L^2(\mathcal{O})} = \int_{\mathcal{O}} f(x)g(x)\,dx. \qquad \diamond$$

6.2.1 *Orthogonal bases in Hilbert spaces*

In this section we introduce and discuss briefly the notions of orthogonality and orthogonal bases in Hilbert spaces, which will be useful in the sequel. We only consider the case of countable orthogonal bases, which is enough for separable Hilbert spaces (see Theorem 6.20 below), mainly considered in this book.

Definition 6.17. *Let H be a Hilbert space. We say that two vectors x and y in H are orthogonal, denoted by $x \perp y$, if $(x,y)_H = 0$. Moreover, if M is a linear subspace of H, we denote by $M^\perp$ the linear subspace defined by*

$$M^\perp = \{\, x \in H \big| x \perp y, \ \forall\, y \in M \,\},$$

called the orthogonal complement of M.

The notion of a basis usually given for a finite-dimensional vector space can be extended to inner product spaces by replacing the finite sums by converging sequences in H.

Definition 6.18. *Let H be a Hilbert space. We say that the sequence $\{w_n\} \in H \setminus \{0\}$ is an orthogonal basis (or a complete orthogonal system) of H if*

$$(w_n, w_m)_H = 0, \qquad \forall n, m \in \mathbb{N},\ n \neq m, \tag{6.12}$$

and for every $x \in H$ there exists a sequence $\{\lambda_n(x)\} \subset \mathbb{R}$, such that

$$x = \sum_{k=1}^{\infty} \lambda_k(x) w_k, \tag{6.13}$$

where

$$\sum_{k=1}^{\infty} \lambda_k(x) w_k = \lim_{n\to\infty} \sum_{k=1}^{n} \lambda_k(x) w_k \quad (\textit{in } H).$$

Furthermore, if

$$\|w_n\|_H = 1, \qquad \forall n \in \mathbb{N}, \tag{6.14}$$

the sequence $\{w_n\}$ is called an orthonormal basis (or a complete orthonormal system).

If $\{w_n\}$ is an orthogonal basis of H, for any $n \in \mathbb{N}$ we denote by W_n the n-dimensional subspace of H spanned by $w_1, \ldots, w_n$.

For every $n \in \mathbb{N}$, the linear map defined by

$$P_n \ : x \in H \longmapsto P_n(x) \doteq \sum_{k=1}^{n} \lambda_k(x)\, w_k \in W_n, \tag{6.15}$$

is called the orthogonal projection on W_n.

Remark 6.19. *Let us explain why if $\{w_n\}$ is an orthogonal basis of H, then W_n is an n-dimensional subspace of H. This is due to the fact that for every $n \in \mathbb{N}$ the set $\{w_1, \ldots, w_n\}$ is a basis of W_n, since by definition W_n is the set of all linear combinations of $w_1, \ldots, w_n$ and moreover these vectors are linearly independent. Indeed, if $\sum_1^n a_i w_i = 0$ for some real numbers $a_1, \ldots, a_n$, then $0 = (\sum_1^n a_i w_i, w_k)_H = a_k \|w_k\|^2_H$, for every $k = 1, \ldots, n$. This implies that $a_k = 0$ for every $k = 1, \ldots, n$, since the vectors w_k are different from the zero vector.*

We have the following important result (see for instance, [43]):

Theorem 6.20. *A separable Hilbert space admits at least an orthonormal basis.*

Remark 6.21. *If $\{w_n\} \in H$ is an orthogonal basis, it is straightforward to deduce from* (6.11)–(6.14) *that for every $n \in \mathbb{N}$,*

$$\|P_n(x)\|_H^2 = \sum_{k=1}^{n} \lambda_k^2(x) \|w_k\|_H^2 . \tag{6.16}$$

The next proposition states the most important properties of orthonormal bases of Hilbert spaces. With the notation from Definition 6.18, we have

Proposition 6.22. *Let H be a Hilbert space which admits an orthonormal basis $\{w_n\}$. Then*

(1) For every $x \in H$, the sequence $\{\lambda_n(x)\} \in \mathbb{R}$ is unique and one has $\lambda_n(x) = (x, w_n)_H$ for any $n \in \mathbb{N}$. Consequently, for all $x \in H$,

$$x = \sum_{k=1}^{\infty} (x, w_k)_H \, w_k, \tag{6.17}$$

and

$$\lim_{n\to\infty} P_n(x) = x. \tag{6.18}$$

Moreover,

$$\|P_n(x)\|_H^2 \leq \|x\|_H^2 = \sum_{k=1}^{\infty} (x, w_k)_H^2. \tag{6.19}$$

(2) Let $x \in H$ *and* $n \in \mathbb{N}$. *If* $y_n = x - P_n(x)$, *then* y_n *belongs to* $W_n^\perp$ *and*

$$\|y_n\|_H^2 = \sum_{k=n+1}^{\infty} (x - P_n(x), w_k)_H^2 = \sum_{k=n+1}^{\infty} (x, w_k)_H^2 \leq \|x\|_H^2, \tag{6.20}$$

so that $H = W_n + W_n^\perp$, *for every* $n \in \mathbb{N}$.

Proof. Let $x \in H$. Multiplying (6.13) by w_n and using (6.12) and (6.14) give

$$(x, w_n)_H = \sum_{k=1}^{\infty} \lambda_k(x)(w_k, w_n)_H = \lambda_n(x), \quad \forall\, n \in \mathbb{N}, \tag{6.21}$$

which shows (6.17). According to (6.14) and (6.15), this implies

$$P_n(x) = \sum_{k=1}^{n} (x, w_k)_H\, w_k, \tag{6.22}$$

which due to (6.17), yields (6.18).

Also, from (6.12), (6.14), (6.16), and (6.21), we deduce that for every $n \in \mathbb{N}$,

$$\|P_n(x)\|_H^2 = \sum_{k=1}^{n} (x, w_k)_H^2. \tag{6.23}$$

Consequently, from (6.3) of Remark 6.2 we have

$$\|x\|_H^2 = \lim_{n\to\infty} \|P_n(x)\|_H^2 = \lim_{n\to\infty} \sum_{k=1}^{n} (x, w_k)_H^2,$$

whence (6.19).

Finally, if $x \in H$ then $y_n = x - P_n(x)$ belongs to $W_n^\perp$ for every $n \in \mathbb{N}$ by (6.15) and Definition 6.17. A similar argument as that used above gives

$$\|y_n\|_H^2 = \left\| \sum_{k=n+1}^{\infty} (x, w_k)_H\, w_k \right\|_H^2 = \lim_{m\to\infty} \left\| \sum_{k=n+1}^{m} (x, w_k)_H\, w_k \right\|_H^2$$

$$= \lim_{m\to\infty} \sum_{k=n+1}^{m} (x, w_k)_H^2,$$

which concludes the proof of (6.20). □

Observe that if $\{w_n\} \subseteq H$ is an orthogonal basis of H, then

$$\left\{ \frac{w_n}{\|w_n\|_H} \right\} \text{ is an orthonormal basis of } H,$$

and by Proposition 6.22, we have,

Corollary 6.23. *Let H be a Hilbert space which admits an orthogonal basis $\{w_n\}$.*
(1) For every $x \in H$, the sequence $\{\lambda_n(x)\} \in \mathbb{R}$ is unique and

$$\lambda_n(x) = \frac{(x, w_n)_H}{\|w_n\|_H} \quad \textit{for any} \quad n \in \mathbb{N},$$

that is (see (6.15)),

$$P_n(x) = \sum_{k=1}^{n} \frac{(x, w_k)_H}{\|w_k\|^2_H} w_k \quad \textit{for any} \quad n \in \mathbb{N}. \tag{6.24}$$

Consequently, for all $x \in H$,

$$x = \sum_{k=1}^{\infty} \frac{(x, w_k)_H}{\|w_k\|^2_H} w_k, \tag{6.25}$$

and

$$\lim_{n \to \infty} P_n(x) = x. \tag{6.26}$$

Moreover,

$$\|P_n(x)\|_H^2 \leq \|x\|_H^2 = \sum_{k=1}^{\infty} \frac{(x, w_k)_H^2}{\|w_k\|^2_H}. \tag{6.27}$$

(2) For every $x \in H$ and $n \in \mathbb{N}$, $y_n = x - P_n(x)$ belongs to $W_n^{\perp}$ and

$$\|y_n\|_H^2 = \sum_{k=n+1}^{\infty} \frac{(y_n, w_k)_H^2}{\|w_k\|^2_H}. \tag{6.28}$$

Remark 6.24. *Let us observe that (6.27) implies that P_n belongs to $\mathcal{L}(H, H)$, that is, P_n is a linear continuous map from H into itself (see Definition 6.4).*

The results above are of interest when H is an infinite dimensional vector space. Indeed, if H admits an orthogonal basis, then any element can be approached by a sequence of functions which belong to finite dimensional subspaces. This fact is widely used in partial differential equations. In particular, this is an essential tool for the evolution problems studied in Chapter 10, as well as in the numerical analysis of partial differential equations using the finite element method.

We end this section by recalling the well-known Riesz Representation Theorem which characterizes the dual of a Hilbert space.

Theorem 6.25 (Riesz Representation Theorem). *Let H be a Hilbert space. For every $F \in H'$, there exists a unique $\tau(F) \in H$ such that*

$$\langle F, v\rangle_{H',H} = (\tau(F), v)_H, \quad \forall v \in H.$$

Moreover, the linear map

$$\tau : F \in H' \mapsto \tau(F) \in H,$$

is a (bijective) isometry, that is,

$$\|F\|_{H'} = \|\tau(F)\|_H, \quad \forall F \in H'.$$

6.3 L^p-spaces and their properties

We give here a brief review of an important class of Banach spaces, the L^p-spaces. In the following, $\mathcal{O}$ is an open set in $\mathbb{R}^N$.

Definition 6.26. *Let $p \in \mathbb{R}$ with $1 \leq p \leq +\infty$. We define the space $L^p(\mathcal{O})$ by*

$$L^p(\mathcal{O}) = \Big\{ f : \mathcal{O} \to \mathbb{R} \mid f \text{ measurable and } \int_{\mathcal{O}} |f(x)|^p \, dx < +\infty \Big\},$$

for $1 \leq p < +\infty$, and the space $L^\infty(\mathcal{O})$ by

$$L^\infty(\mathcal{O}) = \Big\{ f : \mathcal{O} \to \mathbb{R} \mid f \text{ measurable and there exists } C \in \mathbb{R} \text{ such that } |f| \leq C \text{ a.e. on } \mathcal{O} \Big\}.$$

For any $p \in [1, +\infty)$, we define the space $L^p_{loc}(\mathcal{O})$ by

$$L^p_{loc}(\mathcal{O}) = \big\{ f \mid f \in L^p(\omega) \text{ for any open bounded set } \omega \text{ with } \overline{\omega} \subset \mathcal{O} \big\}.$$

Remark 6.27. *In the theory of L^p-spaces, one does not distinguish between two functions which are equal almost everywhere (see Section 1.2). That is, if f_1 and f_2 are equal on $\mathcal{O}$, except on a set with Lebesgue measure zero, they define the same element F of $L^p(\mathcal{O})$. This means that an element F of $L^p(\mathcal{O})$ is an equivalence class with respect to the equivalence relation "equal almost everywhere." We can identify F with any of its equivalence class representatives.*

Every L^p-space can be endowed with a norm; it becomes a Banach space with it. This property is stated in the next result.

Proposition 6.28. *Let $p \in \mathbb{R}$ with $1 \leq p \leq +\infty$. Then $L^p(\mathcal{O})$ is a Banach space for the norm*

$$\|f\|_{L^p(\mathcal{O})} = \begin{cases} \left[\int_{\mathcal{O}} |f(x)|^p \, dx\right]^{\frac{1}{p}} & \text{if } p < +\infty, \\ \inf\left\{C \middle| \; |f| \leq C \ \text{ a.e. on } \mathcal{O}\right\} & \text{if } p = +\infty. \end{cases}$$

If $p = 2$, the space $L^2(\mathcal{O})$ is a Hilbert space for the scalar product

$$(f, g)_{L^2(\mathcal{O})} = \int_{\mathcal{O}} f(x)\, g(x)\, dx. \tag{6.29}$$

Remark 6.29.

(1) If $h \in L^\infty(\mathcal{O})$, its norm $\|h\|_{L^\infty(\mathcal{O})}$ is called the essential supremum of the function h. It is easy to verify that if h is a bounded continuous function on an open subset $\mathcal{O}$ of $\mathbb{R}^N$, then $\|h\|_{L^\infty(\mathcal{O})} = \sup_{\mathcal{O}} |h|$.

(2) According to Definition 6.1, a sequence $\{f_n\}$ in $L^p(\mathcal{O})$ converges to $f \in L^p(\mathcal{O})$ if $\|f_n - f\|_{L^p(\mathcal{O})} \to 0$.

6.3.1 *The Hölder inequality and applications*

Recall that for $1 \leq p \leq +\infty$, p' is said to be the conjugate of p if

$$\begin{cases} \dfrac{1}{p'} = 1 - \dfrac{1}{p} & \text{if } 1 < p < +\infty, \\ p' = 1 & \text{if } p = +\infty, \\ p' = +\infty & \text{if } p = 1. \end{cases} \tag{6.30}$$

Theorem 6.30 (Hölder Inequality). *Let $1 \leq p \leq +\infty$ and p' be its conjugate. Then for any $f \in L^p(\mathcal{O})$ and $g \in L^{p'}(\mathcal{O})$*

$$\int_{\mathcal{O}} |f(x)\, g(x)| \, dx \leq \|f\|_{L^p(\mathcal{O})} \|g\|_{L^{p'}(\mathcal{O})}. \tag{6.31}$$

For $p = 2$, this inequality reduces to the Cauchy–Schwarz inequality in the Hilbert space $L^2(\mathcal{O})$ (see Proposition 6.15), that is

$$\int_{\mathcal{O}} |f(x)\, g(x)| \, dx \leq \|f\|_{L^2(\mathcal{O})} \|g\|_{L^2(\mathcal{O})}. \tag{6.32}$$

We have the following consequences of the Hölder inequality:

Corollary 6.31. *Let $1 \le p \le +\infty$ and p' be its conjugate. If f_n converges to some f in $L^p(\mathcal{O})$, and $g \in L^{p'}(\mathcal{O})$, then*

$$\lim_{n\to\infty} \int_{\mathcal{O}} f_n(x)\, g(x)\, dx = \int_{\mathcal{O}} f(x)\, g(x)\, dx.$$

Proof. Using (6.31), we obtain

$$\left| \int_{\mathcal{O}} \big(f_n(x)\, g(x) - f(x) g(x)\big)\, dx \right| \le \|f_n - f\|_{L^p(\mathcal{O})} \|g\|_{L^{p'}(\mathcal{O})},$$

and the right-hand side approaches zero by assumption. □

Remark 6.32. *Observe that if $1 \le p < +\infty$, $\alpha \in \mathbb{R}^+$, and the function f is in $L^{\alpha p}(\mathcal{O})$, then $|f|^\alpha$ is in $L^p(\mathcal{O})$ and*

$$\||f|^\alpha\|_{L^p(\mathcal{O})} = \|f\|^\alpha_{L^{\alpha p}(\mathcal{O})}. \tag{6.33}$$

Corollary 6.33. *Let $1 \le p < q \le +\infty$. If Ω is a bounded open set in $\mathbb{R}^N$, then*

$$L^q(\Omega) \subseteq L^p(\Omega), \tag{6.34}$$

and for every $f \in L^q(\Omega)$,

$$\|f\|_{L^p(\Omega)} \le c\|f\|_{L^q(\Omega)}, \tag{6.35}$$

where the constant c depends on $|\Omega|$, p, and q.

Proof. The case $q = +\infty$ is obvious. For $q < +\infty$, we apply (6.31) to the functions $f \in L^{\frac{q}{p}}(\Omega)$ and $g \equiv 1 \in L^{(\frac{q}{p})'}(\Omega)$. Hence, using Remark 6.32 we have

$$\int_\Omega |f(x)|^p\, dx \le \||f|^p\|_{L^{\frac{q}{p}}(\Omega)} \|1\|_{L^{(\frac{q}{p})'}(\Omega)} = \|f\|^p_{L^q(\Omega)} |\Omega|^{\frac{q-p}{q}},$$

which gives the result for $c = |\Omega|^{\frac{q-p}{pq}}$. □

Corollary 6.34. *Let $k \ge 2$ and $g_1, \dots, g_k$, be functions such that $g_i \in L^{p_i}(\mathcal{O})$, with $p_i \in (1, +\infty)$ for $i = 1, \dots, k$, and*

$$\frac{1}{p} \doteq \frac{1}{p_1} + \dots + \frac{1}{p_k} \le 1.$$

Then the product $g = g_1 g_2 \cdots g_k$ belongs to $L^p(\mathcal{O})$. Moreover,

$$\|g\|_{L^p(\mathcal{O})} \le \prod_{i=1}^k \|g_i\|_{L^{p_i}(\mathcal{O})}. \tag{6.36}$$

Proof. The proof is done by induction on k. For $k = 2$, let

$$\frac{1}{p} \doteq \frac{1}{p_1} + \frac{1}{p_2} \leq 1,$$

and $g = g_1 g_2$ with $g_i \in L^{p_i}(\mathcal{O})$, $i = 1, 2$.

By Remark 6.32 (written with $\alpha = p$ and $\frac{p_i}{p}$ instead of p), the function $|g_i|^p$ belongs to $L^{\frac{p_i}{p}}(\mathcal{O})$, $i = 1, 2$.

Using (6.33) and the Hölder inequality (6.31) written for $q = \frac{p_1}{p}$ and $q' = \frac{p_2}{p}$, we have the following inequality which proves (6.36) for $k = 2$:

$$\begin{aligned}\|g\|^p_{L^p(\mathcal{O})} = \||g|^p\|_{L^1(\mathcal{O})} &\leq \left\||g_1|^p\right\|_{L^{\frac{p_1}{p}}(\mathcal{O})} \left\||g_2|^p\right\|_{L^{\frac{p_2}{p}}(\mathcal{O})} \\ &= \|g_1\|^p_{L^{p_1}(\mathcal{O})} \|g_2\|^p_{L^{p_2}(\mathcal{O})}.\end{aligned}$$

Suppose now that (6.36) holds for some $k \geq 2$ and consider $k+1$ functions $g_1, \ldots, g_{k+1}$ such that $g_i \in L^{p_i}(\mathcal{O})$, with $p_i \in (1, +\infty)$ for $i = 1, \ldots, k+1$, and

$$\frac{1}{p} \doteq \frac{1}{p_1} + \ldots + \frac{1}{p_{k+1}} \leq 1.$$

Let p'_{k+1} be the conjugate of p_{k+1}, defined in (6.30), and set

$$q_i = \frac{p_i}{p'_{k+1}} \quad \text{for } i = 1, \ldots, k,$$

which gives

$$\frac{1}{q} \doteq \frac{1}{q_1} + \ldots + \frac{1}{q_k} = \left(1 - \frac{1}{p_{k+1}}\right)^{-1} \sum_{i=1}^{k} \frac{1}{p_i} \leq 1.$$

The induction assumption applied to the functions $f_i = |g_i|^{p'_{k+1}} \in L^{q_i}(\mathcal{O})$ implies that $\prod_{i=1}^k |g_i|^{p'_{k+1}}$ is in $L^q(\mathcal{O})$ with

$$\left\| \prod_{i=1}^{k} |g_i|^{p'_{k+1}} \right\|_{L^q(\mathcal{O})} \leq \prod_{i=1}^{k} \left\| |g_i|^{p'_{k+1}} \right\|_{L^{q_i}(\mathcal{O})}. \tag{6.37}$$

On the other hand, since by assumption

$$\frac{1}{p_{k+1}} + \frac{1}{q p'_{k+1}} = \frac{1}{p_{k+1}} + \sum_{i=1}^{k} \frac{1}{p_i} \leq 1,$$

we can apply the result proved above for $k = 2$ to the functions $g_{k+1} \in L^{p_{k+1}}(\mathcal{O})$ and $\prod_{i=1}^k g_i \in L^{q p'_{k+1}}(\mathcal{O})$, to obtain

$$\begin{aligned}\left\| \prod_{i=1}^{k+1} g_i \right\|_{L^p(\mathcal{O})} &\leq \|g_{k+1}\|_{L^{p_{k+1}}(\mathcal{O})} \left\| \prod_{i=1}^{k} g_i \right\|_{L^{q p'_{k+1}}(\mathcal{O})} \\ &= \|g_{k+1}\|_{L^{p_{k+1}}(\mathcal{O})} \left\| \prod_{i=1}^{k} |g_i|^{p'_{k+1}} \right\|^{\frac{1}{p'_{k+1}}}_{L^q(\mathcal{O})},\end{aligned}$$

where we used (6.33) with $\alpha = p'_{k+1}$ and $p = q$. This, in view of (6.37), leads to

$$\begin{aligned}\Big\|\prod_{i=1}^{k+1} g_i\Big\|_{L^p(\mathcal{O})} &\le \big\|g_{k+1}\big\|_{L^{p_{k+1}}(\mathcal{O})} \prod_{i=1}^{k} \Big\||g_i|^{p'_{k+1}}\Big\|_{L^{q_i}(\mathcal{O})}^{\frac{1}{p'_{k+1}}} \\ &= \big\|g_{k+1}\big\|_{L^{p_{k+1}}(\mathcal{O})} \prod_{i=1}^{k} \|g_i\|_{L^{p_i}(\mathcal{O})},\end{aligned}$$

which proves the desired inequality for $k+1$. This ends the proof. □

We shall also make use of the following inequality:

Theorem 6.35 (Young's Inequality). *Let $a, b \ge 0$. Then*

$$ab \le \frac{1}{p}a^p + \frac{1}{p'}b^{p'}, \tag{6.38}$$

where $\frac{1}{p} + \frac{1}{p'} = 1$ and $p \in (1, +\infty)$.

Proof. If $a = 0$ or $b = 0$, inequality (6.38) obviously holds.

Suppose $a > 0$ and $b > 0$. By the convexity of the function e^x on $\mathbb{R}$, for every $t \in [0,1]$ and x, y in $\mathbb{R}$,

$$e^{tx+(1-t)y} \le te^x + (1-t)e^y. \tag{6.39}$$

Since

$$\ln(ab) = \ln a + \ln b = \frac{\ln(a^p)}{p} + \frac{\ln(b^{p'})}{p'},$$

choosing $t = 1/p$ (which implies $1 - t = 1/p'$), $x = \ln a^p$ and $y = \ln b^{p'}$ in (6.39), yields

$$\begin{aligned}ab = e^{\ln(ab)} = \exp\Big(\frac{\ln(a^p)}{p} + \frac{\ln(b^{p'})}{p'}\Big) &\le \frac{\exp\big(\ln(a^p)\big)}{p} + \frac{\exp\big(\ln(b^{p'})\big)}{p'} \\ &= \frac{a^p}{p} + \frac{b^{p'}}{p'}.\end{aligned}$$

This gives the claimed inequality. □

Theorem 6.36 (Interpolation Inequality Lemma). *Let $1 \leq p < r < +\infty$. If f is in $L^p(\mathcal{O}) \cap L^r(\mathcal{O})$, then f is in $L^q(\mathcal{O})$ for every $q \in [p, r]$. Moreover,*

$$\|f\|_{L^q(\mathcal{O})} \leq \|f\|^{\alpha}_{L^p(\mathcal{O})} \|f\|^{1-\alpha}_{L^r(\mathcal{O})}, \tag{6.40}$$

where $\alpha \in [0, 1]$ is given by

$$\frac{1}{q} = \frac{\alpha}{p} + \frac{1-\alpha}{r}. \tag{6.41}$$

Proof. Observe first that (6.41) implies $\alpha = \dfrac{p}{q} \dfrac{(r-q)}{(r-p)}$, so $\alpha \in [0, 1]$. Moreover,

$$\frac{\alpha q}{p} + \frac{q(1-\alpha)}{r} = 1.$$

Since

$$\|f\|^q_{L^q(\mathcal{O})} = \int_{\mathcal{O}} |f(x)|^q \, dx = \int_{\mathcal{O}} |f(x)|^{\alpha q} \, |f(x)|^{(1-\alpha)q} \, dx,$$

in view of (6.33), Hölder's inequality (6.31) applied to the functions

$$|f|^{\alpha q} \in L^{\frac{p}{\alpha q}}(\mathcal{O}) \quad \text{and} \quad |f|^{(1-\alpha)q} \in L^{\frac{r}{(1-\alpha)q}}(\mathcal{O}),$$

gives

$$\begin{aligned} \|f\|^q_{L^q(\Omega)} &\leq \left\| |f|^{\alpha q} \right\|_{L^{\frac{p}{\alpha q}}(\mathcal{O})} \left\| |f|^{(1-\alpha)q} \right\|_{L^{\frac{r}{(1-\alpha)q}}(\mathcal{O})} \\ &= \|f\|^{\alpha q}_{L^p(\mathcal{O})} \|f\|^{(1-\alpha)q}_{L^r(\mathcal{O})}, \end{aligned}$$

which implies (6.40). □

6.3.2 *Main properties of L^p-spaces*

We recall below Fatou's Lemma, Lebesgue's Dominated Convergence Theorem, and Fubini's Theorem in the L^1-setting.

Theorem 6.37 (Fatou's Lemma). *Let $\{f_n\}$ be a bounded sequence of positive functions in $L^1(\mathcal{O})$ and let*

$$f(x) = \liminf_{n \to \infty} f_n(x) \quad \text{a.e. in } \mathcal{O}.$$

Then $f \in L^1(\mathcal{O})$ and

$$\|f\|_{L^1(\mathcal{O})} \leq \liminf_{n \to \infty} \|f_n\|_{L^1(\mathcal{O})}.$$

Theorem 6.38 (Lebesgue's Dominated Convergence Theorem). *Let $\{f_n\}$ be a sequence in $L^1(\mathcal{O})$ satisfying*

$$f_n(x) \to f(x) \quad \text{a.e. in } \mathcal{O}.$$

If there exists a function $h \in L^1(\mathcal{O})$ such that for every n,

$$|f_n(x)| \leq h(x) \quad \text{a.e. in } \mathcal{O},$$

then $f \in L^1(\mathcal{O})$ and f_n converges to f in $L^1(\mathcal{O})$, that is,

$$\|f_n - f\|_{L^1(\mathcal{O})} \to 0.$$

In particular,

$$\int_{\mathcal{O}} f_n(x)\,dx \to \int_{\mathcal{O}} f(x)\,dx.$$

Theorem 6.39 (Fubini's Theorem). *Let $X \subseteq \mathbb{R}^N$ and $Y \subseteq \mathbb{R}^M$ be measurable sets and $f \in L^1(X \times Y)$. Define the mappings*

$$f^x : y \in Y \longmapsto f(x,y),$$

$$f^y : x \in X \longmapsto f(x,y).$$

Then $f^x \in L^1(Y)$ for a.e. $x \in X$ and $f^y \in L^1(X)$ for a.e. $y \in Y$. Moreover,

$$\int_{X\times Y} f\,dx\,dy = \int_X \left(\int_Y f^x\,dy\right) dx = \int_Y \left(\int_X f^y\,dx\right) dy.$$

It can be shown that in general, a sequence which converges in $L^p(\mathcal{O})$ does not converge almost everywhere in $\mathcal{O}$. Nevertheless, we have

Theorem 6.40. *Let $p \in \mathbb{R}$ with $1 \leq p \leq +\infty$, and $\{f_n\}$ be a sequence in $L^p(\mathcal{O})$ such that*

$$f_n \to f \quad \text{in } L^p(\mathcal{O}),$$

for some $f \in L^p(\mathcal{O})$. Then there exists a subsequence $\{f_{n_k}\}$ such that

$$f_{n_k}(x) \to f(x) \quad \text{a.e. in } \mathcal{O}.$$

A relevant property of Banach spaces is the separability.

Definition 6.41. *We say that the Banach space E is separable if there exists a set, at most countable, which is dense in E.*

Proposition 6.42. *The space $L^p(\mathcal{O})$ is separable for $1 \leq p < +\infty$.*

6.3.3 *Dual of L^p-spaces*

Let us end this section with the characterization of the dual of L^p-spaces.

Theorem 6.43 (Green Representation Theorem). *Let $p \in \mathbb{R}$ with $1 \leq p < +\infty$ and p' be its conjugate. If $f \in [L^p(\mathcal{O})]'$, then there exists a unique $g_f \in L^{p'}(\mathcal{O})$ such that*

$$\langle f, \varphi \rangle_{[L^p(\mathcal{O})]', L^p(\mathcal{O})} = \int_{\mathcal{O}} g_f(x)\, \varphi(x)\, dx, \quad \forall \varphi \in L^p(\mathcal{O}).$$

Moreover, the map $f \in [L^p(\mathcal{O})]' \to g_f \in L^{p'}(\mathcal{O})$ is a (bijective) isometry, that is,

$$\|g_f\|_{L^{p'}(\mathcal{O})} = \|f\|_{[L^p(\mathcal{O})]'} \cdot$$

Remark 6.44.
(1) For $p = 2$, the result is an immediate consequence of the Riesz Representation Theorem (Theorem 6.25).
(2) The representation theorem implies that one can identify f with g_f, so for $1 \leq p < +\infty$, the space $[L^p(\mathcal{O})]'$ can be identified with $L^{p'}(\mathcal{O})$.

In particular, $[L^1(\mathcal{O})]' = L^\infty(\mathcal{O})$. Also, $L^1(\mathcal{O}) \subsetneq [L^\infty(\mathcal{O})]'$. Indeed, every $f \in L^1(\mathcal{O})$ can be identified with the map

$$\Psi_f \, : \, \varphi \in L^\infty(\mathcal{O}) \mapsto \int_{\mathcal{O}} f(x)\, \varphi(x)\, dx,$$

which is an element of $[L^\infty(\mathcal{O})]'$. However, $[L^\infty(\mathcal{O})]' \neq L^1(\mathcal{O})$ (we refer the reader to [3], [4] for the proof of this result). It follows that $L^p(\mathcal{O})$ is reflexive for $1 < p < +\infty$, while the spaces $L^1(\mathcal{O})$ and $L^\infty(\mathcal{O})$ are not reflexive.

6.4 Density in L^p-spaces

In this section, we prove that $\mathcal{D}(\mathcal{O})$ (see Definition 6.46 below) is dense in $L^p(\mathcal{O})$. To do so, we first consider the case when $\mathcal{O} = \mathbb{R}^N$ and use the notion of convolution (see Definition 6.52). Based on this result, we prove the density for the case $\mathcal{O} \subset \mathbb{R}^N$.

Density properties are basic tools in the theory of L^p and Sobolev spaces. In particular, the Sobolev Embedding Theorems (Chapter 8), rely on density results.

Definition 6.45. *For any continuous function $\varphi : \mathcal{O} \to \mathbb{R}$, the support of φ, denoted supp φ, is defined by*

$$supp\ \varphi = \overline{\{x \in \mathcal{O} |\ \varphi(x) \neq 0\}} \cap \mathcal{O}.$$

Definition 6.46. *We define $\mathcal{D}(\mathcal{O})$ as the set*

$$\mathcal{D}(\mathcal{O}) = \big\{v : \mathcal{O} \to \mathbb{R} \mid v \text{ is infinitely differentiable on } \mathcal{O} \text{ and } \text{supp } v \text{ is a compact set in } \mathbb{R}^N \text{ contained in } \mathcal{O}\big\},$$

and denote by $C_c^0(\mathcal{O})$, the set of continuous functions $\varphi : \mathcal{O} \to \mathbb{R}$, whose support is a compact set in $\mathbb{R}^N$ contained in $\mathcal{O}$.

Remark 6.47.
(1) The property that supp v is a compact set in $\mathbb{R}^N$ contained in $\mathcal{O}$ means that supp v is a bounded set and

$$\text{supp } v = \overline{\{x \in \mathcal{O} \mid \varphi(x) \neq 0\}}.$$

(2) We can construct functions in $\mathcal{D}(\mathcal{O})$ having an arbitrarily small support. For any $a > 0$, the function φ defined on $\mathbb{R}^N$ by

$$\varphi(x) = \begin{cases} e^{-\frac{a^2}{a^2 - |x|^2}} & \text{if } |x| < a, \\ 0 & \text{if } |x| \geq a, \end{cases} \tag{6.42}$$

is in $C^\infty(\mathbb{R}^N)$ and its support is the closed ball centered at the origin and of radius a.
(3) In the literature, $\mathcal{D}(\mathcal{O})$ is also denoted $C_0^\infty(\mathcal{O})$.

We are now concerned with the notion of support of a function in $L^p(\mathcal{O})$. Definition 6.45 cannot be adopted for this space, since a function in $L^p(\mathcal{O})$ is defined only almost everywhere. An appropriate definition is given below and reduces to Definition 6.45 when the function is continuous. It is justified by the following property:

Proposition 6.48. *Let $1 \leq p \leq +\infty$ and $f \in L^p(\mathcal{O})$. If*

$$\mathcal{A} = \big\{A \subseteq \mathcal{O} \mid A \text{ open in } \mathcal{O} \text{ and } f = 0 \text{ a.e. in } A\big\},$$

and

$$\omega = \bigcup_{A \in \mathcal{A}} A, \tag{6.43}$$

then $f = 0$ almost everywhere on ω.

Proof. For every $n \in \mathbb{N}$, introduce the compact set

$$K_n = \left\{x \in \omega \mid d(x, \partial\omega) \geq \frac{1}{n}\right\} \bigcap \overline{B(0,n)},$$

which is clearly contained in ω. We have

$$\omega = \bigcup_{n \in \mathbb{N}} K_n.$$

Using (6.43) it follows that for any $n \in \mathbb{N}$, there exists $h(n) \in \mathbb{N}$ and $A_1^n, \ldots, A_{h(n)}^n$ in $\mathcal{A}$ such that

$$K_n \subset \bigcup_{i=1}^{h(n)} A_i^n.$$

Hence

$$\omega = \bigcup_{n \in \mathbb{N}} \bigcup_{i=1}^{h(n)} A_i^n,$$

which is a countable union. By construction, $f = 0$ almost everywhere on each A_i^n except on a set of zero (Lebesgue) measure. Since a countable union of sets of zero measure still has a zero measure, this implies that $f = 0$ almost everywhere on ω. $\square$

Definition 6.49. *Let $1 \leq p \leq +\infty$. For any function f in $L^p(\mathcal{O})$, the support of f, denoted supp f, is the closed subset of $\mathcal{O}$ defined by*

$$\mathit{supp}\ f = \mathcal{O} \setminus \omega,$$

where ω is given by (6.43) *in Proposition 6.48.*

We denote by $L_c^p(\mathcal{O})$ the set of functions in $L^p(\mathcal{O})$ whose support is a compact subset of $\mathbb{R}^N$ contained in $\mathcal{O}$.

Remark 6.50. *If f is continuous, Definition 6.49 reduces to Definition 6.45. Indeed, in this case*

$$\mathcal{A} = \big\{ A \subseteq \mathcal{O} \mid A \text{ open in } \mathcal{O} \text{ and } f = 0 \text{ on } A \big\},$$

so that

$$\omega = \bigcup_{A \in \mathcal{A}} A = \mathit{int}\big\{ x \in \mathcal{O} \mid f(x) = 0 \big\},$$

and thus,

$$\mathcal{O} \setminus \omega = \mathcal{O} \cap \omega^c = \mathcal{O} \cap \overline{(\{x \in \mathcal{O} \mid f(x) = 0\})^c} = \mathcal{O} \cap \overline{\{x \in \mathcal{O} \mid f(x) \neq 0\}},$$

where B^c and int B denote the complement with respect to $\mathcal{O}$ and the interior of B, respectively, for any $B \subseteq \mathcal{O}$.

We admit the following density result, which is a classical result in measure theory.

Theorem 6.51. *The set $C_c^0(\mathbb{R}^N)$ introduced in Definition 6.46 is dense in $L^p(\mathbb{R}^N)$ for $1 \leq p < +\infty$.*

6.4.1 *Convolution and mollifiers*

The result stated in Theorem 6.51 is not sufficient here, because we need to approach L^p by more regular functions having continuous derivatives. To do that, we introduce the notion of convolution, which is adapted to this aim. Let us recall its definition and main properties. (For proofs, see [3], [4], [16] or [34].)

Definition 6.52. *Let $1 \leq p \leq +\infty$, $f \in L^1(\mathbb{R}^N)$, and $g \in L^p(\mathbb{R}^N)$. The function $f \star g$ given by*

$$(f \star g)(x) = \int_{\mathbb{R}^N} f(x-y)g(y)\, dy \qquad \text{a.e. for } x \in \mathbb{R}^N, \tag{6.44}$$

is called the convolution of f and g.

The convolution is well-defined, since the definition is justified by the following proposition:

Proposition 6.53. *Let $1 \leq p \leq +\infty$, $f \in L^1(\mathbb{R}^N)$, and $g \in L^p(\mathbb{R}^N)$. Then the function $v(y) = f(x-y)g(y)$ is integrable on $\mathbb{R}^N$ for a.e. $x \in \mathbb{R}^N$. Moreover, if*

$$(f \star g)(x) = \int_{\mathbb{R}^N} f(x-y)g(y)\, dy \qquad \text{a.e. for } x \in \mathbb{R}^N,$$

then $f \star g \in L^p(\mathbb{R}^N)$ and

$$\|f \star g\|_{L^p(\mathbb{R}^N)} \leq \|f\|_{L^1(\mathbb{R}^N)} \|g\|_{L^p(\mathbb{R}^N)}. \tag{6.45}$$

We also have the following important properties of the convolution:

Proposition 6.54.
(1) **(Commutativity)** *Let f and g be in $L^1(\mathbb{R}^N)$. The convolution of f and g is commutative, that is, $f \star g = g \star f$.*
(2) **(Differentiation Rule)** *Let $1 \leq p \leq +\infty$, $f \in \mathcal{D}(\mathbb{R}^N)$, and $g \in L^p(\mathbb{R}^N)$. Then $f \star g \in C^\infty(\mathbb{R}^N)$ and for any $i = 1, \ldots, N$,*

$$\frac{\partial}{\partial x_i}(f \star g) = \frac{\partial f}{\partial x_i} \star g. \tag{6.46}$$

Furthermore, if g belongs to $L^p_c(\mathbb{R}^N)$, then $f \star g \in \mathcal{D}(\mathbb{R}^N)$ and

$$\text{supp } (f \star g) \subset \text{supp } f + \text{supp } g,$$

where

$$\text{supp } f + \text{supp } g = \{x + y \,|\, x \in \text{supp } f, y \in \text{supp } g\}.$$

Remark 6.55. *Let us point out that if only one of the two functions has a compact support, the convolution $f \star g$ does not necessarily have a compact support.*

We now introduce the notion of a mollifier, which allows us to construct regularizing sequences for L^p functions in the sense of the definition below:

Definition 6.56. *Let $B(0,1)$ be the open unit ball in $\mathbb{R}^N$ centered at the origin, and $\rho \in \mathcal{D}(\mathbb{R}^N)$ a positive function such that*

$$\text{supp } \rho \subset \overline{B(0,1)} \quad \text{and} \quad \|\rho\|_{L^1(\mathbb{R}^N)} = 1.$$

For any $n \in \mathbb{N}$, the function

$$\rho_n(x) = n^N \rho(nx), \qquad x \in \mathbb{R}^N,$$

is called the n-mollifier of ρ and the sequence $\{\rho_n\}$ is called a regularizing sequence.

Remark 6.57.

(1) An example of a function satisfying the assumptions of Definition 6.56 is given by

$$\rho = \frac{\varphi}{\|\varphi\|_{L^1(\mathbb{R}^N)}},$$

where φ is the function defined in (6.42) for $a = 1$.

(2) It is easy to verify that the n-mollifier given in Definition 6.56 is a positive function in $\mathcal{D}(\mathbb{R}^N)$ such that

$$\text{supp } \rho_n \subset \overline{B\big(0, \frac{1}{n}\big)} \quad \text{and} \quad \|\rho_n\|_{L^1(\mathbb{R}^N)} = 1.$$

6.4.2 Density results in $L^p(\mathbb{R}^N)$

The following important result states in particular, that for $1 \leq p < +\infty$ the set $C^\infty(\mathbb{R}^N)$ is dense in $L^p(\mathbb{R}^N)$.

Theorem 6.58. *Let $1 \leq p < +\infty$ and $f \in L^p(\mathbb{R}^N)$. For any regularizing sequence $\{\rho_n\}$ given by Definition 6.56, we have*

$$\rho_n \star f \to f \qquad \text{in } L^p(\mathbb{R}^N).$$

Proof. Suppose first that $f \in C_c^0(\mathbb{R}^N)$ and let $\{\rho_n\}$ be a regularizing sequence. By definition,

$$\begin{aligned}(\rho_n \star f)(x) - f(x) &= \int_{\mathbb{R}^N} \rho_n(x-y) f(y)\, dy - f(x) \\ &= \int_{\mathbb{R}^N} \rho_n(x-y)[f(y) - f(x)]\, dy,\end{aligned} \tag{6.47}$$

for every $x \in \mathbb{R}^N$, since by a change of variable and using Remark 6.57

$$\int_{\mathbb{R}^N} \rho_n(x-y)dy = \int_{\mathbb{R}^N} \rho_n(z)dz = 1. \tag{6.48}$$

On the other hand, using estimate (6.45) of Proposition (6.53) we have

$$\begin{aligned}\left|\int_{\mathbb{R}^N} \rho_n(x-y)[f(y)-f(x)]\,dy\right| &= \left|\int_{\|x-y\|\le\frac{1}{n}} \rho_n(x-y)[f(y)-f(x)]\,dy\right| \\ &\le \left(\int_{\mathbb{R}^N} \rho_n(x-y)\,dy\right) \max_{\|x-y\|\le\frac{1}{n}} |f(y)-f(x)| \\ &= \max_{\|x-y\|\le\frac{1}{n}} |f(y)-f(x)|,\end{aligned}$$

which, together with (6.47), prove that

$$\max_{\text{supp } f} |(\rho_n \star f)(x) - f(x)| \to 0, \quad \text{as } n \to \infty,$$

since a function in $C_c^0(\mathbb{R}^N)$ is uniformly continuous. This implies the result for functions in $C_c^0(\mathbb{R}^N)$.

Suppose now that $f \in L^p(\mathbb{R}^N)$. By Theorem 6.51, for every $\eta > 0$ there exists a function $\varphi_\eta \in C_c^0(\mathbb{R}^N)$, such that

$$\|\varphi_\eta - f\|_{L^p(\mathbb{R}^N)} \le \eta.$$

Using again estimate (6.45) and (6.48), we have

$$\begin{aligned}\|\rho_n \star f - f\|_{L^p(\mathbb{R}^N)} &\le \|\rho_n \star f - \rho_n \star \varphi_\eta\|_{L^p(\mathbb{R}^N)} + \|\rho_n \star \varphi_\eta - \varphi_\eta\|_{L^p(\mathbb{R}^N)} \\ &\quad + \|\varphi_\eta - f\|_{L^p(\mathbb{R}^N)} \le 2\eta + \|\rho_n \star \varphi_\eta - \varphi_\eta\|_{L^p(\mathbb{R}^N)}.\end{aligned}$$

By the first part of the proof, we can pass to the limit as $n \to \infty$ to obtain

$$\lim_{n\to\infty} \|\rho_n \star f - f\|_{L^p(\mathbb{R}^N)} = 0,$$

since η is arbitrary, which ends the proof. □

We are now ready to prove the following density result:

Theorem 6.59. *The set $\mathcal{D}(\mathbb{R}^N)$ is dense in $L^p(\mathbb{R}^N)$ for $1 \le p < +\infty$.*

Proof. Let $1 \le p < +\infty$. First by a truncation argument we approximate any element of $L^p(\mathbb{R}^N)$ by functions with compact support, then, using a regularizing sequence, we approximate these functions by functions in $\mathcal{D}(\mathbb{R}^N)$.

From Theorem (6.51) it follows that there exists a sequence $\{v_n\}$ in $C_c^0(\mathbb{R}^N)$ (hence in $L_c^p(\mathbb{R}^N)$), such that

$$v_n \to v \quad \text{in } L^p(\mathbb{R}^N). \tag{6.49}$$

This implies that for every $\varepsilon > 0$ there exists $n_0 \in \mathbb{N}$ such that

$$\|v_{n_0} - v\|_{L^p(\mathbb{R}^N)} < \frac{\varepsilon}{2}. \tag{6.50}$$

Let now $\{\rho_n\}$ be a regularizing sequence given by Definition 6.56. Since v_{n_0} is in $L_c^p(\mathbb{R}^N)$, in view of (2) of Proposition 6.54, the function $\rho_n \star v_{n_0}$ is in $\mathcal{D}(\mathbb{R}^N)$. Moreover, by Theorem 6.58 we have

$$\rho_n \star v_{n_0} \to v_{n_0} \quad \text{in } L^p(\mathbb{R}^N). \tag{6.51}$$

Consequently, there exists $n_\varepsilon \in \mathbb{N}$ such that $v_\varepsilon = \rho_{n_\varepsilon} \star v_{n_0}$ satisfies

$$\|v_\varepsilon - v_{n_0}\|_{L^p(\mathbb{R}^N)} < \frac{\varepsilon}{2}. \tag{6.52}$$

Therefore, according to (6.50) and (6.52),

$$\| v - v_\varepsilon\|_{L^p(\mathbb{R}^N)} \leq \| v - v_{n_0}\|_{L^p(\mathbb{R}^N)} + \| v_{n_0} - v_\varepsilon\|_{L^p(\mathbb{R}^N)} < \varepsilon,$$

proving the density of $\mathcal{D}(\mathbb{R}^N)$ in $L^p(\mathbb{R}^N)$. □

6.4.3 *Density results in $L^p(\mathcal{O})$*

We now show a result similar to Theorem 6.59 for functions defined on an open subset $\mathcal{O}$ of $\mathbb{R}^N$. To do this, we need the following lemma:

Theorem 6.60 (Urysohn's Lemma). *Let $\mathcal{O}$ be an open subset of $\mathbb{R}^N$ and K a compact set contained in $\mathcal{O}$. Then there exists a function $\psi \in \mathcal{D}(\mathcal{O})$ such that*

$$\psi \equiv 1 \quad \text{on} \quad K \quad \text{and} \quad 0 \leq \psi(x) \leq 1 \;\; \text{for every } x \in \mathcal{O}. \tag{6.53}$$

The existence of a function in $C_c^0(\mathcal{O})$ satisfying (6.53) is a classical lemma in measure theory. The construction of a function in $\mathcal{D}(\mathcal{O})$ can be done by convolution arguments. For the proof of this lemma, which is rather technical, we refer the reader to [34], [3], [4], and [16].

An interesting consequence of Urysohn's Lemma is the following result:

Corollary 6.61. *For any open subset $\mathcal{O}$ of $\mathbb{R}^N$, there exists a sequence $\{\psi_n\}$ in $\mathcal{D}(\mathcal{O})$ such that*

$$0 \leq \psi_n(x) \leq 1 \quad \text{and} \quad \psi_n(x) \to 1 \;\; \text{for every } x \in \mathcal{O}. \tag{6.54}$$

Proof. For every $n \in \mathbb{N}$, introduce the compact set

$$K_n = \left\{ x \in \mathcal{O} \mid d(x, \partial\mathcal{O}) \geq \frac{1}{n} \right\} \bigcap \overline{B(0,n)},$$

which is clearly contained in $\mathcal{O}$. The sequence $\{K_n\}$ is increasing and

$$\bigcup_{n \in \mathbb{N}} K_n = \mathcal{O}.$$

For every $n \in \mathbb{N}$, let ψ_n be the function given by Urysohn's Lemma for $K = K_n$. By construction, the sequence $\{\psi_n\}$ has the desired properties since for any fixed x in $\mathcal{O}$, $\psi(x) = 1$ for n sufficiently large. □

We now turn to an important result in the study of Sobolev spaces that will be treated in the next chapter.

Theorem 6.62. *Let $1 \leq p < +\infty$. Then $\mathcal{D}(\mathcal{O})$ is dense in $L^p(\mathcal{O})$ for $\mathcal{O} \subset \mathbb{R}^N$.*

Proof. Let v be in $L^p(\mathcal{O})$ and $\widetilde{v}$ its zero extension

$$\widetilde{v}(x) = \begin{cases} v(x) & \text{if } x \in \mathcal{O}, \\ 0 & \text{otherwise.} \end{cases}$$

Since $\widetilde{v}$ is in $L^p(\mathbb{R}^N)$, Theorem 6.59 provides the existence of a sequence $\{v_n\}$ in $\mathcal{D}(\mathbb{R}^N)$ such that

$$v_n \to \widetilde{v} \quad \text{in } L^p(\mathbb{R}^N).$$

In particular,

$$v_n \to v \quad \text{in } L^p(\mathcal{O}), \tag{6.55}$$

since

$$\lim_{n\to\infty} \|v_n - v\|_{L^p(\mathcal{O})} = \lim_{n\to\infty} \|v_n - \widetilde{v}\|_{L^p(\mathcal{O})} \leq \lim_{n\to\infty} \|v_n - \widetilde{v}\|_{L^p(\mathbb{R}^N)} = 0.$$

Consider now the sequence $\{\psi_n\} \subset \mathcal{D}(\mathcal{O})$ given by Corollary 6.61. As the sequence $\{\psi_n v_n\}$ is also in $\mathcal{D}(\mathcal{O})$, to prove the result it is sufficient to show that

$$\psi_n v_n \to v \quad \text{in } L^p(\mathcal{O}). \tag{6.56}$$

We have

$$\|\psi_n v_n - v\|_{L^p(\mathcal{O})} \leq \|\psi_n v_n - \psi_n v\|_{L^p(\mathcal{O})} + \|\psi_n v - v\|_{L^p(\mathcal{O})} \cdot$$

Using (6.54) and (6.55) gives

$$\begin{aligned}\lim_{n\to\infty} \|\psi_n v_n - \psi_n v\|^p_{L^p(\mathcal{O})} &= \lim_{n\to\infty} \int_{\mathcal{O}} [\psi_n(x)]^p |v_n(x) - v(x)|^p dx \\ &\le \lim_{n\to\infty} \int_{\mathcal{O}} |v_n(x) - v(x)|^p \, dx = 0.\end{aligned} \tag{6.57}$$

On the other hand, by (6.54)

$$|\psi_n - 1|^p \, |v|^p \le |v|^p \qquad \text{a.e. in } \mathcal{O},$$

$$|\psi_n(x) - 1|^p \, |(v(x)|^p \to 0 \quad \text{a.e. in } \mathcal{O},$$

and by the Lebesgue Dominated Convergence Theorem (Thm 6.38),

$$\lim_{n\to\infty} \|\psi_n v - v\|_{L^p(\mathcal{O})} = 0.$$

This, together with (6.57), implies (6.56) and completes the proof. □

Remark 6.63. *It should be noted that this density result does not hold for* $p = +\infty$*, that is,* $C^0(\mathcal{O})$ *is not dense in* $L^\infty(\mathcal{O})$*. Indeed, if* $\{f_n\}$ *is a sequence in* $C^0(\mathcal{O})$ *that converges to* f *in* $L^\infty(\mathcal{O})$*, then it converges uniformly. Since uniform convergence preserves the continuity at the limit,* f *must be continuous. However, a function in* $L^\infty(\mathcal{O})$ *is not necessarily continuous.*

We end this section with a well-known result for functions in $L^1_{loc}(\mathcal{O})$.

Theorem 6.64. *If* $f \in L^1_{loc}(\mathcal{O})$ *such that*

$$\int_{\mathcal{O}} f(x)\, \varphi(x)\, dx = 0 \quad \textit{for every} \quad \varphi \in \mathcal{D}(\mathcal{O}),$$

then

$$f(x) = 0 \quad \textit{a.e for } x \in \mathcal{O}.$$

Observe that if $f \in L^1_{loc}(\mathcal{O})$ and $\varphi \in \mathcal{D}(\mathcal{O})$, then $\int_{\mathcal{O}} f(x)\varphi(x)\, dx$ is well-defined and finite.

Remark 6.65. *If* $f \in L^2(\mathcal{O})$*, this result is a simple consequence of the density result above and of the fact that the zero function is the only function orthogonal to all elements of* $L^2(\mathcal{O})$*. The proof in the general case is more delicate.*

6.5 Weak and weak* convergence

Let $\{x_n\}$ be a sequence in a real Banach space E with norm $\|\cdot\|_E$. Strong convergence in E, which is the usual notion of convergence, is defined in terms of this norm (refer to Definition 6.1). We can define other notions of convergence in E. Specifically, we can use the dual of E to define convergence in a "weak" sense.

6.5.1 *Weak convergence*

We first discuss the weak convergence in Banach spaces.

Definition 6.66. *Let $\{x_n\}$ be a sequence in E. Then $\{x_n\}$ is said to converge weakly to x in E, denoted by*

$$x_n \rightharpoonup x \quad \textit{weakly in } E,$$

if

$$\lim_n \langle x', x_n \rangle_{E',E} \;\to\; \langle x', x \rangle_{E',E}, \quad \forall\, x' \in E'.$$

We call x the weak limit of x_n.

Note that if the weak limit of a sequence in E exists, it is unique. The proof of this statement makes use of the Hahn–Banach Theorem (see [43]). How are the strong and weak limits related? In general, strong convergence implies weak convergence, but the converse is not true. However, if E is a finite dimensional space, the two notions of convergence are equivalent.

Theorem 6.67.
(1) If $x_n \to x$ strongly in E, then $x_n \rightharpoonup x$ weakly in E.
(2) If $\dim E = N < +\infty$, then strong and weak convergence are equivalent.

Proof.
(1) Let $\{x_n\}$ be a sequence in E and suppose $x_n \to x$ strongly in E. By Proposition 6.11, for any $x' \in E'$,

$$\lim_{n\to\infty} \left| \langle x', x_n \rangle_{E',E} - \langle x', x \rangle_{E',E} \right| \le \lim_{n\to\infty} \|x'\|_{E'} \|x_n - x\|_E = 0.$$

Hence $x_n \rightharpoonup x$ weakly in E.

(2) Let $\dim E = N < +\infty$ and $\{x_n\}$ a sequence in E. We only need to show that weak convergence in E implies strong convergence. Suppose $x_n \rightharpoonup x$ weakly in E. Consider a basis $\{e_1, \ldots, e_N\}$ of E with

$\|e_i\|_E = 1$ for $i = 1, \dots, N$. Let $y \in E$. There exist $y_1, \dots, y_N \in \mathbb{R}$, uniquely determined such that

$$y = \sum_{i=1}^{N} y_i e_i.$$

For $i = 1, \dots, N$, define the mappings $f_i : E \longmapsto \mathbb{R}$ by

$$f_i : y \;\mapsto\; y_i \in \mathbb{R}.$$

Then $f_i \in E'$ for $i = 1, \dots, N$. Since $x_n \rightharpoonup x$ weakly in E,

$$\lim_{n\to\infty} \langle f_i, x_n - x\rangle_{E', E} = 0, \qquad i = 1, \dots, N.$$

Consequently,

$$\begin{aligned}\|x_n - x\|_E &= \Big\| \sum_{i=1}^{N} e_i \langle f_i, x_n - x\rangle_{E', E} \Big\|_E \leq \sum_{i=1}^{N} \Big\| e_i \langle f_i, x_n - x\rangle_{E', E} \Big\|_E \\ &= \sum_{i=1}^{N} \Big| \langle f_i, x_n - x\rangle_{E', E} \Big|,\end{aligned}$$

which gives the strong convergence of x_n to x. □

Remark 6.68. *In view of the definition of weak convergence and the Riesz Representation Theorem 6.25 (see also Remark 6.44), weak convergence in the L^p-spaces signifies:*

(1) If $\{u_n\}$ is a sequence in $L^p(\mathcal{O})$ with $1 < p < +\infty$, then

$$u_n \rightharpoonup u \quad \textit{weakly in } L^p(\mathcal{O}),$$

is equivalent to

$$\int_{\mathcal{O}} u_n \, \varphi \, dx \longrightarrow \int_{\mathcal{O}} u \, \varphi \, dx, \qquad \forall\, \varphi \in L^{p'}(\mathcal{O}),$$

where p' is the conjugate of p.

(2) If $\{u_n\}$ is a sequence in $L^1(\mathcal{O})$, then

$$u_n \rightharpoonup u \quad \textit{weakly in } L^1(\mathcal{O}),$$

is equivalent to

$$\int_{\mathcal{O}} u_n \, \varphi \, dx \longrightarrow \int_{\mathcal{O}} u \, \varphi \, dx, \qquad \forall\, \varphi \in L^{\infty}(\mathcal{O}).$$

We now present some properties of weak convergence. We refer the reader to [43] for their proofs. The following result is a particular case of the Banach–Steinhaus Theorem:

Theorem 6.69. *Let $\{x_n\}$ be a sequence weakly convergent to x in E. Then*
(1) $\{x_n\}$ is a bounded sequence in E.
(2) The norm on E is lower semi-continuous with respect to the weak convergence.

Remark 6.70. *The lower semi-continuity of the norm implies that if*

$$f_n \rightharpoonup f \quad \text{weakly in } L^p(\mathcal{O}),$$

for $1 \leq p < +\infty$, then

$$\int_{\mathcal{O}} |f|^p \, dx \leq \liminf_{n\to\infty} \int_{\mathcal{O}} |f_n|^p \, dx.$$

Theorem 6.71. *Let Λ be a linear continuous map from a reflexive Banach space E to a reflexive Banach space F. If*

$$x_n \rightharpoonup x \quad \text{weakly in } E,$$

then

$$\Lambda(x_n) \rightharpoonup \Lambda(x) \quad \text{weakly in } F.$$

Recall that in $\mathbb{R}^N$, bounded sequences have (strongly) convergent subsequences. A similar important property holds for weak convergence but in reflexive Banach spaces.

Theorem 6.72 (Eberlein–Šmuljan). *Let E be a reflexive Banach space. If $\{x_n\}$ is a bounded sequence in E, then*
(1) $\{x_n\}$ has a weakly convergent subsequence. That is, there exists a subsequence $\{x_{n_k}\}$ of $\{x_n\}$ and $x \in E$ such that

$$x_{n_k} \rightharpoonup x \quad \text{weakly in } E.$$

(2) If each weakly convergent subsequence of $\{x_n\}$ has the same weak limit x, then

$$x_n \rightharpoonup x \quad \text{weakly in } E.$$

Remark 6.73. *The above theorem, together with Remark 6.68, implies that for $1 < p < +\infty$, any bounded sequence in $L^p(\mathcal{O})$ has a weakly convergent subsequence.*

The next theorem allows us to pass to the limit in products of "weak-strong" convergent sequences.

Proposition 6.74. *Let $\{x_n\}$ and $\{y_n\}$ be sequences in E and its dual E', respectively, such that*

$$\begin{cases} x_n \rightharpoonup x & \text{weakly in } E, \\ y_n \to y & \text{strongly in } E', \end{cases} \quad \text{or} \quad \begin{cases} x_n \to x & \text{strongly in } E, \\ y_n \rightharpoonup y & \text{weakly in } E'. \end{cases}$$

Then

$$\lim_{n\to\infty} \langle y_n, x_n\rangle_{E',E} = \langle y, x\rangle_{E',E}.$$

Proof. From Proposition 6.11 and Theorem 6.69, in the first case we have

$$\begin{aligned} &\lim_{n\to\infty} \left|\langle y_n, x_n\rangle_{E',E} - \langle y, x\rangle_{E',E}\right| \\ &\qquad = \lim_{n\to\infty} \left|\langle y_n - y, x_n\rangle_{E',E} + \langle y, x_n - x\rangle_{E',E}\right| \\ &\qquad \le \lim_{n\to\infty} \|y_n - y\|_{E'} \|x_n\|_E + \lim_{n\to\infty} \left|\langle y, x_n - x\rangle_{E',E}\right|, \end{aligned}$$

which tends to 0 and proves our assertion. The proof is similar in the second case, due to Remark 6.12. □

6.5.2 *Weak* convergence*

There are still instances where weak convergence may be difficult to investigate. One needs a knowledge of the dual space to study weak convergence and in cases when the dual is too large or complicated, weak convergence is difficult to verify. In such cases, we may use the weak* convergence, which is defined below.

Definition 6.75. *Let F be a Banach space and F' its dual. A sequence $\{x_n\}$ in F' is said to converge weakly* to x, convergence denoted*

$$x_n \rightharpoonup x \quad \text{weakly}^* \text{ in } F',$$

if

$$\langle x_n, y\rangle_{F',F} \to \langle x, y\rangle_{F',F}, \quad \forall y \in F. \tag{6.58}$$

We call x the weak limit of $\{x_n\}$.*

It is easy to verify the following basic properties of weak* convergence:

Theorem 6.76.
(1) If the weak limit of $\{x_n\}$ exists, it is unique.*

(2) If F is a Banach space, then any weakly convergent sequence in F' is also weakly convergent.*
(3) If F is a reflexive Banach space, then weak convergence and weak convergence are equivalent.*

Remark 6.77. *As mentioned in Remark 6.44, the space $[L^\infty(\mathcal{O})]'$ is different from $L^1(\mathcal{O})$ and has a complicated structure that renders weak convergence almost impossible to check. On the other hand, the dual of $L^1(\mathcal{O})$ is given by*

$$[L^1(\mathcal{O})]' = L^\infty(\mathcal{O}),$$

which has well-known properties. It is why the notion of weak convergence is more convenient for the case of $L^\infty(\mathcal{O})$, and can be characterized as follows:*
(1) $u_n \rightharpoonup u$ weakly in $L^\infty(\mathcal{O})$ is equivalent to*

$$\int_{\mathcal{O}} u_n \, \varphi \, dx \to \int_{\mathcal{O}} u \, \varphi \, dx, \quad \forall \, \varphi \in L^1(\mathcal{O}).$$

(2) If $u_n \rightharpoonup x$ weakly in $L^\infty(\mathcal{O})$, then $u_n \rightharpoonup u$ weakly in any $L^p(\mathcal{O})$ with $1 \leq p < +\infty$.*

The second characterization is a consequence of (6.31), the Corollary to Hölder's Theorem and Theorem 6.72. Since $L^1(\mathcal{O})$ is not reflexive, the weak convergence and the weak convergence in $L^\infty(\mathcal{O})$ are not equivalent.*

Finally, we present the analogues of Theorems 6.69 and 6.72. Weakly* convergent sequences have almost similar properties as weakly convergent sequences.

Theorem 6.78. *Let F be a Banach space and F' its dual. If $\{x_n\}$ is a sequence weakly* convergent to x in F', then*
(1) $\{x_n\}$ is a bounded sequence in F'.
(2) The norm on F' is lower semi-continuous with respect to the weak convergence.*

Theorem 6.79. *Let F be a separable Banach space. If $\{x_n\}$ is a bounded sequence in F', then*
(1) $\{x_n\}$ has a weakly convergent subsequence.*
(2) If each weakly convergent subsequence of $\{x_n\}$ has the same weak* limit x, then $x_n \rightharpoonup x$ weakly* in F'.*

Remark 6.80. *The above theorems imply that if* $f_n \rightharpoonup f$ *weakly* in* $L^\infty(\mathcal{O})$, *then*

$$\sup_{\mathcal{O}} |f|^p \leq \liminf_{n\to\infty} \big(\sup_{\mathcal{O}} |f_n|^p \big),$$

and from any bounded sequence in $L^\infty(\mathcal{O})$ *one can extract a weakly* convergent subsequence.*

Chapter 7

The Sobolev Spaces $W^{1,p}$

As mentioned in the introduction to Chapter 6, Sobolev spaces are the appropriate functional settings for finding "weak" solutions of a partial differential equation when a classical solution does not exist. In these spaces, differentiation has to be understood in the "weak" (or distributional) sense. The aim of this chapter is precisely to introduce these notions.

7.1 A motivation

To find a "weak" solution, a variational formulation of the problem is introduced. For simplicity, we illustrate this in the one-dimensional case. We refer the reader to Proposition 9.11 for the general case.

Consider the following one-dimensional problem in the interval $(0,1)$:

$$\begin{cases} -u''(x) + c(x)u(x) = f(x), & \forall x \in (0,1), \\ u(0) = u(1) = 0, \end{cases} \tag{7.1}$$

where $c \in C^0([0,1])$, $c \geq 0$, and $f \in C^0([0,1])$.

Let us introduce the space

$$V = \{v \in C^1([0,1]) \mid v(0) = v(1) = 0\}. \tag{7.2}$$

Since $\mathcal{D}(0,1) \subset V$, by Theorem 6.62, V is dense in $L^2(0,1)$. Define the differential operator

$$A : C^2([0,1]) \cap V \to C^0([0,1]),$$

given by

$$(Au)(x) = -u''(x) + c(x)u(x) \quad \text{for } x \in [0,1].$$

Theorem 7.1. *Let $c \in C^0([0,1])$, $c \geq 0$ and $f \in C^0([0,1])$. Then a function $u \in C^2([0,1]) \cap V$ is a solution of (7.1) if and only if u is a solution of the following problem:*

$$\int_0^1 (u'\,v' + c\,u\,v)\,dx = \int_0^1 f\,v\,dx, \qquad \forall v \in V. \tag{7.3}$$

Proof. Let $u \in C^2([0,1]) \cap V$ be a solution of (7.1). Multiplying the equation in (7.1) by $v \in V$ and integrating on $[0, 1]$, we obtain

$$\int_0^1 (-u'' + c\,u)v\,dx = \int_0^1 f\,v\,dx, \qquad \forall v \in V. \tag{7.4}$$

Integrating by parts and recalling that $v \in V$, we have

$$\int_0^1 -u''\,v\,dx = \int_0^1 u'\,v'\,dx - u'\,v\Big|_0^1 = \int_0^1 u'\,v'\,dx. \tag{7.5}$$

Equations (7.4) and (7.5) imply that u is a solution of (7.3).

Conversely, let $u \in C^2([0,1]) \cap V$ be a solution of (7.3). By (7.5),

$$\int_0^1 (-u'' + c\,u - f)\,v\,dx = 0, \qquad \forall v \in V.$$

Since V is dense in $L^2(0,1)$, we have

$$\int_0^1 (-u'' + c\,u - f)\,v\,dx = 0, \qquad \forall v \in L^2(0,1).$$

This means that the function $\phi = -u'' + cu - f$ is an element of $L^2(0,1)$ satisfying

$$(\phi, v) = 0, \qquad \forall v \in L^2(0,1).$$

It can be easily verified that the zero function is the only function orthogonal to all elements of $L^2(0,1)$ and hence

$$0 = \phi = -u'' + cu - f \qquad \text{a.e. in } (0,1).$$

Since $\phi \in C^0([0,1])$, it follows that $\phi(x) = 0$ for every x in $(0,1)$. This implies that u satisfies the equation in (7.1). Moreover, $u \in V$, so u satisfies the boundary conditions in (7.1). This completes the proof. □

Remark 7.2. *Problem (7.3) is called a variational formulation of problem (7.1). The fact that u is a solution of (7.3) means that the equality is satisfied by every v belonging to the given space V. The function v is called a test function.*

Observe now that (7.3) makes sense even if f, u, v, u', and v' are only in $L^2(0,1)$ and not in spaces of continuous functions. This implies that if we take the closure H of V with respect to the norm

$$\|u\| = \int_0^1 \left(u^2 + (u')^2\right) dx,$$

then the following problem:

$$\int_0^1 (u'\,v' + c\,u\,v)\,dx = \int_0^1 fvdx, \qquad \forall v \in H, \tag{7.6}$$

makes sense also for $u \in H$. Therefore one can try to find a function in H, a bigger space than $C^2([0,1]) \cap V$, satisfying (7.6). This is the idea behind variational formulations. In this case, u is called a *weak solution* of (7.1) and (7.6) is called the *variational formulation* of (7.1). If the solutions and the data are smooth, then u is a solution in the usual (classical) sense. H is an example of a Sobolev space which will be discussed in this chapter.

7.2 Distributions

We first give a brief review of distributions and their properties. We begin with the notion of convergence for sequences in $\mathcal{D}(\mathcal{O})$, the space introduced in Definition 6.46. For notations, we refer to Section 1.2.

Definition 7.3. *A sequence $\{\varphi_n\}$ in $\mathcal{D}(\mathcal{O})$ is said to converge to an element φ in $\mathcal{D}(\mathcal{O})$ if the following are satisfied:*

(1) there exists a compact set $K \subset \mathcal{O}$ such that supp $\varphi \subset K$ and for any $n \in \mathbb{N}$, supp $\varphi_n \subset K$,

(2) for any $\alpha \in \mathbb{N}^N$, $\partial^\alpha \varphi_n$ converges uniformly to $\partial^\alpha \varphi$ on K.

Remark 7.4. *The space $\mathcal{D}(\mathcal{O})$ is not a metric space and the above definition does not induce a topology on $\mathcal{D}(\mathcal{O})$. However, a suitable topology $\mathcal{T}$ on this space may be defined such that the convergence of sequences in $\mathcal{T}$ is exactly that given in Definition 7.3. This topology has a complicated structure and is not needed for our purpose. We refer the reader to [35] for more details.*

Definition 7.5. *A distribution on $\mathcal{O}$ is a map $T : \mathcal{D}(\mathcal{O}) \to \mathbb{R}$ such that*

(1) T is linear,

(2) if $\varphi_n \to \varphi$ in $\mathcal{D}(\mathcal{O})$, then $T(\varphi_n) \to T(\varphi)$ in $\mathbb{R}$.

We denote by $\mathcal{D}'(\mathcal{O})$ the set of distributions on $\mathcal{O}$.

Remark 7.6.

(1) A distribution on $\mathcal{O}$ is simply a linear functional on $\mathcal{D}(\mathcal{O})$ continuous on sequences in $\mathcal{D}(\mathcal{O})$. The usual notation for a distribution T is

$$T(\varphi) = \langle T, \varphi \rangle_{\mathcal{D}'(\mathcal{O}),\mathcal{D}(\mathcal{O})}, \quad \forall \varphi \in \mathcal{D}(\mathcal{O}). \tag{7.7}$$

Note that this is due to the fact that $\mathcal{D}'(\mathcal{O})$ is the dual of $\mathcal{D}(\mathcal{O})$ with respect to the topology $\mathcal{T}$ mentioned in Remark 7.4.

(2) By linearity, condition (2) in the above definition is equivalent to

$$\varphi_n \to 0 \Rightarrow T(\varphi_n) \to 0. \tag{7.8}$$

Example 7.7. Let $x_0 \in \mathbb{R}^N$. The mapping defined by

$$\delta_{x_0}(\varphi) = \varphi(x_0), \quad \forall \varphi \in \mathcal{D}(\mathbb{R}^N),$$

has the following properties:

(1) *(Linearity)* For $\alpha_1, \alpha_2 \in \mathbb{R}$ and $\varphi_1, \varphi_2 \in \mathcal{D}(\mathbb{R}^N)$

$$\begin{aligned}\delta_{x_0}(\alpha_1\varphi_1 + \alpha_2\varphi_2) &= (\alpha_1\varphi_1 + \alpha_2\varphi_2)(x_0) = \alpha_1\varphi_1(x_0) + \alpha_2\varphi_2(x_0) \\ &= \alpha_1\delta_{x_0}(\varphi_1) + \alpha_2\delta_{x_0}(\varphi_2).\end{aligned}$$

(2) *(Continuity on sequences)* Suppose $\varphi_n \to \varphi$ in $\mathcal{D}(\mathbb{R}^N)$. Then there exists a compact set $K \subset \mathbb{R}^N$ such that supp $\varphi_n \subset K$ for all n and φ_n converges uniformly to φ on K. If $x_0 \notin K$, then clearly,

$$\delta_{x_0}(\varphi_n) = \varphi_n(x_0) = 0 = \varphi(x_0) = \delta_{x_0}(\varphi).$$

On the other hand, if $x_0 \in K$, then by uniform convergence,

$$\delta_{x_0}(\varphi_n) = \varphi_n(x_0) \to \varphi(x_0) = \delta_{x_0}(\varphi).$$

Hence δ_{x_0} is a distribution on $\mathbb{R}^N$, that is, $\delta_{x_0} \in \mathcal{D}'(\mathbb{R}^N)$. This distribution is called the Dirac function (or mass) at the point x_0. ◇

7.2.1 *Regular distributions*

In this section we show how L^1-functions can be identified with distributions.

Proposition 7.8. *Let $f \in L^1_{loc}(\mathcal{O})$ and define the mapping T_f on $\mathcal{D}(\mathcal{O})$ by*

$$\langle T_f, \varphi \rangle_{\mathcal{D}'(\mathcal{O}),\mathcal{D}(\mathcal{O})} = \int_{\mathcal{O}} f\varphi \, dx. \tag{7.9}$$

Then T_f is a distribution on $\mathcal{O}$. Moreover, the linear map

$$L : f \in L^1_{loc}(\mathcal{O}) \longmapsto T_f \in \mathcal{D}'(\mathcal{O}),$$

is one-to-one.

Proof. The linearity of T_f follows from the linearity of integration. If the sequence $\{\varphi_n\}$ converges to φ in $\mathcal{D}(\mathcal{O})$, then for a compact set K such that supp $\varphi \subset K$ and supp $\varphi_n \subset K$ for all n,

$$\left| \langle T_f, \varphi_n \rangle_{\mathcal{D}'(\mathcal{O}),\mathcal{D}(\mathcal{O})} - \langle T_f, \varphi \rangle_{\mathcal{D}'(\mathcal{O}),\mathcal{D}(\mathcal{O})} \right| = \left| \int_{\mathcal{O}} f(\varphi_n - \varphi)\, dx \right|$$

$$\leq \|f\|_{L^1(K)} \max_K |\varphi_n - \varphi| \to 0.$$

It follows that T_f is continuous on sequences and therefore, it is a distribution on $\mathcal{O}$.

On the other hand, the mapping L is clearly linear and moreover, $T_f = 0$ if and only if $f = 0$ by Theorem 6.64. This proves the last assertion. □

Remark 7.9. *As a consequence of Proposition 7.8 we can consider $L^1_{loc}(\mathcal{O})$ as a subspace of $\mathcal{D}'(\mathcal{O})$ by identifying f with T_f. In particular, this also holds for $f \in L^p(\mathcal{O})$, since from the inclusion (6.34) in Corollary 6.33, we have $L^p(\mathcal{O}) \subset L^1_{loc}(\mathcal{O})$.*

With the identification done in this remark, we have the following definition:

Definition 7.10. *A distribution T is in $L^1_{loc}(\mathcal{O})$ (respectively in $L^p(\mathcal{O})$), if there exists $f \in L^1_{loc}(\mathcal{O})$ (respectively in $L^p(\mathcal{O})$), such that $T = T_f$ where T_f is given by (7.9). In this case, T is called a regular distribution.*

There exist distributions which are not regular as stated below.

Proposition 7.11. *The Dirac function at a point $x_0 \in \mathbb{R}$ (see Example 7.7) is not a regular distribution.*

Proof. We prove the result by contradiction. Without loss of generality, assume that $x_0 = 0$. Suppose there is a function $v \in L^1_{loc}(\mathbb{R})$ such that

$$\varphi(0) = \delta_0(\varphi) = \int_{\mathbb{R}} v\, \varphi\, dx, \qquad \forall \varphi \in \mathcal{D}(\mathbb{R}). \tag{7.10}$$

In view of Urysohn's Lemma (Thm 6.60), for every $n \in \mathbb{N}$ there exists $\psi_n \in \mathcal{D}\big((-2/n, 2/n)\big)$ such that for $x \in (-2/n, 2/n)$

$$\psi_n(x) \equiv 1 \text{ on } \Big(-\frac{1}{n}, \frac{1}{n}\Big) \quad \text{and} \quad 0 \leq \psi_n(x) \leq 1.$$

Since $\psi_n \to 0$ on $\mathbb{R} \setminus \{0\}$ (hence almost everywhere on $\mathbb{R}$), by the Lebesgue Dominated Convergence Theorem (Thm 6.38),

$$\lim_{n\to\infty} \int_{\mathbb{R}} v\, \psi_n\, dx = 0.$$

This convergence leads to a contradiction, since by our choice of ψ_n and from (7.10), we have

$$1 = \psi_n(0) = \delta_0(\psi_n) = \int_{\mathbb{R}} v\,\psi_n\,dx, \quad \forall n \in \mathbb{N}.$$

This ends the proof. □

7.2.2 *Convergence in the sense of distributions*

We now define the notion of convergence in the space $\mathcal{D}'(\mathcal{O})$ which we refer to as convergence in the sense of distributions.

Definition 7.12. *Let $\{T_n\}$ be a sequence in $\mathcal{D}'(\mathcal{O})$. We say that $\{T_n\}$ converges (in the sense of distributions) to an element $T \in \mathcal{D}'(\mathcal{O})$, and we denote this convergence as,*

$$T_n \longrightarrow T \quad \text{in } \mathcal{D}'(\mathcal{O}),$$

if for all $\varphi \in \mathcal{D}(\mathcal{O})$,

$$\langle T_n, \varphi\rangle_{\mathcal{D}'(\mathcal{O}),\mathcal{D}(\mathcal{O})} \longrightarrow \langle T, \varphi\rangle_{\mathcal{D}'(\mathcal{O}),\mathcal{D}(\mathcal{O})}. \tag{7.11}$$

Example 7.13. Let $\{f_n\}$ be a sequence in $L^1(\mathcal{O})$ such that $f_n \longrightarrow f$ in $L^1(\mathcal{O})$. Then by Remark 7.9

$$f_n \longrightarrow f \quad \text{in } \mathcal{D}'(\mathcal{O}).$$

This convergence in $\mathcal{D}'(\mathcal{O})$ holds since for every $\varphi \in \mathcal{D}(\mathcal{O})$,

$$\begin{aligned}\Big|\langle f_n, \varphi\rangle_{\mathcal{D}'(\mathcal{O}),\mathcal{D}(\mathcal{O})} - \langle f, \varphi\rangle_{\mathcal{D}'(\mathcal{O}),\mathcal{D}(\mathcal{O})}\Big| &= \Big|\int_{\mathcal{O}} f_n\varphi dx - \int_{\mathcal{O}} f\varphi dx\Big| \\ &\leq \int_{\mathcal{O}} |f_n - f|\,|\varphi| dx \leq \max_{\mathcal{O}} |\varphi| \int_{\mathcal{O}} |f_n - f| \to 0,\end{aligned}$$

by the convergence of f_n in $L^1(\mathcal{O})$. Hence

$$\langle T_{f_n}, \varphi\rangle_{\mathcal{D}'(\mathcal{O}),\mathcal{D}(\mathcal{O})} \longrightarrow \langle T_f, \varphi\rangle_{\mathcal{D}'(\mathcal{O}),\mathcal{D}(\mathcal{O})}.$$ ◇

7.2.3 *Derivative of a distribution*

We are now ready to define the derivative of a distribution, which is the basis of the notion of weak derivative for L^p functions. This definition is the foundation of Sobolev spaces.

Definition 7.14. *Let $T \in \mathcal{D}'(\mathcal{O})$. The derivative of T with respect to x_i for any $i = 1, \ldots, N$, denoted $\dfrac{\partial T}{\partial x_i}$, is defined as*

$$\Big\langle \frac{\partial T}{\partial x_i}, \varphi\Big\rangle_{\mathcal{D}'(\mathcal{O}),\mathcal{D}(\mathcal{O})} = -\Big\langle T, \frac{\partial \varphi}{\partial x_i}\Big\rangle_{\mathcal{D}'(\mathcal{O}),\mathcal{D}(\mathcal{O})}, \tag{7.12}$$

for all $\varphi \in \mathcal{D}(\mathcal{O})$.

Since $\dfrac{\partial \varphi}{\partial x_i}$ is in $\mathcal{D}(\mathcal{O})$ and T is a distribution, one can easily verify the following proposition:

Proposition 7.15. *If $T \in \mathcal{D}'(\mathcal{O})$, then*

$$\frac{\partial T}{\partial x_i} \in \mathcal{D}'(\mathcal{O}) \quad \textit{for } i = 1, \ldots, N.$$

That is, the derivative of a distribution is also a distribution.

Moreover, if T_n converges to T in the sense of distributions, then $\dfrac{\partial T_n}{\partial x_i}$ converges to $\dfrac{\partial T}{\partial x_i}$ in the sense of distributions for $i = 1, \ldots, N$.

Remark 7.16. *In particular, by Remark 7.9 and Proposition 7.15, it follows that a function in $L^1_{loc}(\mathcal{O})$ has derivatives (in the distributional sense) of any order.*

Example 7.17.

(1) The derivative of the Dirac function δ_{x_0} defined in Example 7.7 is given by

$$\left\langle \frac{\partial \delta_{x_0}}{\partial x_i}, \varphi \right\rangle_{\mathcal{D}'(\mathbb{R}^N), \mathcal{D}(\mathbb{R}^N)} = -\frac{\partial \varphi}{\partial x_i}(x_0), \quad \forall \varphi \in \mathcal{D}(\mathbb{R}^N).$$

(2) Let H be the Heaviside function on $\mathbb{R}$ defined by

$$H(x) = \begin{cases} 1 & \text{if} \quad x \geq 0, \\ 0 & \text{if} \quad x < 0. \end{cases}$$

Note that $H \in L^1_{loc}(\mathbb{R})$ and we can identify it with the distribution T_H, given by (7.9). Hence the derivative of H with respect to x in the sense of distributions is

$$\left\langle \frac{dH}{dx}, \varphi \right\rangle_{\mathcal{D}'(\mathbb{R}), \mathcal{D}(\mathbb{R})} = -\left\langle H, \frac{d\varphi}{dx} \right\rangle_{\mathcal{D}'(\mathbb{R}), \mathcal{D}(\mathbb{R})} = -\int_{\mathbb{R}} H \frac{d\varphi}{dx}\, dx$$

$$= -\int_0^{+\infty} \frac{d\varphi}{dx}\, dx = \varphi(0), \quad \forall \varphi \in \mathcal{D}(\mathbb{R}),$$

which is simply δ_0, the Dirac mass at 0.

(3) Consider the functions u and v defined on (a, b) as follows:

$$u(x) = \begin{cases} x & \text{if} \quad x \in (a, c], \\ c & \text{if} \quad x \in (c, b), \end{cases}$$

and

$$v(x) = \begin{cases} 1 & \text{if} \quad x \in (a, c], \\ 0 & \text{if} \quad x \in (c, b). \end{cases}$$

It is easily seen that v is the derivative of u in the sense of distributions. ◇

Remark 7.18.

(1) For $x \neq 0$, the (usual) derivative of the Heaviside function is defined and $H'(x) = 0$ a.e. on $\mathbb{R}$. For this particular function, the usual derivative and the derivative in the sense of distributions do not coincide. This is also the case for Example 7.17 (3) above. This occurs whenever the (usual) derivative is only almost everywhere defined. However, if the classical derivative exists for all x and is continuous, the two notions of derivatives coincide, as showed in the next remark.

(2) If f and its derivatives in the sense of distributions $\dfrac{\partial f}{\partial x_i}$ are in $L^1_{loc}(\mathcal{O})$ for $i = 1, \dots, N$, then from Remark 7.9 we have

$$\int_{\mathcal{O}} \frac{\partial f}{\partial x_i}\, \varphi \, dx = -\int_{\mathcal{O}} f \frac{\partial \varphi}{\partial x_i}\, dx, \quad \forall \varphi \in \mathcal{D}(\mathcal{O}). \tag{7.13}$$

By Remark 7.9 and Definition 7.14, for any φ in $\mathcal{D}(\mathcal{O})$

$$\begin{aligned}\int_{\mathcal{O}} f \frac{\partial \varphi}{\partial x_i} dx = \Big\langle f, \frac{\partial \varphi}{\partial x_i} \Big\rangle_{\mathcal{D}'(\mathcal{O}),\mathcal{D}(\mathcal{O})} &= -\Big\langle \frac{\partial f}{\partial x_i}, \varphi \Big\rangle_{\mathcal{D}'(\mathcal{O}),\mathcal{D}(\mathcal{O})} \\ &= -\int_{\mathcal{O}} \frac{\partial f}{\partial x_i}\, \varphi \, dx.\end{aligned}$$

Thus, if $f \in C^1(\mathcal{O})$, its derivatives in the sense of distributions coincide with the usual partial derivatives. In this case, by Green's formula,

$$\int_{\mathcal{O}} f \frac{\partial \varphi}{\partial x_i}\, dx = -\int_{\mathcal{O}} f_{x_i}\, \varphi \, dx,$$

where f_{x_i} denotes the usual derivative. Hence,

$$\int_{\mathcal{O}} f_{x_i}\, \varphi \, dx = \int_{\mathcal{O}} \frac{\partial f}{\partial x_i}\, \varphi \, dx, \quad \forall \varphi \in \mathcal{D}(\mathcal{O}),$$

for every $i = 1, \dots, N$, and by Theorem 6.64,

$$f_{x_i} = \frac{\partial f}{\partial x_i} \quad \text{for every } i = 1, \dots, N.$$

7.3 The space $W^{1,p}$ and its properties

Definition 7.19. *Let $1 \leq p \leq +\infty$. We define the Sobolev space $W^{1,p}(\mathcal{O})$ as*

$$W^{1,p}(\mathcal{O}) = \left\{ u \in L^p(\mathcal{O}) \,\Big|\, \frac{\partial u}{\partial x_i} \in L^p(\mathcal{O}),\ i = 1, \dots, N \right\},$$

where the derivatives are taken in the sense of distributions, together with the associated norm

$$\|u\|_{W^{1,p}(\mathcal{O})} = \|u\|_{L^p(\mathcal{O})} + \sum_{i=1}^{N} \left\| \frac{\partial u}{\partial x_i} \right\|_{L^p(\mathcal{O})}. \tag{7.14}$$

In particular, for $p = 2$, we denote $W^{1,2}(\mathcal{O})$ by $H^1(\mathcal{O})$.

Remark 7.20. *If $\{u_n\}$ is a sequence in the space $W^{1,p}(\mathcal{O})$ such that u_n converges to u in $W^{1,p}(\mathcal{O})$, this means that $\|u_n - u\|_{W^{1,p}(\mathcal{O})} \to 0$, by Definition 6.1.*

In view of Definition 7.19, this is equivalent to the fact that $\{u_n\}$ and $\left\{\frac{\partial u_n}{\partial x_i}\right\}$ converge in $L^p(\mathcal{O})$ to u and $\frac{\partial u}{\partial x_i}$ (for $i = 1, \ldots, N$), respectively.

Proposition 7.21. *Let $1 \le p < +\infty$. The norm in $W^{1,p}(\mathcal{O})$ given by (7.14) is equivalent to*

$$\|u\| = \left(\|u\|^p_{L^p(\mathcal{O})} + \|\nabla u\|^p_{L^p(\mathcal{O})}\right)^{\frac{1}{p}}, \tag{7.15}$$

where, using the definition of ∇u given by (1.6) in Chapter 1, we set

$$\|\nabla u\|_{L^p(\mathcal{O})} = \left(\sum_{i=1}^{N} \left\| \frac{\partial u}{\partial x_i} \right\|^p_{L^p(\mathcal{O})}\right)^{\frac{1}{p}}.$$

To prove the equivalence of the two norms (7.14) and (7.15), we need the following result:

Lemma 7.22. *Let $1 \le p < +\infty$, $m \in \mathbb{N}$, $m \ge 2$, and $a_i \in \mathbb{R}_0^+$ for $i = 1 \ldots, m$. Then*

$$\sum_{i=1}^{m} a_i^p \le \left(\sum_{i=1}^{m} a_i\right)^p \le m^{p-1} \sum_{i=1}^{m} a_i^p. \tag{7.16}$$

Proof. We prove both inequalities in (7.16) by induction on m. Observe that the first one is true for $m = 2$, that is,

$$a_1^p + a_2^p \le (a_1 + a_2)^p. \tag{7.17}$$

Indeed, for a fixed $a_1 \in \mathbb{R}_0^+$, the function defined on $\mathbb{R}_0^+$ by

$$h(a_2) = (a_1 + a_2)^p - a_1^p - a_2^p,$$

is nonnegative, since it increases on $\mathbb{R}_0^+$ and $h(0) = 0$.

Assume now that the inequality

$$\sum_{i=1}^{m} a_i^p \leq \Big(\sum_{i=1}^{m} a_i\Big)^p, \tag{7.18}$$

holds for some $m \geq 2$. By (7.17) we have

$$\sum_{i=1}^{m+1} a_i^p = \Big(\sum_{i=1}^{m} a_i^p\Big) + a_{m+1}^p \leq \Big(\sum_{i=1}^{m} a_i\Big)^p + a_{m+1}^p \leq \Big(\sum_{i=1}^{m} a_i + a_{m+1}\Big)^p,$$

which proves that (7.18) is true for $m+1$, and this gives the first inequality in (7.16).

To prove the second inequality, which is a particular case of Jensen's inequality, we use the convexity of the function x^p on $\mathbb{R}_0^+$ for $1 \leq p < +\infty$. For every $t \in [0,1]$ and $x,\ y \in \mathbb{R}_0^+$,

$$\big(tx + (1-t)y\big)^p \leq tx^p + (1-t)y^p. \tag{7.19}$$

Choosing $t = 1/2$, $x = 2a_1$, and $y = 2a_2$ in (7.19) yields

$$\big(a_1 + a_2\big)^p \leq \frac{1}{2}\big(2a_1\big)^p + \frac{1}{2}\big(2a_2\big)^p = 2^{p-1}\big(a_1^p + a_2^p\big), \tag{7.20}$$

which proves the inequality for $m = 2$.

Suppose now that for some $m \geq 2$, one has

$$\Big(\sum_{i=1}^{m} a_i\Big)^p \leq m^{p-1} \sum_{i=1}^{m} a_i^p, \tag{7.21}$$

for any $a_1, \ldots a_m$ in $\mathbb{R}_0^+$. We have to show that it is true for $m+1$, which is equivalent to showing that for any $a_1, \ldots, a_{m+1}$ in $\mathbb{R}_0^+$,

$$\Big(\sum_{i=1}^{m+1} \frac{a_i}{m+1}\Big)^p \leq \frac{1}{m+1} \sum_{i=1}^{m+1} a_i^p. \tag{7.22}$$

In order to do so, let us choose in (7.19)

$$x = \sum_{i=1}^{m} \frac{a_i}{m}, \quad y = a_{m+1}, \quad \text{and} \quad t = \frac{m}{m+1}.$$

This implies $1 - t = \dfrac{1}{m+1}$.

By (7.21) with $\frac{a_i}{m}$ instead of a_i, we obtain

$$\begin{aligned}\Big(\sum_{i=1}^{m+1}\frac{a_i}{m+1}\Big)^p &= \Big(\frac{m}{m+1}\sum_{i=1}^{m}\frac{a_i}{m}+\frac{a_{m+1}}{m+1}\Big)^p\\ &\le \frac{m}{m+1}\Big(\sum_{i=1}^{m}\frac{a_i}{m}\Big)^p+\frac{a_{m+1}^p}{m+1}\\ &\le \frac{m}{m+1}m^{p-1}\sum_{i=1}^{m}\Big(\frac{a_i}{m}\Big)^p+\frac{a_{m+1}^p}{m+1}\\ &= \frac{1}{m+1}\sum_{i=1}^{m+1}a_i^p.\end{aligned}$$

This shows (7.22) and proves (7.16). □

Proof of Proposition 7.21. Consider the inequalities in (7.16). Choosing

$$m = N+1,\quad a_{N+1} = \|u\|_{L^p(\mathcal{O})},\quad a_i = \Big\|\frac{\partial u}{\partial x_i}\Big\|_{L^p(\mathcal{O})}\quad \text{for } i = 1,\dots,N,$$

we obtain

$$\|u\|^p \le \|u\|^p_{W^{1,p}(\mathcal{O})} \le (N+1)^{p-1}\|u\|^p.$$

By (6.4), the two norms are equivalent. □

Remark 7.23. *Since the equivalence of the two norms depends only on p and N, and not on $\mathcal{O}$, any of the two norms can be used. From now on, we simply denote the norm in $W^{1,p}(\mathcal{O})$ by $\|u\|_{W^{1,p}(\mathcal{O})}$ without specifying which norm is used. The reader can also easily verify that (7.14) and (7.15) are indeed norms.*

Proposition 7.24.
(1) Let $1 \le p \le +\infty$. The space $W^{1,p}(\mathcal{O})$ is a Banach space for the norm $\|u\|_{W^{1,p}(\mathcal{O})}$.

(2) The space $H^1(\mathcal{O})$ (that is $W^{1,2}(\mathcal{O})$) is a Hilbert space for the scalar product defined by

$$(v,w)_{H^1(\mathcal{O})} = (v,w)_{L^2(\mathcal{O})} + \sum_{i=1}^{N}\Big(\frac{\partial v}{\partial x_i},\frac{\partial w}{\partial x_i}\Big)_{L^2(\mathcal{O})}, \tag{7.23}$$

for all $v, w \in H^1(\mathcal{O})$.

Proof. (1) We first show that $W^{1,p}(\mathcal{O})$ is complete. Let $\{u_n\}$ be a Cauchy sequence in $W^{1,p}(\mathcal{O})$. Then for any $\varepsilon > 0$, there exists $n_0 \in \mathbb{N}$ such that for all $m > n \geq n_0$,

$$\|u_n - u_m\|_{L^p(\mathcal{O})} + \sum_{i=1}^{N} \left\| \frac{\partial(u_n - u_m)}{\partial x_i} \right\|_{L^p(\mathcal{O})} = \|u_n - u_m\|_{W^{1,p}(\mathcal{O})} < \varepsilon.$$

It follows that $\{u_n\}$ and $\left\{\dfrac{\partial u_n}{\partial x_i}\right\}$ (for $i = 1, \ldots, N$) are also Cauchy sequences in $L^p(\mathcal{O})$. Since $L^p(\mathcal{O})$ is complete, the sequences $\{u_n\}$ and $\left\{\dfrac{\partial u_n}{\partial x_i}\right\}$ are both convergent in $L^p(\mathcal{O})$. Hence there exist u and v_i in $L^p(\mathcal{O})$ for $i = 1, \ldots, N$, such that

$$u_n \to u \quad \text{and} \quad \frac{\partial u_n}{\partial x_i} \to v_i \quad \text{in } L^p(\mathcal{O}). \tag{7.24}$$

We now prove that $v_i = \dfrac{\partial u}{\partial x_i}$ in the sense of distributions and then the convergence of $\{u_n\}$ in $W^{1,p}(\mathcal{O})$ will follow from Remark 7.20.

From Remark 7.18 one has

$$\int_{\mathcal{O}} \frac{\partial u_n}{\partial x_i} \varphi \, dx = -\int_{\mathcal{O}} u_n \frac{\partial \varphi}{\partial x_i} \, dx, \quad \forall \varphi \in \mathcal{D}(\mathcal{O}).$$

Using Corollary 6.31 and convergence (7.24), we can pass to the limit in this inequality to obtain,

$$\int_{\mathcal{O}} v_i \, \varphi \, dx = -\int_{\mathcal{O}} u \frac{\partial \varphi}{\partial x_i} \, dx, \quad \forall \varphi \in \mathcal{D}(\mathcal{O}),$$

which proves that $v_i = \dfrac{\partial u}{\partial x_i}$ in the sense of distributions. Therefore $u_n \to u$ in $W^{1,p}(\mathcal{O})$ which implies that $W^{1,p}(\mathcal{O})$ is complete.

(2) Clearly, (7.23) is a scalar product on $H^1(\mathcal{O})$ and the norm induced by this product is simply the norm given by (7.15). By statement (1) proved previously and Proposition 7.21, the space $H^1(\mathcal{O})$ is a Banach space with respect to this norm. So, $H^1(\mathcal{O})$ is a Hilbert space with respect to the scalar product (7.23). This completes our proof. □

Proposition 7.25. *The space $W^{1,p}(\mathcal{O})$ is separable for $1 \leq p < +\infty$ and reflexive for $1 < p < +\infty$.*

Proof. From Proposition 6.42 and Remark 6.44, $L^p(\mathcal{O})$ is separable for $1 \leq p < +\infty$ and reflexive for $1 < p < +\infty$. The same is true for

$[L^p(\mathcal{O})]^{N+1}$, endowed with the norm given by (6.5). Since a closed subspace of a separable (respectively, reflexive) Banach space is separable (respectively, reflexive), it suffices to show that $W^{1,p}(\mathcal{O})$ can be identified with a closed subspace of $[L^p(\mathcal{O})]^{N+1}$.

Consider the linear map $T : W^{1,p}(\mathcal{O}) \longmapsto [L^p(\mathcal{O})]^{N+1}$ defined by

$$T(u) = \Big(u, \frac{\partial u}{\partial x_1}, \dots, \frac{\partial u}{\partial x_N}\Big).$$

Recalling the definition of the norms in $W^{1,p}(\mathcal{O})$ and $[L^p(\mathcal{O})]^{N+1}$, we have

$$\|T(u)\|_{[L^p(\mathcal{O})]^{N+1}} = \|u\|_{W^{1,p}(\mathcal{O})},$$

which implies that T is an isometry. Consequently, $W^{1,p}(\mathcal{O})$ can be identified with $T(W^{1,p}(\mathcal{O}))$, which is a closed subspace of $[L^p(\mathcal{O})]^{N+1}$. Indeed, the same argument used in the proof of Proposition 7.24 shows that a limit in $[L^p(\mathcal{O})]^{N+1}$ of a sequence in $T(W^{1,p}(\mathcal{O}))$ belong to $T(W^{1,p}(\mathcal{O}))$. It follows that $W^{1,p}(\mathcal{O})$ is also separable for $1 \leq p < +\infty$ and reflexive for $1 < p < +\infty$. □

The following proposition states two important and basic properties of functions in $W^{1,p}$:

Proposition 7.26. *Let $1 \leq p < +\infty$.*
(1) For every open subset $\mathcal{O}_1$ of $\mathcal{O}$,

$$u \in W^{1,p}(\mathcal{O}) \implies u|_{\mathcal{O}_1} \in W^{1,p}(\mathcal{O}_1).$$

(2) If $\psi \in \mathcal{D}(\mathcal{O})$ and $u \in W^{1,p}(\mathcal{O})$, then $\psi u \in W^{1,p}(\mathcal{O})$ and

$$\frac{\partial(\psi u)}{\partial x_i} = \psi \frac{\partial u}{\partial x_i} + u \frac{\partial \psi}{\partial x_i}, \qquad \forall\, i = 1, \dots, N. \tag{7.25}$$

Proof. (1) Let u be in $W^{1,p}(\mathcal{O})$. Obviously, the restriction $u|_{\mathcal{O}_1}$ of u to $\mathcal{O}_1$ is in $L^p(\mathcal{O}_1)$. To prove the assertion, we only need to show that $\dfrac{\partial}{\partial x_i}(u|_{\mathcal{O}_1})$ is also in $L^p(\mathcal{O}_1)$. By Definition 7.19 it is enough to show that

$$\frac{\partial u}{\partial x_i}\Big|_{\mathcal{O}_1} = \frac{\partial}{\partial x_i}(u|_{\mathcal{O}_1}) \qquad \text{for } i = 1, \dots, N. \tag{7.26}$$

From Definition 7.14 and Remark 7.18, for $i = 1, \dots, N$,

$$\Big\langle \frac{\partial}{\partial x_i}(u|_{\mathcal{O}_1}), \varphi \Big\rangle_{\mathcal{D}'(\mathcal{O}_1), \mathcal{D}(\mathcal{O}_1)} = -\int_{\mathcal{O}_1} u|_{\mathcal{O}_1} \frac{\partial \varphi}{\partial x_i}\, dx, \qquad \forall\, \varphi \in \mathcal{D}(\mathcal{O}_1).$$

On the other hand, since $\mathcal{D}(\mathcal{O}_1) \subset \mathcal{D}(\mathcal{O})$,

$$\int_{\mathcal{O}_1} u|_{\mathcal{O}_1} \frac{\partial \varphi}{\partial x_i}\, dx = \int_{\mathcal{O}} u \frac{\partial \varphi}{\partial x_i}\, dx = -\int_{\mathcal{O}} \frac{\partial u}{\partial x_i} \varphi\, dx = -\int_{\mathcal{O}_1} \frac{\partial u}{\partial x_i}\Big|_{\mathcal{O}_1} \varphi\, dx,$$

for every φ in $\mathcal{D}(\mathcal{O}_1)$. Consequently,

$$\Big\langle \frac{\partial}{\partial x_i}(u|_{\mathcal{O}_1}) - \frac{\partial u}{\partial x_i}\Big|_{\mathcal{O}_1}, \varphi \Big\rangle_{\mathcal{D}'(\mathcal{O}_1),\mathcal{D}(\mathcal{O}_1)} = 0, \qquad \forall\, \varphi \in \mathcal{D}(\mathcal{O}_1),$$

for $i = 1, \ldots, N$, which in view of Proposition 6.64 proves (7.26).

(2) Clearly, ψu belongs to $L^p(\mathcal{O})$. Since

$$\psi \frac{\partial u}{\partial x_i} + u \frac{\partial \psi}{\partial x_i} \in L^p(\mathcal{O}), \qquad \text{for } i = 1, \ldots, N,$$

it suffices to prove the differentiation formula given by (7.25) to show that ψu belongs to $W^{1,p}(\mathcal{O})$.

Let φ be in $\mathcal{D}(\mathcal{O})$. Then $\psi\varphi$ is also in $\mathcal{D}(\mathcal{O})$. By Definition 7.14 and Remark 7.18, we have

$$\begin{aligned}
\Big\langle \frac{\partial(\psi u)}{\partial x_i}, \varphi \Big\rangle_{\mathcal{D}'(\mathcal{O}),\mathcal{D}(\mathcal{O})} &= -\int_{\mathcal{O}} \psi u \frac{\partial \varphi}{\partial x_i} dx = \int_{\mathcal{O}} \frac{\partial u}{\partial x_i} \psi\varphi \, dx + \int_{\mathcal{O}} u \frac{\partial \psi}{\partial x_i} \varphi \, dx \\
&= \Big\langle \psi \frac{\partial u}{\partial x_i}, \varphi \Big\rangle_{\mathcal{D}'(\mathcal{O}),\mathcal{D}(\mathcal{O})} + \Big\langle u \frac{\partial \psi}{\partial x_i}, \varphi \Big\rangle_{\mathcal{D}'(\mathcal{O}),\mathcal{D}(\mathcal{O})} \\
&= \Big\langle \psi \frac{\partial u}{\partial x_i} + u \frac{\partial \psi}{\partial x_i}, \varphi \Big\rangle_{\mathcal{D}'(\mathcal{O}),\mathcal{D}(\mathcal{O})}
\end{aligned}$$

for $i = 1, \ldots, N$. This, by definition, proves (7.25). □

Remark 7.27. *More generally, for $1 \leq p \leq +\infty$ and $m \in \mathbb{N}$, the Sobolev space $W^{m,p}(\mathcal{O})$ is defined as*

$$W^{m,p}(\mathcal{O}) = \big\{ u \,\big|\, u \in L^p(\mathcal{O}), \partial^\alpha u \in L^p(\mathcal{O}), \ \alpha \in \mathbb{N}^N, \ 0 < |\alpha| \leq m \big\},$$

together with the associated norm

$$\|u\|_{W^{m,p}(\mathcal{O})} = \|u\|_{L^p(\mathcal{O})} + \sum_{0<|\alpha|\leq m} \|\partial^\alpha u\|_{L^p(\mathcal{O})}.$$

For $p = 2$ we set

$$H^m(\mathcal{O}) = W^{m,2}(\mathcal{O}).$$

Let us mention that most of the results stated in this section for $W^{1,p}(\mathcal{O})$ are true for the spaces $W^{m,p}(\mathcal{O})$ with $m > 1$.

We consider here only the case $m = 1$, which suffices for our study of second order partial differential equations.

7.3.1 *Density in Sobolev spaces on $\mathbb{R}^N$*

One essential result that makes Sobolev spaces convenient for weak solutions is the density of $\mathcal{D}(\mathbb{R}^N)$ in $W^{1,p}(\mathbb{R}^N)$. To prove it, we need the following lemma:

Lemma 7.28. *Let $1 \le p \le +\infty$. If $\varphi \in \mathcal{D}(\mathbb{R}^N)$ and $u \in W^{1,p}(\mathbb{R}^N)$, then $\varphi \star u$ is in $C^\infty(\mathbb{R}^N)$ and*

$$\frac{\partial}{\partial x_i}(\varphi \star u) = \varphi \star \frac{\partial u}{\partial x_i}, \qquad \forall\, i = 1, \ldots, N.$$

Proof. Let us first show that for every $i = 1, \ldots, N$,

$$\begin{aligned} \frac{\partial}{\partial x_i}(\varphi \star u)(x) &= \frac{\partial}{\partial x_i} \int_{\mathbb{R}^N} \varphi(x-y)\, u(y)\, dy \\ &= \int_{\mathbb{R}^N} \frac{\partial \psi}{\partial x_i}(x,y)\, u(y)\, dy, \end{aligned} \tag{7.27}$$

where we set $\psi(x,y) = \varphi(x-y)$.

The first equality follows simply from the definition of a convolution (Definition 6.52). The second one comes from Definition 7.14, Remark 7.18, and Fubini's Theorem (Thm 6.39), since for $i = 1, \ldots, N$ and $v \in \mathcal{D}(\mathbb{R}^N)$,

$$\begin{aligned} &\Big\langle \frac{\partial}{\partial x_i}\Big(\int_{\mathbb{R}^N} \psi(\,\cdot\,, y)\, u(y)\, dy\Big), v\Big\rangle_{\mathcal{D}'(\mathbb{R}^N),\mathcal{D}(\mathbb{R}^N)} \\ &\qquad = -\int_{\mathbb{R}^N} \Big(\int_{\mathbb{R}^N} \psi(x,y)\, u(y)\, dy\Big) \frac{\partial v}{\partial x_i}(x)\, dx \\ &\qquad = -\int_{\mathbb{R}^N} u(y) \Big(\int_{\mathbb{R}^N} \psi(x,y)\, \frac{\partial v}{\partial x_i}(x)\, dx\Big)\, dy \\ &\qquad = \int_{\mathbb{R}^N} \Big(\int_{\mathbb{R}^N} \frac{\partial \psi}{\partial x_i}(x,y)\, u(y)\, dy\Big) v(x)\, dx \\ &\qquad = \Big\langle \Big(\int_{\mathbb{R}^N} \frac{\partial \psi}{\partial x_i}(\,\cdot\,, y)\, u(y)\, dy\Big), v\Big\rangle_{\mathcal{D}'(\mathbb{R}^N),\mathcal{D}(\mathbb{R}^N)}, \end{aligned}$$

which proves (7.27). On the other hand, as

$$\frac{\partial \psi}{\partial x_i}(x,y) = -\frac{\partial \psi}{\partial y_i}(x,y),$$

using (7.27) and again Remark 7.18, gives

$$\begin{aligned} \frac{\partial}{\partial x_i}(\varphi \star u)(x) &= -\int_{\mathbb{R}^N} \frac{\partial \psi}{\partial y_i}(x,y)\, u(y)\, dy \\ &= \int_{\mathbb{R}^N} \psi(x,y)\, \frac{\partial u}{\partial y_i}(y)\, dy \\ &= \int_{\mathbb{R}^N} \varphi(x-y)\, \frac{\partial u}{\partial y_i}(y)\, dy = \Big(\varphi \star \frac{\partial u}{\partial x_i}\Big)(x). \end{aligned}$$

This ends the proof. $\square$

We now state and prove the density of $\mathcal{D}(\mathbb{R}^N)$ in $W^{1,p}(\mathbb{R}^N)$. As in the proof of Theorem 6.59, we proceed first by truncation and then by regularization. To begin with, we approximate elements of $W^{1,p}(\mathbb{R}^N)$ by functions which are in $W^{1,p}(\mathbb{R}^N)$ and have compact support. Afterwards, we approximate these functions in $W^{1,p}(\mathbb{R}^N)$ by functions in $\mathcal{D}(\mathbb{R}^N)$ using a regularizing sequence.

Theorem 7.29. *Let $1 \leq p < +\infty$. Then $\mathcal{D}(\mathbb{R}^N)$ is dense in $W^{1,p}(\mathbb{R}^N)$.*

Proof. The proof is done in two steps.

Step 1 (Truncation). Let $1 \leq p < +\infty$ and $v \in W^{1,p}(\mathbb{R}^N)$. We now establish the existence of a sequence $\{v_n\}$ in $W^{1,p}(\mathbb{R}^N)$ with compact support and converging to v.

Urysohn's Lemma (Thm 6.60) provides the existence of a (truncation) function ζ in $\mathcal{D}(\mathbb{R}^N)$ such that

$$0 \leq \zeta(x) \leq 1,$$

and satisfying

$$\zeta(x) = \begin{cases} 1 & \text{for } \|x\| < 1, \\ 0 & \text{for } \|x\| \geq 2. \end{cases}$$

For $n \in \mathbb{N}$ and $x \in \mathbb{R}^N$, define

$$\zeta_n(x) = \zeta\Big(\frac{x}{n}\Big),$$

which is still in $\mathcal{D}(\mathbb{R}^N)$ since supp $\zeta_n \subset \overline{B(0,2n)}$. We also introduce the function

$$v_n = \zeta_n v, \qquad \forall\, n \in \mathbb{N},$$

which has compact support (in the sense of Definition 6.49) and by Proposition 7.26, belongs to $W^{1,p}(\mathbb{R}^N)$. Let us show that

$$v_n \to v \quad \text{in } W^{1,p}(\mathbb{R}^N). \tag{7.28}$$

As $\zeta_n = 1$ for $\|x\| < n$,

$$\begin{aligned} \|v_n - v\|^p_{L^p(\mathbb{R}^N)} = \int_{\mathbb{R}^N} |\zeta_n - 1|^p \, |v|^p \, dx &= \int_{\|x\| \geq n} |\zeta_n - 1|^p \, |v|^p \, dx \\ &\leq \int_{\|x\| \geq n} |v|^p \, dx, \end{aligned}$$

and the last integral tends to zero as $n \to \infty$, since $v \in L^p(\mathbb{R}^N)$. Thus

$$v_n \to v \quad \text{in } L^p(\mathbb{R}^N).$$

In view of Remark 7.20, it only remains to prove that

$$\frac{\partial v_n}{\partial x_i} \to \frac{\partial v}{\partial x_i} \quad \text{in } L^p(\mathbb{R}^N), \qquad \forall\, i = 1, \dots, N, \tag{7.29}$$

where, due to Proposition 7.26,

$$\frac{\partial v_n}{\partial x_i} = \zeta_n \frac{\partial v}{\partial x_i} + v\,\frac{\partial \zeta_n}{\partial x_i}, \qquad \forall\, i = 1, \dots, N. \tag{7.30}$$

We have

$$\begin{aligned}\int_{\mathbb{R}^N} \left| v(x)\,\frac{\partial \zeta_n}{\partial x_i}(x) \right|^p dx &= \frac{1}{n^p} \int_{\mathbb{R}^N} \left| v(x)\,\frac{\partial \zeta}{\partial x_i}\left(\frac{x}{n}\right) \right|^p dx \\ &\le \frac{1}{n^p} \max_{\mathbb{R}^N} \left| \frac{\partial \zeta}{\partial x_i} \right|^p \int_{\mathbb{R}^N} |v|^p\, dx.\end{aligned}$$

This implies that for $i = 1, \dots, N$,

$$v\frac{\partial \zeta_n}{\partial x_i} \to 0 \quad \text{in } L^p(\mathbb{R}^N). \tag{7.31}$$

On the other hand, from the properties of the function ζ,

$$\left| \zeta_n \frac{\partial v}{\partial x_i} - \frac{\partial v}{\partial x_i} \right|^p \le \left| \frac{\partial v}{\partial x_i} \right|^p,$$

and

$$\left| \zeta_n \frac{\partial v}{\partial x_i} - \frac{\partial v}{\partial x_i} \right|^p \to \left| \zeta(0) \frac{\partial v}{\partial x_i} - \frac{\partial v}{\partial x_i} \right|^p = 0 \qquad \text{a.e. in } \mathbb{R}^N.$$

By the Lebesgue Dominated Convergence Theorem (Thm 6.38) we have

$$\int_{\mathbb{R}^N} \left| \zeta_n \frac{\partial v}{\partial x_i} - \frac{\partial v}{\partial x_i} \right|^p dx \to 0 \qquad \text{as } n \to \infty.$$

This together with (7.30) and (7.31), yield (7.29), thereby proving (7.28).

Step 2 (Regularization). From (7.28), it follows that for every $\varepsilon > 0$, there exists $n_0 \in \mathbb{N}$ such that

$$\|v_{n_0} - v\|_{W^{1,p}(\mathbb{R}^N)} < \frac{\varepsilon}{2}, \tag{7.32}$$

where v_{n_0} is in $W^{1,p}(\mathbb{R}^N)$ and has compact support.

Let us now prove that if $\{\rho_n\}$ is a regularizing sequence given by Definition 6.56, then

$$\rho_n \star v_{n_0} \to v_{n_0} \quad \text{in } W^{1,p}(\mathbb{R}^N). \tag{7.33}$$

By Theorem 6.58,

$$\rho_n \star v_{n_0} \to v_{n_0} \quad \text{in } L^p(\mathbb{R}^N).$$

Moreover, in view of Lemma 7.28 and again by Theorem 6.58,

$$\frac{\partial}{\partial x_i}(\rho_n \star v_{n_0}) = \rho_n \star \frac{\partial v_{n_0}}{\partial x_i} \to \frac{\partial v_{n_0}}{\partial x_i} \quad \text{in } L^p(\mathbb{R}^N), \qquad \forall\, i = 1, \dots, N,$$

which gives (7.33). Consequently, there exists $n_\varepsilon \in \mathbb{N}$ such that $v_\varepsilon = \rho_{n_\varepsilon} \star v_{n_0}$ satisfies

$$\|v_\varepsilon - v_{n_0}\|_{W^{1,p}(\mathbb{R}^N)} < \frac{\varepsilon}{2}.$$

This together with (7.32) imply

$$\|v - v_\varepsilon\|_{W^{1,p}(\mathbb{R}^N)} \le \|v - v_{n_0}\|_{W^{1,p}(\mathbb{R}^N)} + \|v_{n_0} - v_\varepsilon\|_{W^{1,p}(\mathbb{R}^N)} < \varepsilon.$$

To conclude the proof, observe that since v_{n_0} has compact support, by (2) of Proposition 6.54, the function v_ε is in $\mathcal{D}(\mathbb{R}^N)$. □

7.3.2 *Lipschitz-continuous boundaries*

From now on, if not specified, we consider Sobolev spaces on $\mathbb{R}^N$ or on a bounded open subset Ω of $\mathbb{R}^N$. For a density result on $W^{1,p}(\Omega)$ when $N \ge 2$, we need some regularity on the boundary of Ω. We now define a Lipschitz-continuous boundary which has various definitions in the literature. We adopt the definition given in [29].

Recall that a map $f : A \subseteq \mathbb{R}^N \longmapsto \mathbb{R}^N$ ($N \in \mathbb{N}$) is said to be Lipschitz-continuous if there exists $M \in \mathbb{R}^+$ such that

$$\|f(x) - f(y)\| \le M\|x - y\|, \qquad \forall x, y \in A.$$

Definition 7.30. *Let $N \ge 2$ and Ω be a bounded open subset of $\mathbb{R}^N$. A system of local coordinates is defined as follows: we suppose that there exist an integer $m \in \mathbb{N}$, a set of m functions*

$$\psi_i : Q \doteq (-1,1)^{N-1} \to \mathbb{R} \qquad \textit{for } i = 1, \dots, m,$$

and $r \in \mathbb{R}^+$ such that the maps

$$\Psi_i : U \doteq Q \times (-r, r) \longmapsto V_i \doteq \Psi_i(U) \subset \mathbb{R}^N,$$

given by

$$\Psi_i : (y', y_N) \longmapsto (y', y_N + \psi_i(y'))$$

are homeomorphisms and satisfy

$$\begin{aligned} \Gamma_i &\doteq \Psi_i(Q \times \{0\}) \subset \partial\Omega, \\ U_i^+ &\doteq \Psi_i(Q \times (0, r)) \subset \Omega, \\ U_i^- &\doteq \Psi_i(Q \times (-r, 0)) \subset \mathbb{R}^N \setminus \overline{\Omega}, \end{aligned}$$

for every $i = 1, \ldots, m$, and

$$\partial\Omega = \bigcup_{i=1}^{m} \Gamma_i.$$

We say that $\partial\Omega$ is Lipschitz-continuous (or of class C^k, $k \in \mathbb{N}$, respectively) if there exists a system of local coordinates such that ψ_i is Lipschitz-continuous (or of class C^k, respectively) for $i = 1, \ldots, m$.

Remark 7.31.

(1) A Lipschitz-continuous boundary is simply a finite union of graphs of Lipschitz-continuous functions, which overlap since Q is open. In particular, such a boundary satisfies Condition $\mathcal{C}$ introduced in Section 3.5. Hence one can admit not only smooth graphs but also graphs with corners and without cusps (see Figure 2). For instance, polygons in $\mathbb{R}^2$ or polyhedrons in $\mathbb{R}^3$ have Lipschitz-continuous boundaries. This is important in numerical applications such as finite element methods for partial differential equations, since one uses polygons and polyhedra in these methods.

(2) When $k = 1$, the definition of a C^1 boundary given above is equivalent to that in Definition 3.4.

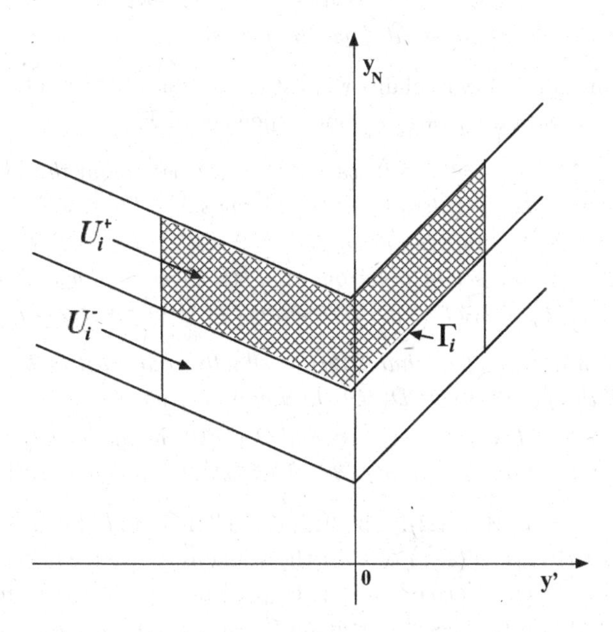

Figure 9. An example of a Lipschitz-continuous boundary

7.3.3 Extension operators and density in $W^{1,p}(\Omega)$

The following result plays an essential role in the study of Sobolev spaces when one deals with functions defined only on a bounded open subset and not on the whole space. We omit its proof and refer the reader to [1] (see also [3], [4], [16]) for the case of an open set with a C^1 boundary. The result in the more difficult case of a Lipschitz-continuous boundary, has been proven in [5] for $1 < p < +\infty$ and in [38] for $p = 1$ and $p = +\infty$ (see also [7]).

Theorem 7.32. *Let $1 \leq p \leq +\infty$. Let $\Omega \subset \mathbb{R}^N$ and suppose that*

- *if $N = 1$, Ω is an open interval in $\mathbb{R}$,*
- *if $N \geq 2$, Ω is either a bounded open subset of $\mathbb{R}^N$ with $\partial\Omega$ Lipschitz-continuous or $\Omega = \mathbb{R}^{N-1} \times \mathbb{R}^+$.*

Then there exists a linear continuous extension operator P from $W^{1,p}(\Omega)$ into $W^{1,p}(\mathbb{R}^N)$ such that

$$
\begin{aligned}
&i)\ \ P(u) = u \quad \text{on } \Omega,\\
&ii)\ \ \|P(u)\|_{L^p(\mathbb{R}^N)} \leq C\|u\|_{L^p(\Omega)},\\
&iii)\ \ \|P(u)\|_{W^{1,p}(\mathbb{R}^N)} \leq C\|u\|_{W^{1,p}(\Omega)},
\end{aligned}
$$

where C is a positive constant depending on Ω and p.

The definition below explains what it means for a function to be of class C^∞ up to the boundary of a bounded open set of $\mathbb{R}^N$.

Definition 7.33. *Let $\Omega \subset \mathbb{R}^N$ be an open set satisfying the assumptions from Theorem 7.32. We denote by $\mathcal{D}(\overline{\Omega})$ the set of the restrictions to $\overline{\Omega}$ of functions in $\mathcal{D}(\mathbb{R}^N)$, and denote by $\mathcal{D}(\mathbb{R}^{N-1} \times \mathbb{R}_0^+)$ the set $\mathcal{D}(\overline{\mathbb{R}^{N-1} \times \mathbb{R}^+})$.*

If $C_c^1(\mathbb{R}^N)$ is the set of functions in $C^1(\mathbb{R}^N)$ with compact support, we denote by $C_c^1(\overline{\Omega})$ the set of the restrictions to $\overline{\Omega}$ of functions in $C_c^1(\mathbb{R}^N)$.

Remark 7.34. *It is clear that $\mathcal{D}(\Omega)$ is strictly contained in $\mathcal{D}(\overline{\Omega})$, since in general, the functions in $\mathcal{D}(\overline{\Omega})$ do not vanish on $\partial\Omega$.*

Theorem 7.35. *Let $1 \leq p < +\infty$ and $\Omega \subset \mathbb{R}^N$ be an open set satisfying the assumptions from Theorem 7.32. Then $\mathcal{D}(\overline{\Omega})$ is dense in $W^{1,p}(\Omega)$.*

Proof. Let u be in $W^{1,p}(\Omega)$. The Extension Theorem (Thm 7.32) gives an extension $P(u) \in W^{1,p}(\mathbb{R}^N)$ of u to the whole of $\mathbb{R}^N$. Therefore Theorem 7.29 provides the existence of a sequence $\{u_n\}$ in $\mathcal{D}(\mathbb{R}^N)$ converging to $P(u)$ in $W^{1,p}(\mathbb{R}^N)$. This implies in particular that the sequence $\{u_n|_{\overline{\Omega}}\} \subset \mathcal{D}(\overline{\Omega})$ converges to $P(u)|_\Omega = u$ in $W^{1,p}(\Omega)$ and proves the claimed result. □

Remark 7.36.

(1) Observe that the density result stated in Theorem 7.29 shows that $W^{1,p}(\mathbb{R}^N)$ is the closure of $\mathcal{D}(\mathbb{R}^N)$ for the norm given by (7.14). This can also be taken as an equivalent definition of $W^{1,p}(\mathbb{R}^N)$. Obviously, in this case one has to prove that the space obtained in this way is the space of the distributions which are in $L^p(\mathbb{R}^N)$ together with their first derivatives.

Similarly, Theorem 7.35 says that if a bounded open subset Ω has a Lipschitz-continuous boundary, then $W^{1,p}(\Omega)$ is the closure of $\mathcal{D}(\overline{\Omega})$ for the norm (7.14). In this case Definition 7.19 is more general, since it can be used for any open subset of $\mathbb{R}^N$.

(2) One has the inclusions

$$\mathcal{D}(\mathbb{R}^N) \subset C^1_c(\mathbb{R}^N) \subset W^{1,p}(\mathbb{R}^N).$$

Therefore the space $W^{1,p}(\mathbb{R}^N)$ is also the closure of $C^1_c(\mathbb{R}^N)$ for the norm (7.14). Similarly, under the assumptions of Theorem 7.35, the space $W^{1,p}(\Omega)$ is, likewise, the closure of $C^1_c(\overline{\Omega})$ for the norm (7.14).

7.3.4 *Chain rule and applications*

The following result shows that the usual chain rule for the derivative of the composition of two differentiable functions is still valid (in the distributional sense) for the composition of a differentiable function with a function belonging to the space $W^{1,p}(\Omega)$.

Theorem 7.37 (Chain Rule). *Let $1 \leq p \leq +\infty$ and $\Omega \subset \mathbb{R}^N$ be an open set satisfying the assumptions from Theorem 7.32. Suppose that G is a function in $C^1(\mathbb{R})$ such that*

$$G(0) = 0 \quad and \quad |G'(s)| \leq M, \quad \forall s \in \mathbb{R}, \tag{7.34}$$

where M is a positive constant. Then for every $u \in W^{1,p}(\Omega)$, the function $G(u)$ belongs to $W^{1,p}(\Omega)$ and (in the distributional sense),

$$\nabla(G(u)) = G'(u)\nabla u. \tag{7.35}$$

Proof. From (7.34) one has

$$|G(s) - G(t)| = \left| \int_t^s G'(x)\, dx \right| \leq M|s-t|, \qquad \forall s, t \in \mathbb{R}. \tag{7.36}$$

Let u be in $W^{1,p}(\Omega)$. Since the composition of measurable functions is still measurable, the function $G(u)$ defined by

$$G(u)(x) = G(u(x)) \qquad \text{a.e. in } \Omega,$$

is measurable. Hence, from (7.34) and (7.36) (written for $s = 0$), the function $G(u)$ belongs to $L^p(\Omega)$, and the function $G'(u)\nabla u$ to $(L^p(\Omega))^N$. So, it suffices to prove (7.35), which will imply that $G(u)$ is in $W^{1,p}(\Omega)$.

To do so, we have to show that

$$\int_\Omega G(u)\,\frac{\partial \varphi}{\partial x_i}\,dx = -\int_\Omega G'(u)\,\frac{\partial u}{\partial x_i}\,\varphi\,dx, \tag{7.37}$$

for every $\varphi \in \mathcal{D}(\Omega)$ and $i = 1, \ldots, N$.

On the other hand, Theorem 7.35 provides the existence of a sequence $\{u_n\}$ in $\mathcal{D}(\overline{\Omega})$ converging to u in $W^{1,p}(\Omega)$. By Remark 7.20 and Theorem 6.40, there exists a subsequence, still denoted $\{u_n\}$, such that

$$u_n(x) \to u(x) \qquad \text{a.e. in } \Omega,$$

which implies

$$G'(u_n) \to G'(u) \qquad \text{a.e. in } \Omega. \tag{7.38}$$

Let us show the convergences

$$\begin{aligned} &G(u_n) \to G(u) && \text{strongly in } L^p(\Omega),\\ &G'(u_n)\nabla u_n \to G'(u)\nabla u && \text{strongly in } (L^p(\Omega))^N. \end{aligned} \tag{7.39}$$

Indeed, from (7.36) one immediately gets the estimate

$$\|G(u_n) - G(u)\|_{L^p(\Omega)} \le M\|u_n - u\|_{L^p(\Omega)}.$$

This implies the first convergence in (7.39), since the right-hand side converges to zero.

For the second convergence, we write

$$\begin{aligned} &\|G'(u_n)\nabla u_n - G'(u)\nabla u\|_{L^p(\Omega)}\\ &\qquad \le \|G'(u_n)(\nabla u_n - \nabla u)\|_{L^p(\Omega)} + \|(G'(u_n) - G'(u))\nabla u\|_{L^p(\Omega)}\\ &\qquad \le M\|\nabla u_n - \nabla u\|_{L^p(\Omega)} + \|(G'(u_n) - G'(u))\nabla u\|_{L^p(\Omega)}. \end{aligned}$$

In view of (7.34) and (7.38), we can apply the Lebesgue Dominated Convergence Theorem (Thm 6.38) to conclude the proof of (7.39).

From Corollary 6.31 and (7.39), we have

$$\begin{aligned} \int_\Omega G(u)\,\frac{\partial \varphi}{\partial x_i}\,dx &= \lim_{n\to\infty}\int_\Omega G(u_n)\,\frac{\partial \varphi}{\partial x_i}\,dx = -\lim_{n\to\infty}\int_\Omega G'(u_n)\,\frac{\partial u_n}{\partial x_i}\varphi\,dx\\ &= -\int_\Omega G'(u)\frac{\partial u}{\partial x_i}\varphi\,dx, \end{aligned}$$

for every $\varphi \in \mathcal{D}(\Omega)$ and $i = 1, \ldots, N$, which proves (7.37). $\square$

An important consequence of this theorem is the next proposition and its corollary, due to G. Stampacchia [36] (see also [37]).

Proposition 7.38. *Let $1 \le p \le +\infty$ and $\Omega \subset \mathbb{R}^N$ as in Theorem 7.32. Let u be in $W^{1,p}(\Omega)$ and define*

$$u_+ = \max\{u, 0\}, \qquad u_- = -\min\{u, 0\}, \qquad \textit{a.e. on } \Omega. \tag{7.40}$$

Then the functions u_+, u_- and $|u|$ belong to $W^{1,p}(\Omega)$ and almost everywhere on Ω, one has

$$\nabla u_+ = \begin{cases} \nabla u & \textit{if } u > 0, \\ 0 & \textit{if } u \le 0, \end{cases} \tag{7.41}$$

$$\nabla u_- = \begin{cases} 0 & \textit{if } u \ge 0, \\ -\nabla u & \textit{if } u < 0, \end{cases} \tag{7.42}$$

and

$$\nabla(|u|) = \begin{cases} \nabla u & \textit{if } u > 0, \\ 0 & \textit{if } u = 0, \\ -\nabla u & \textit{if } u < 0. \end{cases} \tag{7.43}$$

Proof. Let $u \in W^{1,p}(\Omega)$. Since

$$u_- = (-u)_+, \qquad |u| = u_+ + u_-, \tag{7.44}$$

it will be sufficient to prove (7.41).

For every $n \in \mathbb{N}$ and $s \in \mathbb{R}$, define

$$G_n(s) = \begin{cases} \dfrac{\sqrt{1+n^2s^2}-1}{n} & \text{if } s > 0, \\ 0 & \text{if } s \le 0. \end{cases}$$

It is easy to verify that G_n satisfies the assumptions of Theorem 7.37. This implies that for every $n \in \mathbb{N}$, the function $G_n(u)$ belongs to $W^{1,p}(\Omega)$ and

$$\nabla(G_n(u)) = \begin{cases} \dfrac{nu}{\sqrt{1+n^2u^2}}\,\nabla u & \text{if } u > 0, \\ 0 & \text{if } u \le 0. \end{cases}$$

As a consequence, by Definition 7.14 and Remark 7.18,

$$\int_\Omega G_n(u)\,\frac{\partial \varphi}{\partial x_i}\,dx = -\int_{\{x\in\Omega\,|\,u>0\}} \frac{nu}{\sqrt{1+n^2u^2}}\,\frac{\partial u}{\partial x_i}\,\varphi\,dx, \tag{7.45}$$

for $i = 1, \dots, N$ and for every $\varphi \in \mathcal{D}(\Omega)$.

Observe now that

$$G_n(u) \to u_+ \qquad \text{a.e. in } \Omega,$$
$$\frac{n|u|}{\sqrt{1+n^2u^2}} \to 1 \quad \text{a.e. in } \{x \in \Omega \mid u(x) > 0\},$$

and

$$|G_n(u)| \leq |u|, \quad \frac{n|u|}{\sqrt{1+n^2u^2}} \leq 1 \quad \text{a.e. in } \Omega.$$

Therefore by the Lebesgue Dominated Convergence Theorem (Thm 6.38), we can pass to the limit in (7.45) to obtain, for $i = 1, \ldots, N$ and for every $\varphi \in \mathcal{D}(\Omega)$,

$$\int_\Omega u_+ \frac{\partial \varphi}{\partial x_i}\, dx = -\int_{\{x\in\Omega \mid u>0\}} \frac{\partial u}{\partial x_i} \varphi\, dx,$$

which proves (7.41). □

Corollary 7.39. *Let $1 \leq p \leq +\infty$ and $\Omega \subset \mathbb{R}^N$ be an open set satisfying the assumptions from Theorem 7.32. If u is a function in $W^{1,p}(\Omega)$, then its gradient is zero on the level curves, that is, for every $\alpha \in \mathbb{R}$*

$$\nabla u = 0, \quad \text{a.e. in } \{x \in \Omega \mid u(x) = \alpha\}.$$

Proof. Set $v = u - \alpha$. As $v = v_+ - v_-$, the result comes from Proposition 7.38, since $\nabla u = \nabla v = \nabla v_+ - \nabla v_- = 0$ on the set $\{x \in \Omega \mid v = 0\}$. □

The next result extends a well-known property of differentiable functions to functions in $W^{1,p}(\Omega)$.

Theorem 7.40. *Let $1 \leq p < +\infty$ and $\Omega \subset \mathbb{R}^N$ as in Theorem 7.32. If Ω is connected and $u \in W^{1,p}(\Omega)$ such that $\nabla u = 0$ in Ω, then u is constant in Ω.*

Proof. Let P be an extension operator given by Theorem 7.32, and $\{\rho_n\}$ be a regularizing sequence given by Definition 6.56. By Proposition 6.54 and Lemma 7.28, $\rho_n \star P(u)$ belongs to $C^\infty(\mathbb{R}^N)$, and for every $i = 1, \ldots, N$,

$$\frac{\partial}{\partial x_i}(\rho_n \star P(u)) = \rho_n \star \frac{\partial}{\partial x_i}(P(u)),$$

with

$$\text{supp}\Big(\rho_n \star \frac{\partial}{\partial x_i}(P(u))\Big) \subset \text{supp}(\rho_n) + \text{supp}\Big(\frac{\partial}{\partial x_i}(P(u))\Big).$$

Since by assumption, $\nabla u = 0$ on Ω, Remark 6.57 implies that for every connected bounded open set ω such that $\overline{\omega} \subset \Omega$, there is $n(\omega) \in \mathbb{N}$ such that for every $n \geq n(\omega)$ and $i = 1, \ldots, N$,

$$\begin{aligned} \operatorname{supp}\Big(\rho_n \star \frac{\partial}{\partial x_i}(P(u))\Big) &\subset \operatorname{supp}(\rho_n) + \operatorname{supp}\Big(\frac{\partial}{\partial x_i}(P(u))\Big) \\ &\subset \operatorname{supp}(\rho_n) + \big(\mathbb{R}^N \setminus \Omega\big) \subset \mathbb{R}^N \setminus \omega. \end{aligned}$$

This shows that $\nabla\big(\rho_n \star P(u)\big) = 0$ on ω for $n \geq n(\omega)$. As ω is connected, the function $\rho_n \star P(u)$ is constant in ω. That is, for every $n \geq n(\omega)$, there exists $c_n \in \mathbb{R}$ such that $\rho_n \star P(u) = c_n$ in ω.

On the other hand, from Theorem 6.58,

$$\rho_n \star P(u) \to P(u) \quad \text{in } L^p(\mathbb{R}^N),$$

and by Theorem 6.40, there exists a subsequence $\{\rho_{n_k} \star P(u)\}$ such that

$$\big(\rho_{n_k} \star P(u)\big)(x) \to P(u)(x) \quad \text{a.e. in } \mathbb{R}^N,$$

which implies

$$u = \lim_{n\to\infty} c_n, \qquad \text{a.e. in } \omega,$$

so that u is constant on ω. This concludes the proof due to the arbitrariness of ω and the connectedness of Ω. □

7.3.5 *One-dimensional Sobolev inclusions*

We end this section with a relevant result for the case $N = 1$ (see (1.4) for notation).

Theorem 7.41. *Let $1 \leq p < +\infty$. Then*

$$W^{1,p}(\mathbb{R}) \subset C_b^0(\mathbb{R}), \tag{7.46}$$

and for every $u \in W^{1,p}(\mathbb{R})$,

$$\|u\|_{L^\infty(\mathbb{R})} \leq c(p)\|u\|_{W^{1,p}(\mathbb{R})}, \tag{7.47}$$

for some constant $c(p)$ depending only on p.

Moreover, if $p > 1$,

$$|u(x) - u(y)| \leq |x - y|^{\frac{p-1}{p}} \, \|u'\|_{L^p(\mathbb{R})} \quad \text{for every } x,\ y \in \mathbb{R}, \tag{7.48}$$

that is, u is Hölder continuous with exponent $\dfrac{p-1}{p}$ (see Definition 3.22).

Proof. Since $1 \leq p < +\infty$, we can argue by density. We shall prove that there exists a constant $c(p)$ which depends only on p such that, for every u in $\mathcal{D}(\mathbb{R})$, one has

$$|u(x)| \leq c(p)\|u\|_{W^{1,p}(\mathbb{R})}, \quad \forall x \in \mathbb{R}. \tag{7.49}$$

For the case $p = 1$ we have

$$u(x) = \int_{-\infty}^{x} u'(s)\, ds \quad \text{for every } x \in \mathbb{R}.$$

Then,

$$|u(x)| \leq \int_{-\infty}^{+\infty} |u'(s)|\, ds \quad \text{for every } x \in \mathbb{R},$$

which gives (7.49) with $c(p) = 1$, for functions in $\mathcal{D}(\mathbb{R})$.

Now, suppose $p > 1$. For $u \in \mathcal{D}(\mathbb{R})$, introduce the function $w(x) = |u(x)|^p$, which still has a compact support and belongs to $C^1(\mathbb{R})$. One has $w' = p|u|^{p-1}u' \operatorname{sgn}(u)$ where $\operatorname{sgn}(x)$ is the function defined on $\mathbb{R}$ as

$$\operatorname{sgn}(x) = \begin{cases} 1 & \text{if } x > 0, \\ 0 & \text{if } x = 0, \\ -1 & \text{if } x < 0. \end{cases}$$

It follows that

$$w(x) = \int_{-\infty}^{x} p\,|u(s)|^{p-1} u'(s) \operatorname{sgn}(u(s))\, ds, \quad \forall\, x \in \mathbb{R}.$$

This yields, using Hölder's inequality (6.31) and (6.33),

$$|u(x)|^p = |w(x)| \leq p \left\| |u|^{p-1} \right\|_{L^{p'}(\mathbb{R})} \|u'\|_{L^p(\mathbb{R})} = p\, \|u\|^{p-1}_{L^p(\mathbb{R})} \|u'\|_{L^p(\mathbb{R})},$$

where p' is the conjugate of p. By Young's Inequality (6.38), we have

$$|u(x)| \leq p^{\frac{1}{p}} \|u\|^{\frac{1}{p'}}_{L^p(\mathbb{R})} \|u'\|^{\frac{1}{p}}_{L^p(\mathbb{R})} \leq p^{\frac{1}{p}} \left(\frac{\|u\|_{L^p(\mathbb{R})}}{p'} + \frac{\|u'\|_{L^p(\mathbb{R})}}{p} \right),$$

from which we get (7.49) with

$$c(p) = p^{\frac{1}{p}} \max\left\{ \frac{1}{p}, \frac{1}{p'} \right\},$$

and prove (7.47) for functions in $\mathcal{D}(\mathbb{R})$. Observe also that for every $x, y \in \mathbb{R}$ and $u \in \mathcal{D}(\mathbb{R})$,

$$u(x) - u(y) = \int_{y}^{x} u'(t)\, dt.$$

Therefore if $p > 1$, by Hölder's inequality (6.31), one has (7.48) for $x, y \in \mathbb{R}$ and functions in $\mathcal{D}(\mathbb{R})$.

Now, suppose u is a function in $W^{1,p}(\mathbb{R})$. By Theorem 7.29, there exists a sequence $\{u_n\}$ in $\mathcal{D}(\mathbb{R})$ which converges to u in $W^{1,p}(\mathbb{R})$. For $n, m \in \mathbb{N}$, using (7.47) with $u = u_n - u_m$ gives

$$\|u_n - u_m\|_{L^\infty(\mathbb{R})} \leq c(p)\, \|u_n - u_m\|_{W^{1,p}(\mathbb{R})}\,. \tag{7.50}$$

Note that $\{u_n\}$ is a sequence in $C_b^0(\mathbb{R})$ which is complete, and (7.47) shows that $\{u_n\}$ is a Cauchy sequence. This implies the convergence of $\{u_n\}$ to some function $\widehat{u}$ in $C_b^0(\mathbb{R})$.

On the other hand, by Theorem 6.40, there exists a subsequence still denoted $\{u_n\}$, which converges to u almost everywhere. Hence

$$u = \widehat{u} \quad \text{a.e. in } \mathbb{R},$$

and this proves (7.46).

Moreover, using (7.47) and (7.48) for u_n and passing to the limit as $n \to \infty$, gives the desired inequalities for every $u \in W^{1,p}(\mathbb{R})$. □

An immediate consequence of Theorem 7.41 and Theorem 7.32 concerns the space $W^{1,p}(I)$, where I is an interval in $\mathbb{R}$.

Theorem 7.42. *Let $1 \leq p < +\infty$ and I an open interval in $\mathbb{R}$. Then*

$$W^{1,p}(I) \subset C^0(\bar{I}) \cap L^\infty(I), \tag{7.51}$$

and there exists a constant $c(I,p)$, which depends on I and p such that

$$\|u\|_{L^\infty(I)} \leq c(I,p)\|u\|_{W^{1,p}(I)} \qquad \textit{for every } u \in W^{1,p}(I).$$

Remark 7.43.

(1) From the proof of Theorem 7.41, we can see that the inclusion in $C^0(\bar{I})$ stated in Theorem 7.42 means that a function u in $W^{1,p}(I)$ is equal almost everywhere to a continuous function. According to Remark 6.27, this signifies that it has a class representative which is continuous. In the following, u will be identified with its continuous representative. Consequently, the value $u(x)$ makes sense at any point $x \in \bar{I}$. In particular, if $I = (a,b)$, then $u(a)$ and $u(b)$ are well-defined.

(2) The inclusion of $W^{1,p}(\mathbb{R})$ in $C_b^0(\mathbb{R})$ stated in Theorem 7.41 remains true for $p = +\infty$. This will be shown in the next chapter (Remark 8.8).

(3) As we shall also see in the following chapter, Theorem 7.42 fails for $N \geq 2$. This is why in the next section we introduce an appropriate notion for the values on the boundary of an open set Ω.

7.4 The notion of trace

As seen in the previous chapters, partial differential equations are coupled with some boundary conditions prescribing, for instance, the value of the solution on the boundary. As discussed in the beginning of this chapter, introducing a suitable variational formulation allows us to consider a partial differential equation in a weaker sense, that is, in the framework of Sobolev spaces.

What about the boundary conditions in this case? It can be easily seen that a function in $W^{1,p}(\Omega)$ is, in general, not in $C^0(\overline{\Omega})$ (except for $N=1$, see Theorem 7.42). Hence as an element of $L^p(\Omega)$, it is only defined almost everywhere so that its restriction on the boundary makes no sense.

In this section, we consider functions defined on $\Omega \subseteq \mathbb{R}^N$, $N \geq 2$, and introduce the notion of trace which gives a meaning to the restriction of functions in $W^{1,p}(\Omega)$ to $\partial\Omega$, for $\partial\Omega$ sufficiently smooth.

7.4.1 *Integral on Lipschitz-continuous boundaries*

Let us first describe how the notion of integrals over a surface, usually given for boundaries of class C^2, can be extended to Lipschitz-continuous boundaries (see [29] for more details). To do that, we use the following property of Lipschitz-continuous functions:

Theorem 7.44 (Rademacher's Theorem). *Let f be a Lipschitz-continuous function on an open set $A \subset \mathbb{R}^N$. Then f is almost everywhere differentiable. Moreover, its gradient is almost everywhere equal to its gradient in the distributional sense and ∇f belongs to $\left(L^\infty(A)\right)^N$.*

By construction, the surfaces Γ_i appearing in Definition 7.30, which are open in $\partial\Omega$, overlap. To define a surface integral over $\partial\Omega$, we have to use the following result on the partition of unity (see for instance [3], [4]):

Theorem 7.45. *Let F be a compact subset of $\mathbb{R}^N$ $(N \geq 2)$ and $A_1, \ldots, A_m$ open sets in $\mathbb{R}^N$ such that $F \subset \bigcup_{i=1}^m A_i$. Then for $i = 1, \ldots, m$, there exists $\gamma_i \in \mathcal{D}(A_i)$ satisfying*

$$0 \leq \gamma_i \leq 1 \quad \text{on } A_i \quad \text{and} \quad \sum_{i=1}^m \gamma_i(x) = 1 \quad \text{on } F.$$

Definition 7.46. *Let $N \geq 2$. With the notations of Definition 7.30, let Ω be a bounded open subset of $\mathbb{R}^N$ with a Lipschitz-continuous boundary and γ_i the functions given by Theorem 7.45 for the choice $F = \partial\Omega$ and $A_i = V_i$, for $i = 1, \ldots, m$.*

Let u *be a measurable function defined on* $\partial\Omega$. *We say that* u *is integrable on* $\partial\Omega$ *if for* $i = 1, \ldots, m$, *the function*

$$u(y', \psi_i(y'))\, \gamma_i(y', \psi_i(y'))\, \sqrt{1 + |\nabla \psi_i(y')|^2}$$

is integrable on Q. *We define the integral of* u *over* $\partial\Omega$ *by*

$$\int_{\partial\Omega} u(x)\, ds = \sum_{i=1}^{m} \int_{\Gamma_i} u(x)\gamma_i(x)\, ds,$$

where, for $i = 1, \ldots, m$,

$$\int_{\Gamma_i} u(x)\gamma_i(x)\, ds = \int_Q u(y', \psi_i(y'))\, \gamma_i(y', \psi_i(y'))\, \sqrt{1 + |\nabla \psi_i(y')|^2}\, dy'.$$

Remark 7.47. *It can be shown that the definition of the integral of* u *over* $\partial\Omega$ *is independent of the choice of the local coordinates* ψ_i *and of the functions* γ_i *of the partition of unity.*

Definition 7.48. *Let* Ω *be a bounded open subset of* $\mathbb{R}^N$ $(N \geq 2)$ *with a Lipschitz-continuous boundary. We define the space* $L^p(\partial\Omega)$, $1 \leq p < +\infty$, *by*

$$L^p(\partial\Omega) = \Big\{ u : \partial\Omega \to \mathbb{R} \mid u \text{ measurable and } \int_{\partial\Omega} |u(x)|^p\, ds < +\infty \Big\},$$

and the space $L^\infty(\partial\Omega)$ *by*

$$L^\infty(\partial\Omega) = \{ u : \partial\Omega \to \mathbb{R} \mid u \text{ measurable and there exists } C \in \mathbb{R} \text{ with } |u| \leq C, \text{ a.e. on } \partial\Omega \}.$$

We endow these spaces with the norm

$$\|u\|_{L^p(\partial\Omega)} = \begin{cases} \Big[\int_{\partial\Omega} |u(x)|^p\, ds \Big]^{\frac{1}{p}} & \text{if } p < +\infty, \\ \inf \{ C \,|\, |u| \leq C \text{ a.e. on } \partial\Omega \} & \text{if } p = +\infty. \end{cases}$$

It can be shown that $L^p(\partial\Omega)$ is a Banach space for the above norm.

7.4.2 *The Trace Theorem*

We now state the Trace Theorem. For simplicity, we discuss here only the case $p = 2$ which will be needed in the study of variational partial differential equations in Chapter 9. The case $p > 2$ can be proved by similar arguments.

Theorem 7.49 (Trace Theorem). *Let $N \geq 2$.*
(1) There exists a unique linear continuous map, called trace,

$$\gamma : H^1(\mathbb{R}^{N-1} \times \mathbb{R}^+) \longmapsto L^2(\mathbb{R}^{N-1}),$$

such that $\gamma(u) = u|_{\mathbb{R}^{N-1}}$ for any $u \in H^1(\mathbb{R}^{N-1} \times \mathbb{R}^+) \cap C^0(\mathbb{R}^{N-1} \times \mathbb{R}_0^+)$. The function $\gamma(u)$ is called the trace of u on $\mathbb{R}^{N-1}$.
(2) If Ω is a bounded open subset of $\mathbb{R}^N$ such that $\partial\Omega$ is Lipschitz continuous, then there exists a unique linear continuous map

$$\gamma : H^1(\Omega) \longmapsto L^2(\partial\Omega),$$

such that $\gamma(u) = u|_{\partial\Omega}$ for any $u \in H^1(\Omega) \cap C^0(\overline{\Omega})$. The function $\gamma(u)$ is called the trace of u on $\partial\Omega$.

Proof. We only prove the first statement of the theorem. The second statement follows from the first one by using Definition 7.30 and by mapping locally $\partial\Omega$ into $\mathbb{R}^N$ (see [29], [1] or [16] for details).

Let us consider the restriction map γ_0 defined as follows:

$$\gamma_0 : v \in \mathcal{D}(\mathbb{R}^{N-1} \times \mathbb{R}_0^+) \longmapsto v|_{\mathbb{R}^{N-1}},$$

with $\mathcal{D}(\mathbb{R}^{N-1} \times \mathbb{R}_0^+)$ given by Definition 7.33. Clearly, γ_0 is a linear map. Let us prove that

$$\|\gamma_0(v)\|_{L^2(\mathbb{R}^{N-1})} \leq \|v\|_{H^1(\mathbb{R}^{N-1}\times\mathbb{R}^+)}, \qquad \forall v \in \mathcal{D}(\mathbb{R}^{N-1} \times \mathbb{R}_0^+). \tag{7.52}$$

For $x \in R^{N-1} \times \mathbb{R}_0^+$, set $x = (x', x_N)$ where $x' \in \mathbb{R}^{N-1}$. If v belongs to $\mathcal{D}(\mathbb{R}^{N-1} \times \mathbb{R}_0^+)$, we have

$$\begin{aligned}
|v(x',0)|^2 &= -\int_0^{+\infty} \frac{\partial}{\partial x_N}\left(|v(x',x_N)|^2\right) dx_N \\
&= -\int_0^{+\infty} 2v(x',x_N)\frac{\partial v}{\partial x_N}(x',x_N)\, dx_N.
\end{aligned}$$

Using Young's inequality (6.38) with $p = q = 2$, yields

$$|v(x',0)|^2 \leq \int_0^{+\infty} |v(x',x_N)|^2\, dx_N + \int_0^{+\infty} \left|\frac{\partial v}{\partial x_N}(x',x_N)\right|^2 dx_N.$$

By integrating over $\mathbb{R}^{N-1}$ with respect to x' and using Fubini's Theorem (Thm 6.39), we obtain the following inequality which implies (7.52):

$$\int_{\mathbb{R}^{N-1}} |v(x',0)|^2\, dx' \leq \int_{\mathbb{R}^{N-1}\times\mathbb{R}^+} |v(x)|^2\, dx + \int_{\mathbb{R}^{N-1}\times\mathbb{R}^+} \left|\frac{\partial v}{\partial x_N}(x)\right|^2 dx.$$

Now, let u be in $H^1(\mathbb{R}^{N-1} \times \mathbb{R}^+)$. From Theorem 7.35 there exists a sequence $\{u_n\} \in \mathcal{D}(\mathbb{R}^{N-1} \times \mathbb{R}_0^+)$ converging to u in $H^1(\mathbb{R}^{N-1} \times \mathbb{R}^+)$. Inequality (7.52) and the linearity of γ_0 give

$$\|\gamma_0(u_n) - \gamma_0(u_m)\|_{L^2(\mathbb{R}^{N-1})} \leq \|u_n - u_m\|_{H^1(\mathbb{R}^{N-1}\times\mathbb{R}^+)}, \quad \forall m, n \in \mathbb{N}.$$

Therefore $\{\gamma_0(u_n)\}$ is a Cauchy sequence in the complete space $L^2(\mathbb{R}^{N-1})$ which converges to some u_0 in $L^2(\mathbb{R}^{N-1})$. From (7.52), we obtain

$$\|u_0\|_{L^2(\mathbb{R}^{N-1})} \leq \|u\|_{H^1(\mathbb{R}^{N-1}\times\mathbb{R}^+)}.$$

We now show that u_0 does not depend on the sequence $\{u_n\}$. Suppose there exists another sequence $\{w_n\}$ in $\mathcal{D}(\mathbb{R}^{N-1} \times \mathbb{R}_0^+)$ which also converges to u in $H^1(\mathbb{R}^{N-1} \times \mathbb{R}^+)$. This implies that $\{\gamma_0(w_n)\}$ converges in $L^2(\mathbb{R}^{N-1})$ to some u_1 and by (7.52), we have

$$\begin{aligned}\|u_0 - u_1\|_{L^2(\mathbb{R}^{N-1})} &\leq \|u_0 - \gamma_0(u_n)\|_{L^2(\mathbb{R}^{N-1})} + \|\gamma_0(u_n) - \gamma_0(w_n)\|_{L^2(\mathbb{R}^{N-1})} \\ &\quad + \|u_1 - \gamma_0(w_n)\|_{L^2(\mathbb{R}^{N-1})} \\ &\leq \|u_0 - \gamma_0(u_n)\|_{L^2(\mathbb{R}^{N-1})} + \|u_n - w_n\|_{H^1(\mathbb{R}^{N-1}\times\mathbb{R}^+)} \\ &\quad + \|u_1 - \gamma_0(w_n)\|_{L^2(\mathbb{R}^{N-1})},\end{aligned}$$

where we pass to the limit to get $u_1 = u_0$.

Consequently, by setting $\gamma(u) = u_0$ one defines uniquely a linear and continuous map γ from $H^1(\mathbb{R}^{N-1} \times \mathbb{R}^+)$ to $L^2(\mathbb{R}^{N-1})$. By construction, it is the unique linear extension of γ_0 to $H^1(\mathbb{R}^{N-1} \times \mathbb{R}^+)$.

Finally, let $u \in H^1(\mathbb{R}^{N-1} \times \mathbb{R}^+) \cap C^0(\mathbb{R}^{N-1} \times \mathbb{R}_0^+)$. One can find an approximating sequence $\{u_n\}$ such that (see [3], [4]),

$$u_n \to u \quad \text{in } C^0(\mathbb{R}^{N-1} \times \mathbb{R}_0^+).$$

So, $\gamma_0(u_n) = u_n|_{\mathbb{R}^{N-1}}$ converges to $u|_{\mathbb{R}^{N-1}}$ in $C^0(\mathbb{R}^{N-1})$ and to $\gamma(u)$ in $L^2(\mathbb{R}^{N-1})$. This proves that $\gamma(u) = u|_{\mathbb{R}^{N-1}}$. □

Remark 7.50. *One can show by counterexamples that there exist functions in $L^2(\partial\Omega)$ which are not traces of any element of $H^1(\Omega)$, that is, γ is not a surjection.*

Definition 7.51. *Let Ω be either a bounded open subset of $\mathbb{R}^N$ $(N \geq 2)$ such that $\partial\Omega$ is Lipschitz continuous, or $\Omega = \mathbb{R}^{N-1} \times \mathbb{R}^+$. We define*

$$H^{\frac{1}{2}}(\partial\Omega) = \gamma(H^1(\Omega)),$$

where γ is the trace map given by Theorem 7.49.

The following theorem gives the main properties of this space. We omit its proof which is rather technical.

Theorem 7.52. *Let Ω be a bounded open subset of $\mathbb{R}^N$ $(N \geq 2)$ such that $\partial\Omega$ is Lipschitz continuous or $\Omega = \mathbb{R}^{N-1} \times \mathbb{R}^+$.*

Then the following assertions hold:

(1) The set $H^{\frac{1}{2}}(\partial\Omega)$ is a Banach space for the norm defined by

$$\|u\|^2_{H^{\frac{1}{2}}(\partial\Omega)} = \int_{\partial\Omega} |u(x)|^2 \, ds + \int_{\partial\Omega}\int_{\partial\Omega} \frac{|u(x) - u(y)|^2}{|x-y|^{N+1}} \, ds_x \, ds_y.$$

(2) The set $\{u|_{\partial\Omega} \mid u \in C^\infty(\mathbb{R}^N)\}$ is dense in $H^{\frac{1}{2}}(\partial\Omega)$.

(3) The mapping $\gamma : H^1(\Omega) \longmapsto H^{\frac{1}{2}}(\partial\Omega)$ is linear and continuous.

(4) There exists a linear continuous map (called lifting operator)

$$G : g \in H^{\frac{1}{2}}(\partial\Omega) \longmapsto u_g \in H^1(\Omega) \tag{7.53}$$

such that $\gamma(u_g) = g$, for any $g \in H^{\frac{1}{2}}(\partial\Omega)$.

We end this section with a property which extends the usual Green's formula for smooth functions to functions in $H^1(\Omega)$.

Let us recall that if $\partial\Omega$ is Lipschitz continuous, then the unit outward normal vector to Ω is well-defined almost everywhere, that is, except on an $(N-1)$-dimensional subset of $\partial\Omega$ with zero measure (see [29]). Therefore by a density argument and using the definition of the trace, we have a Green-type formula.

Theorem 7.53. *Let Ω be a bounded open subset of $\mathbb{R}^N$ $(N \geq 2)$ such that $\partial\Omega$ is Lipschitz continuous. Then for every $u, v \in H^1(\Omega)$*

$$\int_\Omega u(x) \frac{\partial v}{\partial x_i}(x) \, dx = -\int_\Omega v(x) \frac{\partial u}{\partial x_i}(x) \, dx + \int_{\partial\Omega} \gamma(u)(x) \, \gamma(v)(x) \, n_i(x) \, ds,$$

for $1 \leq i \leq N$, where $n = (n_1, \ldots, n_N)$ denotes the unit outward normal vector to Ω.

In the following, we denote the trace $\gamma(v)$ simply by v.

7.5 The space H_0^1 and its properties

We introduce here the space $H_0^1(\mathcal{O})$, which, as we will see in Chapter 9, is well adapted to treat variational elliptic problems with homogeneous boundary Dirichlet conditions.

Definition 7.54. *Let $1 \leq p \leq +\infty$. We define the Sobolev space $W_0^{1,p}(\mathcal{O})$ as the closure of $\mathcal{D}(\mathcal{O})$ with respect to the norm of $W^{1,p}(\mathcal{O})$. In particular, we set*

$$H_0^1(\mathcal{O}) = W_0^{1,2}(\mathcal{O}).$$

Remark 7.55. *As $\mathcal{D}(\mathcal{O})$ is a linear subspace of the Banach space $W^{1,p}(\mathcal{O})$, it follows from the definition that $W_0^{1,p}(\mathcal{O})$ is a closed subspace of $W^{1,p}(\mathcal{O})$, hence a Banach space itself. Moreover, it can be easily verified that*
(1) $H_0^1(\mathcal{O})$ is a Hilbert space for the scalar product (7.23).
(2) In view of Theorem 7.29 and Definition 7.54, $W_0^{1,p}(\mathbb{R}^N) = W^{1,p}(\mathbb{R}^N)$.

Proposition 7.56. *Let $1 \leq p \leq +\infty$. If u is a function in $W_0^{1,p}(\mathcal{O})$, then its zero extension $\tilde{u}$ defined by*

$$\widetilde{u}(x) = \begin{cases} u(x) & \text{for } x \in \mathcal{O}, \\ 0 & \text{otherwise,} \end{cases}$$

belongs to $W^{1,p}(\mathbb{R}^N)$, and to $W_0^{1,p}(\mathcal{O}_1)$ for every open subset $\mathcal{O}_1$ containing $\mathcal{O}$. Moreover,

$$\|u\|_{W_0^{1,p}(\mathcal{O})} = \|\widetilde{u}\|_{W_0^{1,p}(\mathcal{O}_1)} = \|\widetilde{u}\|_{W^{1,p}(\mathbb{R}^N)} .$$

Proof. It is easily verified that $\widetilde{u}$ and $\widetilde{\dfrac{\partial u}{\partial x_i}}$ are in $L^p(\mathbb{R}^N)$ with

$$\|u\|_{L^p(\mathcal{O})} = \|\widetilde{u}\|_{L^p(\mathbb{R}^N)} \text{ and } \left\|\frac{\partial u}{\partial x_i}\right\|_{L^p(\mathcal{O})} = \left\|\widetilde{\frac{\partial u}{\partial x_i}}\right\|_{L^p(\mathbb{R}^N)} \quad \text{for } i = 1, \ldots, N.$$

Therefore to prove that $\widetilde{u}$ belongs to $W^{1,p}(\mathbb{R}^N)$, it suffices to show that

$$\frac{\partial \widetilde{u}}{\partial x_i} = \widetilde{\frac{\partial u}{\partial x_i}} \qquad \text{for } i = 1, \ldots, N. \tag{7.54}$$

By definition, there exists a sequence $\{u_n\}$ in $\mathcal{D}(\mathcal{O})$ such that

$$u_n \to u \quad \text{strongly in } W_0^{1,p}(\mathcal{O}).$$

So, using Definition 7.14 and Corollary 6.31, for all $\varphi \in \mathcal{D}(\mathbb{R}^N)$,

$$\begin{aligned} \left\langle \frac{\partial \widetilde{u}}{\partial x_i}, \varphi \right\rangle_{\mathcal{D}'(\mathbb{R}^N), \mathcal{D}(\mathbb{R}^N)} &= - \int_{\mathbb{R}^N} \widetilde{u} \, \frac{\partial \varphi}{\partial x_i} \, dx = - \int_{\mathcal{O}} u \, \frac{\partial \varphi}{\partial x_i} \, dx \\ &= - \lim_{n \to \infty} \int_{\mathcal{O}} u_n \, \frac{\partial \varphi}{\partial x_i} \, dx = \lim_{n \to \infty} \int_{\mathcal{O}} \varphi \, \frac{\partial u_n}{\partial x_i} \, dx \\ &= \int_{\mathcal{O}} \varphi \, \frac{\partial u}{\partial x_i} = \int_{\mathbb{R}^N} \varphi \, \widetilde{\frac{\partial u}{\partial x_i}} \, dx = \left\langle \widetilde{\frac{\partial u}{\partial x_i}}, \varphi \right\rangle_{\mathcal{D}'(\mathbb{R}^N), \mathcal{D}(\mathbb{R}^N)}. \end{aligned}$$

To prove that $\widetilde{u} \in W_0^{1,p}(\mathcal{O}_1)$, observe that if $\{u_n\}$ is a sequence in $\mathcal{D}(\mathcal{O})$ converging to u in $W_0^{1,p}(\mathcal{O})$, then $\{\widetilde{u}_n\}$ belongs to $\mathcal{D}(\mathcal{O}_1)$ and from (7.54), it converges to $\widetilde{u}$ in $W_0^{1,p}(\mathcal{O}_1)$. By Definition 7.54, it follows that $\widetilde{u}$ is in $W_0^{1,p}(\mathcal{O}_1)$. The equality of the norms is immediate. □

We also have the following property:

Proposition 7.57. *Let $1 \leq p < +\infty$ and Ω an open interval in $\mathbb{R}$ if $N = 1$ and a bounded open subset of $\mathbb{R}^N$ with a Lipschitz-continuous boundary if $N \geq 2$. If u is a function in $W^{1,p}(\Omega)$ with compact support in Ω, then u belongs to $W_0^{1,p}(\Omega)$.*

Proof. If $u \in W^{1,p}(\Omega)$, then by Theorem 7.35, there exists a sequence $\{u_n\}$ in $\mathcal{D}(\overline{\Omega})$ which converges to u in $W^{1,p}(\Omega)$.

Moreover, if the support of u is a compact set in Ω, by Urysohn's Lemma (Thm 6.60) there exists $\psi \in \mathcal{D}(\Omega)$ such that $\psi \equiv 1$ on supp u. Thus $\{\psi u_n\}$ is a sequence in $\mathcal{D}(\Omega)$ which converges to $\psi u = u$ in $W^{1,p}(\Omega)$, which means that u is in $W_0^{1,p}(\Omega)$. □

Remark 7.58. *More generally, one can prove that the result stated in Proposition 7.57 holds for any open set in $\mathbb{R}^N$.*

To simplify the presentation of the remainder of this section, we treat only the case $p = 2$. It is this case that will be used in our study of second order partial differential equations.

7.5.1 *A characterization of $H_0^1(\Omega)$*

The result below is an interesting characterization of the space $H_0^1(\Omega)$.

Theorem 7.59.
(1) Let $\Omega = (a, b) \subset \mathbb{R}$. Then

$$H_0^1(\Omega) = \{ u \in H^1(\Omega) | \, u(a) = u(b) = 0 \}.$$

(2) For $N \geq 2$, let Ω be a bounded open subset of $\mathbb{R}^N$ such that $\partial\Omega$ is Lipschitz continuous. Then

$$H_0^1(\Omega) = \{ u \in H^1(\Omega) | \, \gamma(u) = 0 \ \text{ on } \partial\Omega \}.$$

Proof. Let Ω be the interval (a, b) if $N = 1$ and a bounded open subset of $\mathbb{R}^N$ with $\partial\Omega$ Lipschitz-continuous if $N \geq 2$.

Let $u \in H_0^1(\Omega)$. By definition, there exists a sequence $\{u_n\}$ in $\mathcal{D}(\Omega)$ converging to u in $H_0^1(\Omega)$. If $N = 1$, by Theorem 7.42 we deduce that $\{u_n\}$

converges uniformly to u in $[a,b]$, so that $u(a) = u(b) = 0$. If $N \geq 2$, the continuity of the trace operator given by Theorem 7.49 implies that $\gamma(u_n)$ converges to $\gamma(u)$ in $L^2(\partial\Omega)$. Therefore $\gamma(u) = 0$, since $\gamma(u_n) \equiv 0$.

Let us prove the converse. For simplicity, we suppose that u belongs to $H^1(\Omega) \cap C^0(\overline{\Omega})$ (which, by Theorem 7.42, is always true for $N = 1$).

Let G be a function in $C^1(\mathbb{R})$ such that

$$G(s) = \begin{cases} 0 & \text{if } |s| \leq 1, \\ s & \text{if } |s| \geq 2, \end{cases} \qquad \text{and } |G(s)| \leq |s|, \quad \forall s \in \mathbb{R}.$$

Define

$$u_n = \frac{1}{n} G(n\,u) \quad \text{for every } n \in \mathbb{N}.$$

As $u \in C^0(\overline{\Omega})$, the set

$$K_n \doteq \left\{ x \in \Omega \middle| \; |u(x)| \geq \frac{1}{n} \right\}$$

is a compact set contained in Ω for every $n \in \mathbb{N}$. Since supp $u_n \subset K_n$, it follows that the support of u_n must also be a compact set contained in Ω. Moreover, G satisfies (7.34) for some $M \in \mathbb{R}$, so that by Theorem 7.37, the function u_n is in $H^1(\Omega)$ and $\nabla u_n = G'(nu)\nabla u$ for every $n \in \mathbb{N}$. In view of Proposition 7.57, the sequence $\{u_n\}$ is in $H_0^1(\Omega)$.

Now, as

$$u_n(x) \to u(x) \quad \text{and} \quad \nabla u_n(x) \to \nabla u(x) \quad \text{a.e. in } \Omega,$$

from the properties of G one has,

$$u_n^2(x) \leq u^2(x) \in L^1(\Omega) \quad \text{a.e. in } \Omega,$$

and

$$|\nabla u_n(x)|^2 \leq M^2 |\nabla u(x)|^2 \in L^1(\Omega) \quad \text{a.e. in } \Omega.$$

By Remark 7.20 and the Lebesgue Dominated Convergence Theorem (Thm 6.38), $\{u_n\}$ converges to u in $H^1(\Omega)$. This concludes the proof, since $H_0^1(\Omega)$ is a closed subspace of $H^1(\Omega)$. □

Remark 7.60.

(1) For $N \geq 2$, suppose that we have $u \in H^1(\Omega)$ and that $\partial\Omega$ is Lipschitz continuous. In view of the Trace Theorem (Thm 7.49), the condition $u = 0$ on $\partial\Omega$ can be replaced by the weaker one, $\gamma(u) = 0$ on $\partial\Omega$.

(2) If $\partial\Omega$ is not Lipschitz continuous, we cannot define the trace on $\partial\Omega$ anymore, but we can replace the condition $\gamma(u) = 0$ by the weaker condition that u belongs to $H_0^1(\Omega)$. This is justified by Theorem 7.59 and plays an essential role when dealing with partial differential equations with a boundary Dirichlet condition, as we shall see in the following.

7.5.2 *The Poincaré inequality in $H_0^1(\Omega)$*

By definition, the space $H_0^1(\Omega)$ is endowed with the H^1-norm. The following result provides an equivalent norm on $H_0^1(\Omega)$ and plays an important role in the study of partial differential equations.

Theorem 7.61 (Poincaré Inequality). *Let Ω be a bounded open set in $\mathbb{R}^N$. Then there exists a constant C_Ω, such that*

$$\|u\|_{L^2(\Omega)} \leq C_\Omega \, \|\nabla u\|_{L^2(\Omega)}, \quad \forall u \in H_0^1(\Omega), \tag{7.55}$$

and C_Ω depends only on the diameter of Ω.

Proof. We first prove the result for the case Ω is a cube in $\mathbb{R}^N$ of the form $\Omega = \left(-\frac{a}{2}, \frac{a}{2}\right)^N$. Let $u \in \mathcal{D}(\Omega)$. Then

$$u(x) = u(x', x_N) = \int_{-\frac{a}{2}}^{x_N} \frac{\partial u}{\partial x_N}(x', t)\, dt, \tag{7.56}$$

where $x' \in \mathbb{R}^{N-1}$.

Applying the Cauchy–Schwarz inequality, we obtain

$$\begin{aligned}\left|\int_{-\frac{a}{2}}^{x_N} \frac{\partial u}{\partial x_N}(x',t)\, dt\right|^2 &\leq \left(\int_{-\frac{a}{2}}^{x_N} \left|\frac{\partial u}{\partial x_N}(x',t)\right| dt\right)^2 \leq \left(\int_{-\frac{a}{2}}^{\frac{a}{2}} \left|\frac{\partial u}{\partial x_N}(x',t)\right| dt\right)^2 \\ &\leq \left[\left(\int_{-\frac{a}{2}}^{\frac{a}{2}} dt\right)^{\frac{1}{2}} \left(\int_{-\frac{a}{2}}^{\frac{a}{2}} \left|\frac{\partial u}{\partial x_N}(x',t)\right|^2 dt\right)^{\frac{1}{2}}\right]^2 \\ &= a \int_{-\frac{a}{2}}^{\frac{a}{2}} \left|\frac{\partial u}{\partial x_N}(x',t)\right|^2 dt.\end{aligned}$$

From (7.56), by integrating this inequality over Ω and using Fubini's Theorem (Thm 6.39), we get

$$\int_\Omega |u|^2\, dx = \int_\Omega |u(x', x_N)|^2\, dx'\, dx_N \leq a^2 \int_\Omega \left|\frac{\partial u}{\partial x_N}\right|^2 dx \leq a^2 \int_\Omega |\nabla u|^2\, dx.$$

Therefore we have

$$\|u\|_{L^2(\Omega)} \leq a \|\nabla u\|_{L^2(\Omega)},$$

for all $u \in \mathcal{D}(\Omega)$, and by density, for all $u \in H_0^1(\Omega)$.

If Ω is an arbitrary bounded open subset, we choose $Q = \left(-\frac{a}{2}, \frac{a}{2}\right)^N$ to be a cube in $\mathbb{R}^N$ such that $\Omega \subset Q$. In this case, a depends on the diameter of Ω. For $u \in H_0^1(\Omega)$, by Proposition 7.56 we have that $\widetilde{u} \in H_0^1(Q)$. Moreover,

$$\|u\|_{L^2(\Omega)} = \|\widetilde{u}\|_{L^2(Q)} \quad \text{and} \quad \|\nabla u\|_{L^2(\Omega)} = \|\nabla \widetilde{u}\|_{L^2(Q)},$$

and the assertion follows. □

Corollary 7.62. *Let Ω be a bounded open set in $\mathbb{R}^N$. Then*

$$\|\nabla u\|_{L^2(\Omega)} \leq \|u\|_{H^1(\Omega)} \leq (1 + C_\Omega)\|\nabla u\|_{L^2(\Omega)}, \quad \forall u \in H_0^1(\Omega),$$

for some constant C_Ω depending only on the diameter of Ω. Moreover,

$$\|u\|_{H_0^1(\Omega)} = \|\nabla u\|_{L^2(\Omega)}, \tag{7.57}$$

defines a norm on $H_0^1(\Omega)$ equivalent to the norm of $H^1(\Omega)$ given by (7.15).

Remark 7.63.

(1) In the following, the space $H_0^1(\Omega)$ is equipped with the norm (7.57).

(2) The inequality does not hold in $H^1(\Omega)$, since if $u \in H^1(\Omega)$ then for any constant $c \in \mathbb{R}$, the function $u+c$ is also in $H^1(\Omega)$. However, $\|u+c\| \to +\infty$ as $c \to +\infty$.

(3) Up to a translation of the coordinate axes, we can always take $a =$ diam(Ω) in the proof of the Poincaré inequality. Then in inequality (7.55), one has $C_\Omega \leq$ diam(Ω). The constant C_Ω is referred hereafter as the Poincaré constant.

7.5.3 *Dual of $H_0^1(\Omega)$*

We end this section with a discussion on $H^{-1}(\Omega)$, the dual space of $H_0^1(\Omega)$.

Definition 7.64. *Let Ω be a bounded open set in $\mathbb{R}^N$, $p \in [1, +\infty]$ and p' its conjugate given by (6.30).*

We define $W^{-1,p}(\Omega)$ as the Banach space

$$W^{-1,p}(\Omega) = \big(W_0^{1,p'}(\Omega)\big)',$$

endowed with the norm (see Definition 6.4)

$$\|F\|_{W^{-1,p}(\Omega)} = \sup_{W_0^{1,p'}(\Omega)\setminus\{0\}} \frac{\left|\langle F, u\rangle_{W^{-1,p}(\Omega), W_0^{1,p'}(\Omega)}\right|}{\|u\|_{W_0^{1,p'}(\Omega)}}. \tag{7.58}$$

In particular, for $p = 2$ we set

$$H^{-1}(\Omega) = \big(H_0^1(\Omega)\big)',$$

so that (7.58) reads

$$\|F\|_{H^{-1}(\Omega)} = \sup_{H_0^1(\Omega)\setminus\{0\}} \frac{\left|\langle F, u\rangle_{H^{-1}(\Omega), H_0^1(\Omega)}\right|}{\|u\|_{H_0^1(\Omega)}}.$$

One can prove the following result, which gives a complete description of this space, stated for simplicity for $p = 2$:

Theorem 7.65. *Let F be in $H^{-1}(\Omega)$. Then there exist $N+1$ functions $f_0, f_1, \dots, f_N$ in $L^2(\Omega)$ such that*

$$F = f_0 + \sum_{i=1}^{N} \frac{\partial f_i}{\partial x_i}, \tag{7.59}$$

in the sense of distributions. Moreover,

$$\|F\|^2_{H^{-1}(\Omega)} = \inf \sum_{i=0}^{N} \|f_i\|^2_{L^2(\Omega)},$$

where the infimum is taken over all the vectors $(f_0, f_1, \dots, f_N)$ in $[L^2(\Omega)]^{N+1}$ for which (7.59) holds.

Conversely, if $(f_0, f_1, \dots, f_N)$ is a vector in $[L^2(\Omega)]^{N+1}$, then (7.59) defines an element F of $H^{-1}(\Omega)$ satisfying

$$\|F\|^2_{H^{-1}(\Omega)} \leq \sum_{i=0}^{N} \|f_i\|^2_{L^2(\Omega)}.$$

We also have the following definition:

Definition 7.66. *Let Ω be a bounded open subset of $\mathbb{R}^N$ such that $\partial\Omega$ is Lipschitz continuous. We denote by $H^{-\frac{1}{2}}(\partial\Omega)$ the Banach space defined by*

$$H^{-\frac{1}{2}}(\partial\Omega) = \left(H^{\frac{1}{2}}(\partial\Omega)\right)',$$

endowed with the norm

$$\|F\|_{H^{-\frac{1}{2}}(\partial\Omega)} = \sup_{H^{\frac{1}{2}}(\partial\Omega)\setminus\{0\}} \frac{\left|\langle F, u\rangle_{H^{-\frac{1}{2}}(\partial\Omega), H^{\frac{1}{2}}(\partial\Omega)}\right|}{\|u\|_{H^{\frac{1}{2}}(\partial\Omega)}}.$$

Remark 7.67.

(1) From Theorem 7.65,

$$L^2(\Omega) \subset H^{-1}(\Omega).$$

Since $H_0^1(\Omega) \subset L^2(\Omega)$, for every $u \in H_0^1(\Omega)$,

$$\langle v, u\rangle_{H^{-1}(\Omega), H_0^1(\Omega)} = \langle v, u\rangle_{L^2(\Omega), L^2(\Omega)} = \int_\Omega u\, v\, dx.$$

Moreover, for every v in $L^2(\Omega)$

$$\|v\|_{H^{-1}(\Omega)} \leq C_\Omega \|v\|_{L^2(\Omega)}, \tag{7.60}$$

where C_Ω is the Poincaré constant given by Theorem 7.61.

(2) Let $N \geq 2$ and Ω a bounded open subset of $\mathbb{R}^N$ such that $\partial\Omega$ is Lipschitz continuous. A similar argument as above shows that if v is in $L^2(\partial\Omega)$, then v is in $H^{-\frac{1}{2}}(\partial\Omega)$ and for every $u \in H^{\frac{1}{2}}(\partial\Omega)$,

$$\langle v, u\rangle_{H^{-\frac{1}{2}}(\partial\Omega), H^{\frac{1}{2}}(\partial\Omega)} = \langle v, u\rangle_{L^2(\partial\Omega), L^2(\partial\Omega)} = \int_{\partial\Omega} u\, v\, ds.$$

(3) It can be proven that $(H^1(\Omega))'$ can be identified with the direct sum $H^{-1}(\Omega) \oplus H^{-\frac{1}{2}}(\partial\Omega)$. Also, $L^2(\Omega) \subset (H^1(\Omega))'$ and if v is in $L^2(\Omega)$,

$$\langle v, u\rangle_{(H^1(\Omega))', H^1(\Omega)} = \langle v, u\rangle_{L^2(\Omega), L^2(\Omega)} = \int_\Omega u\, v\, dx,$$

for all $u \in H^1(\Omega)$. Then

$$\|v\|_{(H^1(\Omega))'} \leq \|v\|_{L^2(\Omega)}. \tag{7.61}$$

Indeed, from the Cauchy–Schwarz Inequality (6.32) and Proposition 6.11,

$$\|v\|_{(H^1(\Omega))'} = \sup_{H^1(\Omega)\setminus\{0\}} \frac{\left|\langle v, u\rangle_{(H^1(\Omega))', H^1(\Omega)}\right|}{\|u\|_{H^1(\Omega)}} = \sup_{H^1(\Omega)\setminus\{0\}} \frac{\left|\int_\Omega v\, u\, dx\right|}{\|u\|_{H^1(\Omega)}}$$

$$\leq \frac{\|v\|_{L^2(\Omega)}\|u\|_{L^2(\Omega)}}{\|u\|_{H^1(\Omega)}} \leq \|v\|_{L^2(\Omega)}.$$

7.6 A characterization of $H^1(\mathbb{R}^N)$

In this section, we present an important result which characterizes the space $H^1(\mathbb{R}^N)$ and gives, using the Fourier transform, a definition equivalent to (7.19). This represents a different approach to Sobolev spaces in the case $p = 2$. We present here this characterization for completeness.

We start by recalling the definition and the main properties of the Fourier transform in L^2 (for proofs and details, see [1], [16], [29], [34]).

Definition 7.68. *Let u be in $L^1(\mathbb{R}^N)$. The functions $\hat{u}$ and $\check{u}$ on $\mathbb{R}^N$ defined by*

$$\hat{u}(\xi) = \frac{1}{(2\pi)^{\frac{N}{2}}} \int_{\mathbb{R}^N} e^{-i\xi\cdot x} u(x)\, dx \quad \text{for a.e. } \xi \in \mathbb{R}^N,$$

and

$$\check{u}(\xi) = \frac{1}{(2\pi)^{\frac{N}{2}}} \int_{\mathbb{R}^N} e^{i\xi\cdot x} u(x)\, dx \quad \text{for a.e. } \xi \in \mathbb{R}^N,$$

are called the Fourier transform of u and the inverse Fourier transform of u, respectively.

Let us point out that in Definition 7.68, the functions $\hat{u}$ and $\check{u}$ are complex-valued. The following result states one of the main properties of the Fourier transform:

Theorem 7.69 (Plancherel's Theorem). *Let u be in $L^1(\mathbb{R}^N) \cap L^2(\mathbb{R}^N)$. Then its Fourier transform $\hat{u}$ and its inverse Fourier transform $\check{u}$ are in $L^2(\mathbb{R}^N)$ and*

$$\|\hat{u}\|_{L^2(\mathbb{R}^N)} = \|\check{u}\|_{L^2(\mathbb{R}^N)} = \|u\|_{L^2(\mathbb{R}^N)}. \tag{7.62}$$

Equation (7.62) is called Parseval's formula. Plancherel's Theorem allows us to define the Fourier transform also for functions in $L^2(\mathbb{R}^N)$.

Proposition 7.70. *Let u be a function in $L^2(\mathbb{R}^N)$. If $\{u_n\}$ is any sequence in $L^1(\mathbb{R}^N) \cap L^2(\mathbb{R}^N)$ which converges to u in $L^2(\mathbb{R}^N)$, then the sequence $\{\hat{u}_n\}$ (respectively $\{\check{u}_n\}$) converges in $L^2(\mathbb{R}^N)$ to some element denoted by $\hat{u}$ (respectively $\check{u}$). Moreover, $\hat{u}$ and $\check{u}$ do not depend on the choice of the sequence $\{u_n\}$.*

Proof. We prove only the result for $\{\hat{u}_n\}$, the proof for $\{\check{u}_n\}$ being similar.

Let u be in $L^2(\mathbb{R}^N)$ and $\{u_n\}$ a sequence in $L^1(\mathbb{R}^N) \cap L^2(\mathbb{R}^N)$ which converges to u in $L^2(\mathbb{R}^N)$. This implies that $\{u_n\}$ is a Cauchy sequence in $L^2(\mathbb{R}^N)$. By Parseval's formula (7.62), the sequence $\{\hat{u}_n\}$ is also a Cauchy sequence in $L^2(\mathbb{R}^N)$, which is complete. Consequently, $\{\hat{u}_n\}$ converges to some $\hat{u}$ in $L^2(\mathbb{R}^N)$.

Now, suppose $\{v_n\} \subset L^1(\mathbb{R}^N) \cap L^2(\mathbb{R}^N)$ is another sequence converging to u in $L^2(\mathbb{R}^N)$ and denote by $\hat{v}$ the limit of $\{\hat{v}_n\}$ in $L^2(\mathbb{R}^N)$. Applying again Parseval's formula, we have

$$\|\hat{u} - \hat{v}\|_{L^2(\mathbb{R}^N)} = \lim_{n\to\infty} \|\hat{u}_n - \hat{v}_n\|_{L^2(\mathbb{R}^N)} = \lim_{n\to\infty} \|u_n - v_n\|_{L^2(\mathbb{R}^N)} = 0.$$

This shows that $\hat{u} = \hat{v}$ and completes the proof. □

In view of the density of $L^1(\mathbb{R}^N) \cap L^2(\mathbb{R}^N)$ in $L^2(\mathbb{R}^N)$ the following definition makes sense:

Definition 7.71. *Let u be in $L^2(\mathbb{R}^N)$. The functions $\hat{u}$ and $\check{u}$ in $L^2(\mathbb{R}^N)$ given by Proposition 7.70 are still called the Fourier transform and the inverse Fourier transform of u, respectively.*

Proposition 7.72. *The following properties hold:*

(1) Parseval's formula (7.62) *is valid also for functions in $L^2(\mathbb{R}^N)$.*

(2) Let u and v be in $L^2(\mathbb{R}^N)$. Then

$$\int_{\mathbb{R}^N} u(x)\, v(x)\, dx = \int_{\mathbb{R}^N} \hat{u}(\xi)\, \overline{\hat{v}}(\xi)\, d\xi. \tag{7.63}$$

(3) For every u in $L^2(\mathbb{R}^N)$,

$$u = \check{\hat{u}} = \widehat{\check{u}}. \tag{7.64}$$

(4) If u is a real-valued function in $H^1(\mathbb{R}^N)$, then

$$\widehat{\nabla u}(\xi) = i\xi\, \hat{u}(\xi), \qquad \textit{for a.e. } \xi \in \mathbb{R}^N. \tag{7.65}$$

Now, we can prove the main result of this section.

Theorem 7.73. *Let u be a real-valued function in $L^2(\mathbb{R}^N)$. Then*

$$u(x) \in H^1(\mathbb{R}^N) \iff (1 + |\xi|)\, \hat{u}(\xi) \in L^2(\mathbb{R}^N).$$

Moreover,

$$\|u\|_{H^1(\mathbb{R}^N)} \leq \|(1 + |\xi|)\, \hat{u}(\xi)\|_{L^2(\mathbb{R}^N)} \leq \sqrt{2}\|u\|_{H^1(\mathbb{R}^N)}, \tag{7.66}$$

for every $u \in H^1(\mathbb{R}^N)$, where the norm is given by (7.15).

Proof. Let u be in $H^1(\mathbb{R}^N)$. By Parseval's formula and (7.65), one has

$$\begin{aligned}
\int_{\mathbb{R}^N} (1 + |\xi|)^2\, |\hat{u}(\xi)|^2\, d\xi &\leq 2 \int_{\mathbb{R}^N} (1 + |\xi|^2)|\hat{u}(\xi)|^2\, d\xi \\
&= 2 \int_{\mathbb{R}^N} (|u(\xi)|^2 + |\nabla u(\xi)|^2\, d\xi \\
&= 2\|u\|^2_{H^1(\mathbb{R}^N)},
\end{aligned}$$

which proves that $(1+|\xi|)\hat{u}(\xi)$ is in $L^2(\mathbb{R}^N)$ and gives the second inequality in (7.66).

To prove the converse, suppose that $(1 + |\xi|)\, \hat{u}(\xi)$ is in $L^2(\mathbb{R}^N)$. This implies, in particular, that $\hat{u}(\xi)$ is in $L^2(\mathbb{R}^N)$, so that by Theorem 7.69, u is in $L^2(\mathbb{R}^N)$.

Now, let φ be a (real-valued) function in $\mathcal{D}(\mathbb{R}^N)$. Using Definition 7.14, (7.63), (7.64), and (7.65), we have

$$\begin{aligned}
\left\langle \frac{\partial u}{\partial x_k}, \varphi \right\rangle_{\mathcal{D}'(\mathbb{R}^N), \mathcal{D}(\mathbb{R}^N)} &= -\int_{\mathbb{R}^N} \frac{\partial \varphi}{\partial x_k}(x) u(x)\, dx = -\int_{\mathbb{R}^N} \widehat{\frac{\partial \varphi}{\partial x_k}}(\xi)\, \overline{\hat{u}}(\xi)\, d\xi \\
&= -\int_{\mathbb{R}^N} i\, \xi_k\, \hat{\varphi}(\xi)\, \overline{\hat{u}}(\xi)\, d\xi = \int_{\mathbb{R}^N} \hat{\varphi}(\xi)\, \overline{i\xi_k \hat{u}}(\xi)\, d\xi \\
&= \int_{\mathbb{R}^N} \hat{\varphi}(x)\, \overline{w_k}(x)\, dx \;=\; \int_{\mathbb{R}^N} \varphi(x)\, \overline{\check{w}_k}(x)\, dx,
\end{aligned}$$

for every $k = 1, \ldots, N$, where we set $w_k(\xi) = i\xi_k \hat{u}(\xi)$. Hence, for each k in $\{1, \ldots, N\}$, $\dfrac{\partial u}{\partial x_k} = \overline{\check{w}_k} \in L^2(\mathbb{R}^N)$, which proves that u is in $H^1(\mathbb{R}^N)$.

Finally, the first inequality in (7.66) follows directly from Parseval's formula (7.62) and (7.65) since

$$\begin{aligned}\int_{\mathbb{R}^N} (|u(\xi)|^2 + |\nabla u(\xi)|^2 \, d\xi &= \int_{\mathbb{R}^N} (1 + |\xi|^2)|\hat{u}(\xi)|^2 \, d\xi \\ &\leq \int_{\mathbb{R}^N} (1 + |\xi|)^2 |\hat{u}(\xi)|^2 \, d\xi,\end{aligned}$$

which ends the proof. □

Remark 7.74. *This result means that*

$$H^1(\mathbb{R}^N) = \left\{ u \middle| \, u \in L^2(\mathbb{R}^N), (1 + |\xi|)\hat{u}(\xi) \in L^2(\mathbb{R}^N) \right\},$$

and

$$\|u\| = \|(1 + |\xi|)\, \hat{u}(\xi)\|_{L^2(\mathbb{R}^N)} \tag{7.67}$$

is a norm on $H^1(\mathbb{R}^N)$ equivalent to the norm given by (7.15).

Chapter 8

Sobolev Embedding Theorems

The Sobolev Embedding Theorems state one of the most interesting properties of the Sobolev spaces. It is precisely the embedding properties of these spaces which make them suitable spaces in the study of solutions of partial differential equations. They show that in any dimension N, functions in $W^{1,p}$ have actually better integrability properties depending on p and N. This is due to the fact that these functions and their derivatives are in L^p.

Moreover, for regular and bounded open sets we also have compact embedding results. Let us emphasize that these results are essential tools when studying partial differential equations. We present their proofs in this chapter; they are technical and require some delicate arguments.

8.1 Continuous embedding theorems

Let us first recall the definition of a continuous embedding.

Definition 8.1. *Let X and Y be Banach spaces such that $X \subset Y$. We say that the embedding (or the inclusion) is continuous if the mapping*

$$i_X : x \in X \to x \in Y$$

is continuous.

Remark 8.2. *With this definition and that of the space $C_b^0(\mathbb{R})$ (see (1.4)), the inclusion*

$$W^{1,p}(\mathbb{R}) \subset C_b^0(\mathbb{R}),$$

is a continuous embedding for $1 \leq p \leq +\infty$.

When $N \geq 2$, we shall consider separately the three cases $p < N$, $p = N$, and $p > N$ for which we have different embedding results.

8.1.1 The case $1 \leq p < N$ for $\mathbb{R}^N$

We first present the Sobolev Embedding Theorem for the case $p \in [1, N)$. For its proof, we need the following lemma:

Lemma 8.3. *Let $N \geq 2$. For $i = 1, 2, \ldots, N$, let f_i be a nonnegative function in $L^{N-1}(\mathbb{R}^{N-1})$ and for $x = (x_1, \ldots, x_N) \in \mathbb{R}^N$, define*

$$\hat{x}_i = (x_1, \ldots, x_{i-1}, x_{i+1}, \ldots, x_N) \in \mathbb{R}^{N-1}, \tag{8.1}$$

for every $i = 1, \ldots, N$. Then

$$\begin{aligned} \int_{\mathbb{R}^N} \prod_{i=1}^{N} f_i(\hat{x}_i) dx &\leq \prod_{i=1}^{N} \left[\int_{\mathbb{R}^{N-1}} [f_i(\hat{x}_i)]^{N-1} d\hat{x}_i \right]^{\frac{1}{N-1}} \\ &= \prod_{i=1}^{N} \|f_i\|_{L^{N-1}(\mathbb{R}^{N-1})}. \end{aligned} \tag{8.2}$$

Proof. We prove by induction on N.

For $N = 2$ the equality in (8.2) holds, since $f_1(\hat{x}_1)$ is independent of x_1 and $f_2(\hat{x}_2)$ is independent of x_2. Hence

$$\int_{R^2} f_1(\hat{x}_1) f_2(\hat{x}_2)\, dx = \left[\int_R f_1(\hat{x}_1)\, d\hat{x}_1 \right] \left[\int_R f_2(\hat{x}_2)\, d\hat{x}_2 \right],$$

by Fubini's Theorem (Thm 6.39).

Let us assume (8.2) holds for some N and let f_i be nonnegative functions in $L^N(\mathbb{R}^N)$ for $i = 1, 2, \ldots, N+1$. Denote by $\xi = (x, x_{N+1})$ the point in $\mathbb{R}^{N+1}$ with $x \in \mathbb{R}^N$ and let

$$f(\xi) = \prod_{i=1}^{N+1} f_i(x_1, \ldots, x_{i-1}, x_{i+1}, \ldots, x_{N+1}). \tag{8.3}$$

We have to prove that the claimed inequality is true for $N+1$, that is,

$$\int_{\mathbb{R}^{N+1}} f(\xi) d\xi \leq \prod_{i=1}^{N+1} \left(\int_{\mathbb{R}^N} [f_i(x)]^N dx \right)^{\frac{1}{N}}. \tag{8.4}$$

Integrating (8.3) over $\mathbb{R}^N$ with respect to x and applying Hölder's inequality (6.31) with $p = N$ and $p' = \frac{N}{N-1}$ yield

$$\begin{aligned} \int_{\mathbb{R}^N} f(x, x_{N+1}) dx &= \int_{\mathbb{R}^N} f_{N+1}(x) \prod_{i=1}^{N} f_i(\hat{x}_i, x_{N+1}) dx \\ &\leq \|f_{N+1}\|_{L^N(\mathbb{R}^N)} \left(\int_{\mathbb{R}^N} \prod_{i=1}^{N} f_i^{\frac{N}{N-1}}(\hat{x}_i, x_{N+1}) dx \right)^{\frac{N-1}{N}}, \end{aligned}$$

where $\hat{x}_i$ is given by (8.1). Consequently, applying the induction on the functions $f_i^{\frac{N}{N-1}}(\,\cdot\,, x_{N+1})$, we have

$$\int_{\mathbb{R}^N} f(x, x_{N+1})dx \leq \|f_{N+1}\|_{L^N(\mathbb{R}^N)} \prod_{i=1}^{N} \left(\int_{R^{N-1}} f_i^N(\hat{x}_i, x_{N+1}) d\hat{x}_i \right)^{\frac{1}{N}}$$

$$= \|f_{N+1}\|_{L^N(\mathbb{R}^N)} \prod_{i=1}^{N} \|f_i(\,\cdot\,, x_{N+1})\|_{L^N(R^{N-1})},$$

for almost every x_{N+1} in $\mathbb{R}$.

Integrating the above inequality over $\mathbb{R}$ with respect to x_{N+1} and applying inequality (6.36) for $k = N$, $p = 1$, $p_i = N$, and

$$g_i = \|f_i(\,\cdot\,, x_{N+1})\|_{L^N(R^{N-1})},$$

we obtain

$$\begin{aligned}
\int_{\mathbb{R}^{N+1}} f(\xi)\, d\xi &\leq \|f_{N+1}\|_{L^N(\mathbb{R}^N)} \int_{\mathbb{R}} \prod_{i=1}^{N} \|f_i(\,\cdot\,, x_{N+1})\|_{L^N(R^{N-1})}\, dx_{N+1} \\
&\leq \|f_{N+1}\|_{L^N(\mathbb{R}^N)} \prod_{i=1}^{N} \left(\int_{\mathbb{R}} \|f_i(\,\cdot\,, x_{N+1})\|^N_{L^N(R^{N-1})}\, dx_{N+1} \right)^{\frac{1}{N}} \\
&= \|f_{N+1}\|_{L^N(\mathbb{R}^N)} \prod_{i=1}^{N} \left(\int_{\mathbb{R}} \int_{\mathbb{R}^{N-1}} f_i^N(x) dx \right)^{\frac{1}{N}} \\
&= \prod_{i=1}^{N+1} \left(\int_{\mathbb{R}^N} f_i^N(x) dx \right)^{\frac{1}{N}}.
\end{aligned}$$

It follows that inequality (8.4) is true, which completes the proof. □

Theorem 8.4. *Let $N \geq 2$. If $1 \leq p < N$ and*

$$\frac{1}{p^*} = \frac{1}{p} - \frac{1}{N}, \tag{8.5}$$

then

$$W^{1,p}(\mathbb{R}^N) \subset L^{p^*}(\mathbb{R}^N), \tag{8.6}$$

and there exists a constant $C = C(p, N)$ such that

$$\|u\|_{L^{p^*}(\mathbb{R}^N)} \leq C\, \|\nabla u\|_{L^p(\mathbb{R}^N)}, \tag{8.7}$$

for all $u \in W^{1,p}(\mathbb{R}^N)$.

Proof. We first prove that (8.7) holds for functions in $\mathcal{D}(\mathbb{R}^N)$, then we argue by density. Notice that $N \geq 2$ since $1 \leq p < N$.

Case 1: $p = 1$, $p^* = \dfrac{N}{N-1}$.

Let $u \in \mathcal{D}(\mathbb{R}^N)$. Since u has compact support, it can be expressed as

$$u(x) = \int_{-\infty}^{x_i} \frac{\partial u}{\partial x_i}(x_1, \ldots, x_{i-1}, t, x_{i+1}, \ldots, x_N)\, dt,$$

for $i = 1, \ldots, N$. It follows that

$$|u(x)| \leq \int_{-\infty}^{\infty} \left| \frac{\partial u}{\partial x_i}(x) \right| dx_i.$$

Using (8.1), if we set

$$f_i(\hat{x}_i) = \left(\int_{-\infty}^{\infty} \left| \frac{\partial u}{\partial x_i}(x) \right| dx_i \right)^{\frac{1}{N-1}},$$

then $f_i \in L^{N-1}(\mathbb{R}^{N-1})$ and

$$|u(x)|^{\frac{1}{N-1}} \leq f_i(\hat{x}_i), \tag{8.8}$$

for $i = 1, \ldots, N$. Hence

$$|u(x)|^{\frac{N}{N-1}} \leq \prod_{i=1}^{N} f_i(\hat{x}_i).$$

By assumption, the functions f_i are nonnegative and independent of the variable x_i. Integrating both sides of the above inequality over $\mathbb{R}^N$ and applying Lemma 8.3 and (8.8), we get

$$\begin{aligned}
\int_{\mathbb{R}^N} |u(x)|^{\frac{N}{N-1}}\, dx &\leq \int_{\mathbb{R}^N} \prod_{i=1}^{N} f_i(\hat{x}_i)\, dx \leq \prod_{i=1}^{N} \left(\int_{\mathbb{R}^{N-1}} (f_i(\hat{x}_i))^{N-1} d\hat{x}_i \right)^{\frac{1}{N-1}} \\
&= \prod_{i=1}^{N} \left(\int_{\mathbb{R}^{N-1}} \left(\int_{-\infty}^{\infty} \left| \frac{\partial u}{\partial x_i}(x) \right| dx_i \right) d\hat{x}_i \right)^{\frac{1}{N-1}} \\
&= \prod_{i=1}^{N} \left(\int_{\mathbb{R}^N} \left| \frac{\partial u}{\partial x_i}(x) \right| dx \right)^{\frac{1}{N-1}}.
\end{aligned}$$

It follows that

$$\left[\int_{\mathbb{R}^N} |u(x)|^{\frac{N}{N-1}} \right]^{\frac{N-1}{N}} \leq \left[\prod_{i=1}^{N} \left(\int_{\mathbb{R}^N} \left| \frac{\partial u}{\partial x_i}(x) \right| dx \right)^{\frac{1}{N-1}} \right]^{\frac{N-1}{N}}. \tag{8.9}$$

Therefore

$$\|u\|_{L^{\frac{N}{N-1}}(\mathbb{R}^N)} \leq \prod_{i=1}^{N} \left\| \frac{\partial u}{\partial x_i} \right\|_{L^1(\mathbb{R}^N)}^{\frac{1}{N}} \leq \prod_{i=1}^{N} \|\nabla u\|_{L^1(\mathbb{R}^N)}^{\frac{1}{N}} = \|\nabla u\|_{L^1(\mathbb{R}^N)},$$

which proves (8.7) for functions in $\mathcal{D}(\mathbb{R}^N)$, for $p = 1$ and with $C = 1$.

Case 2: $1 < p < N$.

Letting $u = |v|^t$ in inequality (8.9) (also true in this case for functions in $\mathcal{D}(\mathbb{R}^N)$) for $t > 1$ and $v \in \mathcal{D}(\mathbb{R}^N)$, we obtain

$$\left[\int_{\mathbb{R}^N} |v|^{t\frac{N}{N-1}}\right]^{\frac{N-1}{N}} \leq \left[\prod_{i=1}^{N}\left(t\int_{\mathbb{R}^N} |v|^{t-1}\left|\frac{\partial v}{\partial x_i}\right| dx\right)^{\frac{1}{N-1}}\right]^{\frac{N-1}{N}}.$$

The left-hand side of the inequality is simply the norm of $|v|^t$ in $L^{\frac{N}{N-1}}(\mathbb{R}^N)$. Using (6.33) and Hölder's inequality (6.31),

$$\begin{aligned}
\|v\|^t_{L^{t\frac{N}{N-1}}(\mathbb{R}^N)} &\leq t\prod_{i=1}^{N}\left(\int_{\mathbb{R}^N} |v|^{t-1}\left|\frac{\partial v}{\partial x_i}\right| dx\right)^{\frac{1}{N}} \\
&\leq t\prod_{i=1}^{N}\left(\||v|^{t-1}\|_{L^{p'}(\mathbb{R}^N)}\left\|\frac{\partial v}{\partial x_i}\right\|_{L^p(\mathbb{R}^N)}\right)^{\frac{1}{N}} \\
&= t\prod_{i=1}^{N}\left(\|v\|^{t-1}_{L^{p'(t-1)}(\mathbb{R}^N)}\left\|\frac{\partial v}{\partial x_i}\right\|_{L^p(\mathbb{R}^N)}\right)^{\frac{1}{N}} \\
&= t\|v\|^{t-1}_{L^{p'(t-1)}(\mathbb{R}^N)}\prod_{i=1}^{N}\left\|\frac{\partial v}{\partial x_i}\right\|^{\frac{1}{N}}_{L^p(\mathbb{R}^N)},
\end{aligned}$$

where p' is the conjugate of p. Consequently,

$$\|v\|^t_{L^{t\frac{N}{N-1}}(\mathbb{R}^N)} \leq t\|v\|^{t-1}_{L^{p'(t-1)}(\mathbb{R}^N)}\|\nabla v\|_{L^p(\mathbb{R}^N)}. \tag{8.10}$$

Since $1 < p < N$,

$$1 - \frac{1}{N} > \frac{1}{p} - \frac{1}{N} = \frac{1}{p^*}.$$

We can choose $t = p^*\dfrac{N-1}{N} > 1$, so that

$$p'(t-1) = p^* = t\frac{N}{N-1}.$$

This implies that inequality (8.10) is equivalent to

$$\|v\|^t_{L^{p^*}(\mathbb{R}^N)} \leq t\|v\|^{t-1}_{L^{p^*}(\mathbb{R}^N)}\|\nabla v\|_{L^p(\mathbb{R}^N)},$$

so that,

$$\|v\|_{L^{p^*}(\mathbb{R}^N)} \leq C\|\nabla v\|_{L^p(\mathbb{R}^N)},$$

with $C = p^*\frac{N-1}{N}$ and proves (8.7) for the case $1 < p < N$ and v in $\mathcal{D}(\mathbb{R}^N)$.

Now, if $1 \leq p < N$ and $u \in W^{1,p}(\mathbb{R}^N)$, then by the density of $\mathcal{D}(\mathbb{R}^N)$ in $W^{1,p}(\mathbb{R}^N)$ (see Theorem 7.29), there exists a sequence $\{u_n\} \in \mathcal{D}(\mathbb{R}^N)$ such that

$$u_n \to u \quad \text{in } W^{1,p}(\mathbb{R}^N).$$

Due to Remark 7.20, this implies that

$$u_n \to u \quad \text{in } L^p(\mathbb{R}^N),$$

and

$$\nabla u_n \to \nabla u \quad \text{in } [L^p(\mathbb{R}^N)]^N.$$

Hence by Theorem 6.40 there exists a subsequence of $\{u_n\}$, still denoted by $\{u_n\}$, such that

$$u_n \to u \qquad \text{a.e. in } \mathbb{R}^N.$$

Letting $v = u_n$ in (8.7) and using Fatou's Lemma (Thm 6.37), we get

$$\begin{aligned}\|u\|_{L^{p^*}(\mathbb{R}^N)} &\leq \liminf_{n\to\infty} \|u_n\|_{L^{p^*}(\mathbb{R}^N)} \leq C \lim_{n\to\infty} \|\nabla u_n\|_{L^p(\mathbb{R}^N)} \\ &= C\|\nabla u\|_{L^p(\mathbb{R}^N)},\end{aligned}$$

where $C = C(p, N)$. This shows that a function u in $W^{1,p}(\mathbb{R}^N)$ also belongs to $L^{p^*}(\mathbb{R}^N)$ and completes the proof of (8.7). □

Corollary 8.5. *Let $N \geq 2$. If $1 \leq p < N$ and $\frac{1}{p^*} = \frac{1}{p} - \frac{1}{N}$, then*

$$W^{1,p}(\mathbb{R}^N) \subset L^q(\mathbb{R}^N), \tag{8.11}$$

with continuous embedding for all $q \in [p, p^*]$. That is, there exists a constant $C = C(p, N)$ such that

$$\|u\|_{L^q(\mathbb{R}^N)} \leq C\|u\|_{W^{1,p}(\mathbb{R}^N)}, \quad \forall u \in W^{1,p}(\mathbb{R}^N).$$

Proof. The embedding holds true for the case $q = p$ by the definition of $W^{1,p}$ and by Theorem 8.4 for $q = p^*$.

Suppose $p < q < p^*$. Then $\frac{1}{q}$ lies between $\frac{1}{p^*}$ and $\frac{1}{p}$, so we can express

$$\frac{1}{q} = \frac{\alpha}{p} + \frac{1-\alpha}{p^*}, \qquad \alpha \in (0, 1).$$

By Theorem 8.4, and the Interpolation Inequality (Lemma 6.36) with $r = p^*$, any function $u \in W^{1,p}(\mathbb{R}^N)$ is in $L^q(\mathbb{R}^N)$ and

$$\begin{aligned}\|u\|_{L^q(\mathbb{R}^N)} &\leq \|u\|^{\alpha}_{L^p(\mathbb{R}^N)} \|u\|^{1-\alpha}_{L^{p^*}(\mathbb{R}^N)} \\ &\leq \alpha\|u\|_{L^p(\mathbb{R}^N)} + (1-\alpha)\|u\|_{L^{p^*}(\mathbb{R}^N)},\end{aligned}$$

where the last inequality is due to Young's Inequality (6.38) with $p = 1/\alpha$ and $p' = 1/(1-\alpha)$. Now, as $\alpha \in (0,1)$, by (8.7)

$$\begin{aligned}\|u\|_{L^q(\mathbb{R}^N)} \leq \|u\|_{L^p(\mathbb{R}^N)} + \|u\|_{L^{p^*}(\mathbb{R}^N)} &\leq \|u\|_{L^p(\mathbb{R}^N)} + C\|\nabla u\|_{L^p(\mathbb{R}^N)} \\ &\leq C\|u\|_{W^{1,p}(\mathbb{R}^N)},\end{aligned}$$

since $C \geq 1$. This ends the proof. □

8.1.2 *The case $p = N$ for $\mathbb{R}^N$*

Theorem 8.6. *Suppose that $N \geq 2$. Then*

$$W^{1,N}(\mathbb{R}^N) \subset L^q(\mathbb{R}^N), \tag{8.12}$$

with continuous embedding for all $q \in [N, +\infty)$.

Proof. Let $u \in \mathcal{D}(\mathbb{R}^N)$. Observe that inequality (8.10) still holds for $p = N$ and $p' = N/(N-1)$. Hence for $t > 1$, we have

$$\|u\|^t_{L^{t\frac{N}{N-1}}(\mathbb{R}^N)} \leq t\|u\|^{t-1}_{L^{\frac{N}{N-1}(t-1)}(\mathbb{R}^N)}\|\nabla u\|_{L^N(\mathbb{R}^N)}.$$

We apply Young's Inequality (6.38) with $p = t/(t-1)$ and $p' = t$ to obtain

$$\begin{aligned}\|u\|_{L^{t\frac{N}{N-1}}(\mathbb{R}^N)} &\leq t^{1/t}\|u\|^{\frac{t-1}{t}}_{L^{(t-1)\frac{N}{N-1}}(\mathbb{R}^N)}\|\nabla u\|^{\frac{1}{t}}_{L^N(\mathbb{R}^N)} \\ &\leq C(t)\Big(\|u\|_{L^{(t-1)\frac{N}{N-1}}(\mathbb{R}^N)} + \|\nabla u\|_{L^N(\mathbb{R}^N)}\Big),\end{aligned}$$

where $C = C(t)$ is an expression dependent on t. Since $t > 1$, we can choose $t = N$ and we now have,

$$\|u\|_{L^{\frac{N^2}{N-1}}(\mathbb{R}^N)} \leq C\big(\|u\|_{L^N(\mathbb{R}^N)} + \|\nabla u\|_{L^N(\mathbb{R}^N)}\big) \leq C\|u\|_{W^{1,N}(\mathbb{R}^N)},$$

from which we get, by (6.40) for $p = N$ and $r = N^2/(N-1)$,

$$\begin{aligned}\|u\|_{L^q(\mathbb{R}^N)} \leq \|u\|^{\alpha}_{L^N(\mathbb{R}^N)}\|u\|^{(1-\alpha)}_{L^{\frac{N^2}{N-1}}(\mathbb{R}^N)} &\leq C\|u\|^{\alpha}_{W^{1,N}(\mathbb{R}^N)}\|u\|^{(1-\alpha)}_{W^{1,N}(\mathbb{R}^N)} \\ &= C\|u\|_{W^{1,N}(\mathbb{R}^N)},\end{aligned}$$

for all $q \in \big[N, N^2/(N-1)\big]$ and $\alpha \in (0,1)$ such that

$$\frac{1}{q} = \frac{\alpha}{N} + (1-\alpha)\frac{N-1}{N^2}.$$

Repeating the process with $t = N+1, \ldots, N+k, \ldots$, gives

$$\|u\|_{L^q(\mathbb{R}^N)} \leq C\|u\|_{W^{1,N}(\mathbb{R}^N)}, \tag{8.13}$$

for all finite $q \geq N$ and u in $\mathcal{D}(\mathbb{R}^N)$, with $C = C(q, N)$.

The same density argument used in the proof of Theorem 8.4 implies $W^{1,p}(\mathbb{R}^N) \subset L^q(\mathbb{R}^N)$ for all $q \in [N, +\infty)$ and (8.13) holds for all u in $W^{1,p}(\mathbb{R}^N)$, which completes the proof. □

8.1.3 *The case $p > N$ for $\mathbb{R}^N$*

In the last case, we have

Theorem 8.7. *Let $1 < N < p < +\infty$.*
(1) There exists a constant $C = C(p, N)$ such that for every $u \in W^{1,p}(\mathbb{R}^N)$ the following inequality, called Morrey's inequality, holds (that is, u is Hölder continuous with exponent $1 - \frac{N}{p}$):

$$|u(x) - u(y)| \leq C|x-y|^{1-\frac{N}{p}} \|\nabla u\|_{L^p(\mathbb{R}^N)} \quad \text{for a.e. } x, y \in \mathbb{R}^N. \tag{8.14}$$

(2) Moreover,

$$W^{1,p}(\mathbb{R}^N) \subset L^\infty(\mathbb{R}^N),$$

with continuous embedding.

Proof. (1) Let $u \in \mathcal{D}(\mathbb{R}^N)$ and Q an open cube in $\mathbb{R}^N$ with sides parallel to the coordinate axes and of length r such that $0 \in Q$. Denote the mean (or average) value of u on Q by $\mathcal{M}_Q(u)$, defined as

$$\mathcal{M}_Q(u) = \frac{1}{|Q|} \int_Q u(x)\, dx,$$

where $|Q| = r^N$, the measure (or volume) of Q. We first show that

$$|\mathcal{M}_Q(u) - u(0)| \leq \frac{r^{1-\frac{N}{p}}}{1-\frac{N}{p}} \|\nabla u\|_{L^p(Q)}. \tag{8.15}$$

Observe that

$$u(x) - u(0) = \int_0^1 \frac{d}{dt}(u(tx))\, dt = \int_0^1 \Big(\sum_{i=1}^N x_i \frac{\partial u}{\partial x_i}(tx) \Big) dt,$$

for $x \in Q$.

Integrating over Q gives

$$\mathcal{M}_Q(u) - u(0) = \frac{1}{|Q|}\sum_{i=1}^{N}\int_Q\int_0^1 x_i\frac{\partial u}{\partial x_i}(tx)\,dt\,dx.$$

Since $x \in Q$, it follows that $|x_i| \leq r$ and hence

$$|\mathcal{M}_Q(u) - u(0)| \leq \frac{1}{|Q|}\sum_{i=1}^{N}\int_Q\int_0^1 |x_i|\left|\frac{\partial u}{\partial x_i}(tx)\right| dt\,dx$$

$$\leq \frac{r}{|Q|}\sum_{i=1}^{N}\int_Q\int_0^1 \left|\frac{\partial u}{\partial x_i}(tx)\right| dt\,dx.$$

By the change of variable $y = tx$ and applying Fubini's Theorem (Thm 6.39) and Hölder's inequality (6.31) (with $f = \left|\frac{\partial u}{\partial x_i}\right|$ and $g = 1$), we have successively

$$|\mathcal{M}_Q(u) - u(0)| \leq \frac{1}{r^{N-1}}\sum_{i=1}^{N}\int_0^1 \frac{1}{t^N}\int_{tQ}\left|\frac{\partial u}{\partial x_i}(y)\right| dy\,dt$$

$$\leq \frac{1}{r^{N-1}}\sum_{i=1}^{N}\int_0^1 \frac{1}{t^N}\left\|\frac{\partial u}{\partial x_i}\right\|_{L^p(tQ)}|tQ|^{\frac{1}{p'}}\,dt$$

$$= \frac{r^{\frac{N}{p'}}}{r^{N-1}}\sum_{i=1}^{N}\int_0^1 \frac{t^{\frac{N}{p'}}}{t^N}\left\|\frac{\partial u}{\partial x_i}\right\|_{L^p(tQ)}dt$$

$$\leq r^{1-\frac{N}{p}}\left(\int_0^1 t^{-\frac{N}{p}}\,dt\right)\sum_{i=1}^{N}\left\|\frac{\partial u}{\partial x_i}\right\|_{L^p(Q)}$$

$$= \frac{r^{1-\frac{N}{p}}}{1-\frac{N}{p}}\|\nabla u\|_{L^p(Q)},$$

since $N < p$, $t \leq 1$ (so that $tQ \subset Q$) and $|tQ| = t^N r^N$. This proves (8.15).

Now suppose that Q is an arbitrary open cube in $\mathbb{R}^N$ of length r, with sides parallel to the coordinate axes, and let $x \in Q$. By translation, the above inequality written for $v(y) = u(y+x)$ with y in the cube $Q-x$ results to

$$|\mathcal{M}_Q(u) - u(x)| \leq \frac{r^{1-\frac{N}{p}}}{1-\frac{N}{p}}\|\nabla u\|_{L^p(Q)}, \tag{8.16}$$

for all $x \in Q$. Hence for any $x, y \in Q$,

$$|u(x) - u(y)| \leq |\mathcal{M}_Q(u) - u(x)| + |\mathcal{M}_Q(u) - u(y)|$$

$$\leq 2\frac{r^{1-\frac{N}{p}}}{1-\frac{N}{p}}\|\nabla u\|_{L^p(Q)}.$$

Notice that for any $x, y \in \mathbb{R}^N$, there exists an open cube Q in $\mathbb{R}^N$ with sides parallel to the coordinate axes and of length $r = 2|x-y|$ such that x and y belong to Q. This implies that for every $x, y \in \mathbb{R}^N$ and $u \in \mathcal{D}(\mathbb{R}^N)$

$$|u(x) - u(y)| \leq \frac{2^{2-\frac{N}{p}} |x-y|^{1-\frac{N}{p}}}{1-\frac{N}{p}} \|\nabla u\|_{L^p(Q)}. \tag{8.17}$$

Now, if u is in $W^{1,p}(\mathbb{R}^N)$, then by the density of $\mathcal{D}(\mathbb{R}^N)$ in $W^{1,p}(\mathbb{R}^N)$ (Theorem 7.29), there exists a sequence $\{u_n\} \in \mathcal{D}(\mathbb{R}^N)$ such that

$$u_n \to u \quad \text{in } W^{1,p}(\mathbb{R}^N),$$

whence

$$\|\nabla u_n\|_{L^p(\mathbb{R}^N)} \to \|\nabla u\|_{L^p(\mathbb{R}^N)}$$

and from Theorem 6.40, there exists a subsequence, still denoted $\{u_n\}$, such that

$$u_n \to u \quad \text{a.e. in } \mathbb{R}^N.$$

In consequence, by (8.17)

$$|u_n(x) - u_n(y)| \leq \frac{2^{2-\frac{N}{p}} |x-y|^{1-\frac{N}{p}}}{1-\frac{N}{p}} \|\nabla u_n\|_{L^p(\mathbb{R}^N)} \quad \text{a.e. for } x, y \text{ in } \mathbb{R}^N.$$

As $n \to \infty$,

$$|u(x) - u(y)| \leq \frac{2^{2-\frac{N}{p}} |x-y|^{1-\frac{N}{p}}}{1-\frac{N}{p}} \|\nabla u\|_{L^p(\mathbb{R}^N)} \quad \text{a.e. for } x, y \text{ in } \mathbb{R}^N,$$

which completes the proof of (1) with

$$C = \frac{2^{2-\frac{N}{p}}}{1-\frac{N}{p}}. \tag{8.18}$$

(2) We prove the assertion for $N < p < +\infty$, the case $p = +\infty$ being trivial. Let $u \in \mathcal{D}(\mathbb{R}^N)$, $x \in \mathbb{R}^N$ and consider an open cube Q containing x with sides parallel to the coordinate axes and of length 1. Since $|Q| = 1$ and $\frac{N}{p} < 1$, it follows from (8.16) and Hölder's inequality that

$$\begin{aligned}|u(x)| &\leq |\mathcal{M}_Q(u)| + |\mathcal{M}_Q(u) - u(x)| \leq \left|\int_Q u(x)\,dx\right| + \frac{1}{1-\frac{N}{p}} \|\nabla u\|_{L^p(Q)} \\ &\leq \|u\|_{L^p(Q)} + \frac{1}{1-\frac{N}{p}} \|\nabla u\|_{L^p(Q)} \leq C\|u\|_{W^{1,p}(Q)} \leq C\|u\|_{W^{1,p}(\mathbb{R}^N)},\end{aligned}$$

where

$$C = \frac{1}{1 - \frac{N}{p}}.$$

The density argument used in the proof of the first assertion shows that this inequality holds also almost everywhere for u in $W^{1,p}(\mathbb{R}^N)$. Hence

$$\|u\|_{L^\infty(\mathbb{R}^N)} \leq C\|u\|_{W^{1,p}(\mathbb{R}^N)},$$

for every $u \in W^{1,p}(\mathbb{R}^N)$, and this ends the proof of the claimed result. □

Remark 8.8.
(1) Inequality (8.14) implies that if $N < p < +\infty$, a function u in $W^{1,p}(\mathbb{R}^N)$ is equal almost everywhere to a bounded and continuous function. As in the one-dimensional case (see Remark 7.43), u can be identified with its continuous representative. This means that with this identification we can state that for $N < p < +\infty$, the inclusion

$$W^{1,p}(\mathbb{R}^N) \subset C_b^0(\mathbb{R}^N), \tag{8.19}$$

is continuous.

(2) The inclusion in (8.19) *is still valid if $p = +\infty$. Indeed, suppose that u is in $W^{1,\infty}(\mathbb{R}^N)$ and for a fixed x in $\mathbb{R}^N$, let B_x be an open ball containing x. By Theorem 6.60, there exists a function $\psi \in \mathcal{D}(\mathbb{R}^N)$ such that $\psi \equiv 1$ on B_x. So, the function ψu has compact support, which implies that it belongs to $W^{1,p}(\mathbb{R}^N)$ for any finite $p > N$ and from (8.14)*

$$|u(x) - u(y)| \leq C|x-y|^{1-\frac{N}{p}}\|\nabla(\psi u)\|_{L^p(\mathbb{R}^N)} \quad \text{a.e. in } B_x.$$

This allows us to conclude as in the previous case.

8.1.4 *Embedding results for open sets*

The above embedding results for $\mathbb{R}^N$ can be extended to the half-space or bounded open sets as follows:

Theorem 8.9. *Let $N \geq 2$ and Ω a bounded open subset of $\mathbb{R}^N$ with a Lipschitz-continuous boundary or $\Omega = \mathbb{R}^{N-1} \times \mathbb{R}^+$.*
(1) If $1 \leq p < N$ and $\dfrac{1}{p^*} = \dfrac{1}{p} - \dfrac{1}{N}$, then

$$W^{1,p}(\Omega) \subset L^q(\Omega), \tag{8.20}$$

with continuous embedding for all $q \in [p, p^*]$.

(2) If $p = N$, then

$$W^{1,N}(\Omega) \subset L^q(\Omega), \tag{8.21}$$

with continuous embedding for all $q \in [N, +\infty)$.
(3) If $N < p \leq +\infty$, then the following inclusion holds true:

$$W^{1,p}(\Omega) \subset L^\infty(\Omega), \tag{8.22}$$

with continuous embedding and (in the sense of Remark 8.8)

$$W^{1,p}(\Omega) \subset C^0(\bar{\Omega}). \tag{8.23}$$

Proof. We only prove (1). The embeddings in (2) and (3) follow by a similar argument, using Theorem 8.6 and Theorem 8.7 (see also Remark 8.8), respectively.

Suppose that $1 \leq p < N$ and $q \in [p, p^*]$. Due to the assumption on Ω, the Extension Theorem (Thm 7.32) applies and shows, together with Corollary 8.5, that there exist some constants c and C, depending on Ω, and p such that

$$\|u\|_{L^q(\Omega))} \leq \|P(u)\|_{L^q(\mathbb{R}^N)} \leq c\|P(u)\|_{W^{1,p}(\mathbb{R}^N)} \leq C\|u\|_{W^{1,p}(\Omega)},$$

where P is an extension operator given by Theorem 7.32. Consequently, (8.20) holds. □

Remark 8.10. *It can be shown by counterexamples (see for instance [1]), that the embeddings given by Theorem 8.9 are optimal. Moreover, they are not true if Ω is not bounded or does not have a Lipschitz continuous boundary, for instance if $\partial\Omega$ has a cusp.*

Let us point out here that Rademacher's Theorem (Thm 7.44) implies that a Lipschitz-continuous function on an open set $\mathcal{O}$ in $\mathbb{R}^N$ is in $W^{1,\infty}(\mathcal{O})$. When the open set is bounded, we have the following converse result that actually is a characterization of the space $W^{1,\infty}(\Omega)$.

Theorem 8.11. *Let $\Omega \subset \mathbb{R}^N$ such that Ω is an open interval in $\mathbb{R}$ if $N = 1$ and if $N \geq 2$, Ω is a bounded open subset of $\mathbb{R}^N$ with a Lipschitz-continuous boundary. Then the functions in $W^{1,\infty}(\Omega)$ are Lipschitz-continuous on Ω.*

Proof. Suppose first that u is a function in $W^{1,\infty}(\mathbb{R}^N)$ with compact support, so that in particular, $u \in W^{1,p}(\mathbb{R}^N)$ for every $p \in (1, +\infty)$.

Let $K = \text{supp}\, u$ and denote by $|K|$ its measure. If $p \in (1, +\infty)$ and $N = 1$, then for every $x, y \in \mathbb{R}$ we have from (7.48)

$$|u(x) - u(y)| \leq |x - y|^{\frac{p-1}{p}} \, |K|^{\frac{1}{p}} \, \|u'\|_{L^\infty(\mathbb{R})}.$$

If $N \geq 2$, from (8.14) and (8.18) we have for every $x, y \in \mathbb{R}^N$,

$$|u(x) - u(y)| \leq \frac{2^{2-\frac{N}{p}}}{1 - \frac{N}{p}} |x - y|^{1-\frac{N}{p}} |K|^{\frac{1}{p}} \|\nabla u\|_{L^\infty(\mathbb{R}^N)}.$$

Passing to the limit in these inequalities as $p \to +\infty$ yields

$$\begin{aligned} &|u(x) - u(y)| \leq |x - y| \, \|u'\|_{L^\infty(\mathbb{R})} && \text{if } N = 1, \\ &|u(x) - u(y)| \leq 4 \, |x - y| \, \|\nabla u\|_{L^\infty(\mathbb{R}^N)} && \text{if } N > 2, \end{aligned} \tag{8.24}$$

for every $x, y \in \mathbb{R}^N$. This proves that u is Lipschitz-continuous on $\mathbb{R}^N$.

Suppose now that v is a function in $W^{1,\infty}(\Omega)$ and let $P(v)$ be its extension to $\mathbb{R}^N$ given by Theorem 7.32. By Urysohn's Lemma (Thm 6.60), there exists ψ in $\mathcal{D}(\mathbb{R}^N)$ such that $\psi \equiv 1$ on $\overline{\Omega}$. It follows that the function $\psi P(v)$ has compact support and in view of Proposition 7.26, it belongs to $W^{1,\infty}(\mathbb{R}^N)$. Therefore we can apply (8.24) to $u = \psi P(v)$. Choosing x and y to be in Ω ends the proof. □

The embeddings stated in Theorem 8.9 are in particular still true for the Sobolev space $W_0^{1,p}(\mathcal{O})$ and in this case, no assumption on the boundary of the open set $\mathcal{O}$ is needed.

Theorem 8.12. *Let $N \geq 2$ and $\mathcal{O}$ be an open subset of $\mathbb{R}^N$.*
(1) If $1 \leq p < N$ and $\dfrac{1}{p^*} = \dfrac{1}{p} - \dfrac{1}{N}$, then

$$W_0^{1,p}(\mathcal{O}) \subset L^q(\mathcal{O}), \tag{8.25}$$

with continuous embedding for all $q \in [p, p^*]$.
(2) If $p = N$, then

$$W_0^{1,N}(\mathcal{O}) \subset L^q(\mathcal{O}), \tag{8.26}$$

with continuous embedding for all $q \in [N, +\infty)$.
(3) If $N < p \leq +\infty$, then the following inclusions hold (in the sense of Remark 8.8):

$$W_0^{1,p}(\mathcal{O}) \subset C^0(\overline{\mathcal{O}}), \tag{8.27}$$

and

$$W_0^{1,p}(\mathcal{O}) \subset L^\infty(\mathcal{O}), \tag{8.28}$$

with continuous embedding.

Proof. The proof is similar to that of Theorem 8.9, but using the zero extension $\tilde{u}$ (see Proposition 7.56) instead of the extension operator P.

Suppose first that $1 \leq p < N$ and $q \in [p, p^*]$. From Proposition 7.56 and the continuous embedding (8.11), it follows that there exists a constant $C = C(p, q, N)$ such that

$$\|u\|_{L^q(\mathcal{O}))} = \|\tilde{u}\|_{L^q(\mathbb{R}^N)} \leq C\|\tilde{u}\|_{W^{1,p}(\mathbb{R}^N)} = C\|u\|_{W_0^{1,p}(\mathcal{O})}.$$

This proves (8.25). The embeddings for the remaining cases follow from a similar argument, using Theorems 8.6 and 8.7 (see also Remark 8.8), respectively. □

8.2 Compact embedding theorems

Let us first define the notions of compact mappings and inclusions:

Definition 8.13. *Let X and Y be two Banach spaces. We say that the mapping $h : X \longmapsto Y$ is compact if h maps bounded sets of X into relatively compact sets in Y. Equivalently, h is compact if for any bounded sequence $\{x_n\}$ of X, the sequence $\{h(x_n)\}$ has a subsequence which converges in Y. If $X \subset Y$, we say that the inclusion is compact if the mapping*

$$i_X : x \in X \longmapsto x \in Y,$$

is compact, that is, if any bounded sequence of X has a subsequence which converges in Y.

Remark 8.14. *A compact linear map $h : X \mapsto Y$ is continuous. Indeed, if B_1 is the unit ball of X centered at the point 0, the compactness implies in particular, that there exists a constant c such that*

$$\|h(w)\|_Y \leq c, \quad \forall\, w \in B_1.$$

By linearity

$$\frac{\|h(x)\|_Y}{\|x\|_X} = \Big\|h\Big(\frac{x}{\|x\|_X}\Big)\Big\|_Y \leq c, \quad \forall\, x \in X.$$

This proves that h is bounded, hence continuous by Theorem 6.7. Let us point out that, as can be shown by counterexamples, this is not true anymore if f is nonlinear.

We also recall the following well-known theorem which gives a compactness criterion in C^0:

Theorem 8.15 (Arzela–Ascoli Compactness Theorem). *Let $\{f_n\}$ be a bounded sequence in $C_b^0(\mathbb{R}^N)$ which is equicontinuous, that is,*

$$\forall\, \varepsilon > 0,\ \exists\, \delta > 0, \quad \|x - y\| < \delta \Longrightarrow |f_n(x) - f_n(y)| < \varepsilon, \quad \forall\ n \in \mathbb{N}.$$

Then $\{f_n\}$ admits a subsequence which is uniformly convergent on any compact subset of $\mathbb{R}^N$.

Remark 8.16. *One can easily check that a bounded sequence of functions in $C_b^1(\mathbb{R}^N)$ (see (1.4)) is also equicontinuous. As a consequence of Theorem 8.15, the inclusion*

$$C_b^1(\mathbb{R}^N) \subset C^0(K),$$

is compact for any compact set K in $\mathbb{R}^N$.

8.2.1 *The cases $N = 1$ and $N \geq 2$ with $p > N$*

We start with the case $N = 1$ for which the result is a simple consequence of the Arzela–Ascoli Theorem (Thm 8.15) and Theorem 7.42.

Theorem 8.17. *Let $1 < p < +\infty$. If I is a bounded open interval in $\mathbb{R}$, then the inclusion $W^{1,p}(I) \subset C^0(\bar{I})$ is compact.*

When $N \geq 2$ and $p > N$, another consequence of Theorem 8.15 is also the result below.

Theorem 8.18. *Let $N \geq 2$ and Ω a bounded open subset of $\mathbb{R}^N$ with a Lipschitz continuous boundary. If $N < p \leq +\infty$, the following inclusion is compact:*

$$W^{1,p}(\Omega) \subset C^0(\overline{\Omega}). \tag{8.29}$$

Proof. Since Ω is bounded, $W^{1,\infty}(\Omega) \subset W^{1,p}(\Omega)$ for any p. It suffices to show the result for $p < +\infty$.

Let $\{u_n\}$ be a bounded sequence in $W^{1,p}(\Omega)$. Let P be an extension operator given by Theorem 7.32. By Theorem 8.7, the sequence $\{P(u_n)\}$ is also bounded on $W^{1,p}(\mathbb{R}^N)$ and $L^\infty(\mathbb{R}^N)$, respectively. Moreover, since (8.14) holds for any function $\{P(u_n)\}$ with a constant independent of n, this sequence is equicontinuous. Therefore by Theorem 8.15, the sequence admits a subsequence $\{P(u_{n_k})\}$ which converges in $C^0(K)$ for any compact set K in $\mathbb{R}^N$. This implies in particular that $\{u_{n_k}\}$ is compact in $C^0(\overline{\Omega})$ and ends the proof. □

Remark 8.19. *The assumptions on Ω, namely, bounded and with a Lipschitz-continuous boundary, are necessary for the compact inclusion to hold (see [1] for counterexamples).*

8.2.2 *The Rellich–Kondrachov Theorem (case $1 \leq p < N$)*

We give the following compactness result which generalizes Theorem 8.15 to Sobolev spaces $W^{1,p}$. It states that a sequence of functions which is bounded in L^p, as well as the sequence of their derivatives, admit a subsequence which converges in L^q for a suitable q. This is one of the deepest and most delicate results on Sobolev Spaces.

Theorem 8.20 (The Rellich–Kondrachov Compactness Theorem). *Let $N \geq 2$ and Ω be a bounded open subset of $\mathbb{R}^N$ with a Lipschitz-continuous boundary. If $1 \leq p < N$, the inclusion*

$$W^{1,p}(\Omega) \subset L^q(\Omega), \tag{8.30}$$

is compact for all $1 \leq q < p^*$, where p^* is given by

$$\frac{1}{p^*} = \frac{1}{p} - \frac{1}{N}.$$

Proof. Since Ω is bounded, by Theorem 8.9 and Corollary 6.33, the inclusion holds true. Therefore we only have to prove that if $\{u_n\}$ is a bounded sequence in $W^{1,p}(\Omega)$, then it has a subsequence converging in $L^q(\Omega)$.

To do this, we use regularization, the Arzela–Ascoli Compactness Theorem, and an adaptation of the classical diagonal argument.

Step 1. We first show that we can assume, without loss of generality, that $\{u_n\} \subset W^{1,p}(\mathbb{R}^N)$ and that there exists a bounded open set $U \subset \mathbb{R}^N$ with a Lipschitz-continuous boundary, and a positive constant c such that

$$\overline{\Omega} \subset U, \quad \text{supp } u_n \subset U, \quad \|u_n\|_{W^{1,p}(U)} \leq c, \quad \forall\, n \in \mathbb{N}. \tag{8.31}$$

To do so, we prove that if $\{u_n\} \subset W^{1,p}(\Omega)$, we can construct a sequence $\{w_n\}$ in $W^{1,p}(\mathbb{R}^N)$ satisfying (8.31) and such that $w_n = u_n$ in Ω.

Let P be an extension operator given by the Extension Theorem (Thm 7.32), then the sequence $\{P(u_n)\}$ is bounded in $W^{1,p}(\mathbb{R}^N)$ by construction.

Suppose U is a bounded open set with a Lipschitz-continuous boundary containing $\overline{\Omega}$ and let $\psi_{\overline{\Omega}}$ be the function given by Urysohn's Lemma (Thm 6.60) for the compact set $\overline{\Omega}$ and the open set U. Define

$$w_n = \psi_{\overline{\Omega}}\, P(u_n), \quad \forall\, n \in \mathbb{N}.$$

Obviously, $\{w_n\}$ is contained in $W^{1,p}(\mathbb{R}^N)$ and $w_n = u_n$ in Ω as $\psi_{\overline{\Omega}} = 1$ in Ω. Since supp $\psi_{\overline{\Omega}} \subset U$, we deduce that

$$\text{supp } w_n \subset U.$$

Moreover,

$$\begin{aligned}\|w_n\|_{W^{1,p}(U)} &= \|\psi_{\overline{\Omega}}\, P(u_n)\|_{L^p(U)} + \left\|\psi_{\overline{\Omega}}\, \nabla\big(P(u_n)\big) + P(u_n)\nabla\psi_{\overline{\Omega}}\right\|_{L^p(U)} \\ &\le \|\, P(u_n)\|_{L^p(U)} + \left\|\nabla\big(P(u_n)\big)\right\|_{L^p(U)} \\ &\quad + \|\nabla\, \psi_{\overline{\Omega}}\, \|_{L^\infty(U)}\, \|P(u_n)\|_{L^p(U)} \le\, c\,\|u_n\|_{W^{1,p}(\Omega)} \le c_1,\end{aligned}$$

by the continuity of the extension operator P, Proposition 7.26 and the boundedness of the sequence $\{u_n\}$ in $W^{1,p}(\Omega)$. Hence we can assume that $\{u_n\}$ satisfies (8.31).

Step 2. Let $\{\rho_h\}$ be a regularizing sequence given by Definition 6.56, and let $1 \le q < p^*$. By Theorem 6.58,

$$u_n^h \doteq \rho_h \star u_n \to u_n \quad \text{in } L^q(\mathbb{R}^N) \text{ as } h \to \infty, \tag{8.32}$$

for every fixed $n \in \mathbb{N}$.

In this step, we show that there exists a positive constant C and a number $\alpha \in (0,1)$ such that

$$\|u_n^h - u_n\|_{L^q(\mathbb{R}^N)} \le \frac{C}{h^\alpha}, \qquad \forall\, n,\ h \in \mathbb{N}, \tag{8.33}$$

which implies that the convergence in (8.32) is uniform with respect to n.

Suppose u_n is in $\mathcal{D}(U)$. By Proposition 6.54(1), the properties of the function ρ, and by the change of variable $y = hz$, we have

$$\begin{aligned}u_n^h(x) - u_n(x) &= (u_n \star \rho_h)(x) - u_n(x) = \int_{\mathbb{R}^N} u_n(x-z)h^N\rho(hz)dz - u_n(x) \\ &= \int_{\mathbb{R}^N} \rho(y)u_n\left(x - \frac{y}{h}\right) dy - u_n(x) \\ &= \int_{B(0,1)} \rho(y)\left[u_n\left(x - \frac{y}{h}\right) - u_n(x)\right] dy \\ &= \int_{B(0,1)} \rho(y) \int_0^1 \frac{d}{dt}\left(u_n\left(x - \frac{ty}{h}\right)\right) dt\, dy \\ &= -\frac{1}{h}\int_{B(0,1)} \rho(y) \int_0^1 \left[\nabla u_n\left(x - \frac{ty}{h}\right)\cdot y\right] dt\, dy.\end{aligned}$$

We integrate this equality over U with respect to x. Performing in the integral over U the following change of variable

$$z = x - \frac{ty}{h},$$

we obtain

$$\begin{aligned}\int_U \left|u_n^h(x) - u_n(x)\right| \, dx &\leq \frac{1}{h} \int_{B(0,1)} \rho(y) \int_0^1 \int_U \left|\left(\nabla u_n\right)\left(x - \frac{ty}{h}\right)\right| \, dx \, dt \, dy \\ &= \frac{1}{h} \int_{B(0,1)} \rho(y) \int_0^1 \int_U \left|(\nabla u_n)(z)\right| \, dz \, dt \, dy \\ &= \frac{1}{h} \int_U \left|(\nabla u_n)(z)\right| \, dz,\end{aligned}$$

since $|y| \leq 1$ and u_n vanishes outside U in view of (8.31).

From the Density Theorem (Thm 7.35), it follows that this inequality also holds when u_n is in $W^{1,p}(U)$. Hence from (8.31) and (6.35),

$$\left\|u_n^h - u_n\right\|_{L^1(U)} \leq \frac{1}{h} \left\|\nabla u_n\right\|_{L^1(U)} \leq \frac{c}{h} \left\|\nabla u_n\right\|_{L^p(U)} \leq \frac{c_1}{h}, \tag{8.34}$$

where c_1 is a positive constant independent of h and n.

Observe now that from of Proposition 6.53 and Remark 6.57, we have

$$\left\|u_n^h - u_n\right\|_{L^{p^\star}(U)} \leq 2 \left\|u_n\right\|_{L^{p^\star}(U)}.$$

By (8.34) and applying (6.40) of the Interpolation Inequality Lemma with $r = p^*$, $1 \leq q < p^\star$, we have

$$\begin{aligned}\left\|u_n^h - u_n\right\|_{L^q(U)} &\leq \left\|u_n^h - u_n\right\|_{L^1(U)}^{\alpha} \left\|u_n^h - u_n\right\|_{L^{p^\star}(U)}^{1-\alpha} \\ &\leq \left(\frac{c_1}{h}\right)^{\alpha} 2^{1-\alpha} \left\|u_n\right\|_{L^{p^\star}(U)}^{1-\alpha},\end{aligned}$$

with $\alpha \in (0,1)$. This implies (8.33), since in view of (8.31) and Theorem 8.9, the sequence $\{u_n\}$ is bounded on $L^{p^\star}(U)$.

Step 3. Let us prove that for every fixed $h \in \mathbb{N}$, the sequence $\{u_n^h\}$ is compact in $C^0(\overline{U})$. To do that, we apply the Arzela–Ascoli theorem (Thm 8.15) for every fixed h. From (8.31) and Proposition 6.54 the sequence $\{u_n^h\}$ is in $\mathcal{D}(\mathbb{R}^N)$ and due to Remark 6.57, it is not restrictive to assume that they are in $\mathcal{D}(U)$ (otherwise we replace U by a bigger bounded open subset). Consequently, from (6.35)

$$\begin{aligned}|u_n^h(x)| &\leq \int_{\mathbb{R}^N} \rho_h(x-y)|u_n(y)|dy \leq \|\rho_h\|_{L^\infty(\mathbb{R}^N)} \|u_n\|_{L^1(\mathbb{R}^N)} \\ &= h^N \|\rho\|_{L^\infty(\mathbb{R}^N)} \|u_n\|_{L^1(U)} \leq ch^N \|u_n\|_{L^p(U)},\end{aligned}$$

for every $x \in \overline{U}$. On the other hand, from Definition 6.56 and using again Proposition 6.54 we have

$$\begin{aligned}|\nabla u_n^h(x)| &\leq \int_{\mathbb{R}^N} |(\nabla \rho_h)(x-y)| \, |u_n(y)|dy \leq \|\nabla \rho_h\|_{L^\infty(\mathbb{R}^N)} \|u_n\|_{L^1(\mathbb{R}^N)} \\ &= h^{N+1} \|\nabla \rho\|_{L^\infty(\mathbb{R}^N)} \|u_n\|_{L^1(U)} \leq ch^{N+1} \|u_n\|_{L^p(U)}.\end{aligned}$$

The two inequalities above, together with (8.31) and Remark 8.16, show that the Arzela–Ascoli Theorem (Thm 8.15) applies. Hence for every $h \in \mathbb{N}$ the sequence $\{u_n^h\}$ is compact in $C^0(\overline{U})$.

Step 4. To conclude, we use here a diagonal-type argument.

We start with $h = 1$. From Step 3 we can extract from $\{u_n^1\}$ a subsequence $\{u^1_{\tau_1(n)}\}$ which uniformly converges in $\overline{U}$, and consequently in $L^q(U)$, since U is bounded. Hence this subsequence is a Cauchy sequence in this space, so that

$$\limsup_{n,m\to\infty} \|u^1_{\tau_1(n)} - u^1_{\tau_1(m)}\|_{L^q(U)} = 0. \tag{8.35}$$

This yields, using (8.33), the following inequality for the subsequence $\{u_{\tau_1(n)}\}$ of $\{u_n\}$:

$$\begin{aligned}\limsup_{n,m\to\infty} \|u_{\tau_1(n)} - u_{\tau_1(m)}\|_{L^q(U)} &\leq \limsup_{n,m\to\infty} \|u_{\tau_1(n)} - u^1_{\tau_1(n)}\|_{L^q(U)} \\ &+ \limsup_{n,m\to\infty} \|u^1_{\tau_1(n)} - u^1_{\tau_1(m)}\|_{L^q(U)} \\ &+ \limsup_{n,m\to\infty} \|u^1_{\tau_1(m)} - u_{\tau_1(m)}\|_{L^q(U)} \leq 2C,\end{aligned} \tag{8.36}$$

where C is given by (8.33).

Let us take now $h = 2$. The same argument used to prove (8.35), applied to $\{u_{\tau_1(n)}\}$ instead of $\{u_n\}$, shows that we can extract from $\{u^2_{\tau_1(n)}\}$ a subsequence $\{u^2_{\tau_2(n)}\}$ such that

$$\limsup_{n,m\to\infty} \|u^2_{\tau_2(n)} - u^2_{\tau_2(m)}\|_{L^q(U)} = 0.$$

Therefore, for the subsequence $\{u_{\tau_2(n)}\}$ of $\{u_{\tau_1(n)}\}$, using (8.33) we obtain

$$\begin{aligned}\limsup_{n,m\to\infty} \|u_{\tau_2(n)} - u_{\tau_2(m)}\|_{L^q(U)} &\leq \limsup_{n,m\to\infty} \|u_{\tau_2(n)} - u^2_{\tau_2(n)}\|_{L^q(U)} \\ &+ \limsup_{n,m\to\infty} \|u^2_{\tau_2(n)} - u^2_{\tau_2(m)}\|_{L^q(U)} + \limsup_{n,m\to\infty} \|u^2_{\tau_2(m)} - u_{\tau_2(m)}\|_{L^q(U)} \\ &\leq \frac{2C}{2^\alpha}.\end{aligned}$$

Taking successively $h = 3, 4, \dots$, we can construct a subsequence $\{u_{\tau_h(n)}\}$ of $\{u_{\tau_{h-1}(n)}\}$ for any h such that

$$\limsup_{n,m\to\infty} \|u_{\tau_h(n)} - u_{\tau_h(m)}\|_{L^q(U)} \leq \frac{2C}{h^\alpha}.$$

This implies that for every fixed $h \in \mathbb{N}$ there exists $n_h \in \mathbb{N}$, such that

$$\|u_{\tau_h(n)} - u_{\tau_h(m)}\|_{L^q(U)} \leq \frac{C_1}{h^\alpha}, \qquad \forall n, m \geq n_h, \tag{8.37}$$

with $C_1 > 2C$.

Consider now the sequence $\{n_h^*\}_{h\in\mathbb{N}}$ defined by

$$n_1^* = n_1, \qquad n_h^* = \max\{n_h, n_{h-1}^* + 1\},\ \forall\, h > 1, \tag{8.38}$$

which by construction is strictly increasing, that is

$$n_h^* < n_{h+1}^*, \quad \forall\, h \in \mathbb{N}. \tag{8.39}$$

Observe that $\{\tau_h(n_h^*)\}_{h\in\mathbb{N}}$ is also a strictly increasing sequence. Indeed, for every $h \in \mathbb{N}$ the sequence $\{\tau_{h+1}(n)\}_{n\in\mathbb{N}}$ is a subsequence of $\{\tau_h(n)\}_{n\in\mathbb{N}}$, which implies using (8.39)

$$\tau_h(n_h^*) \leq \tau_{h+1}(n_h^*) < \tau_{h+1}(n_{h+1}^*), \quad \forall\, h \in \mathbb{N}. \tag{8.40}$$

It follows that the sequence $\{v_h\}_{h\in\mathbb{N}}$ given by

$$v_h = u_{\tau_h(n_h^*)}, \quad \forall\, h \in \mathbb{N}, \tag{8.41}$$

is a subsequence of $\{u_n\}$. For $h, k \in \mathbb{N}$ with $h < k$, let us estimate

$$\|v_h - v_k\|_{L^q(U)} = \|u_{\tau_h(n_h^*)} - u_{\tau_k(n_k^*)}\|_{L^q(U)}. \tag{8.42}$$

Since $\{\tau_k(n)\}_{n\in\mathbb{N}}$ is a subsequence of $\{\tau_h(n)\}_{n\in\mathbb{N}}$, there exists $p \in \mathbb{N}$ such that $\tau_k(n_k^*) = \tau_h(p)$. Moreover, from (8.40) we derive

$$\tau_h(p) = \tau_k(n_k^*) > \tau_h(n_h^*),$$

which implies that $p > n_h^* \geq n_h$ since $\{\tau_h(n)\}_{n\in\mathbb{N}}$ is strictly increasing and (8.38) holds true. Consequently, we can apply (8.37) with $n = n_h^*$ and $m = p$, and from (8.42) we obtain

$$\|u_{\tau_h(n_h^*)} - u_{\tau_k(n_k^*)}\|_{L^q(U)} = \|u_{\tau_h(n_h^*)} - u_{\tau_h(p)}\|_{L^q(U)} \leq \frac{C_1}{h^\alpha},$$

for $h, k \in \mathbb{N}$ with $h < k$.

This proves that $\{v_h\}_{h\in\mathbb{N}}$, which is a subsequence of $\{u_n\}_{n\in\mathbb{N}}$, is a Cauchy sequence in $L^q(U)$, so that it converges in $L^q(U)$. This ends the proof. □

8.2.3 *Compactness for the case* $p = N$

It remains to study the case $p = N$, for which we have

Theorem 8.21. *Let $N \geq 2$ and Ω be a bounded open subset of $\mathbb{R}^N$ with a Lipschitz-continuous boundary. The inclusion*

$$W^{1,N}(\Omega) \subset L^q(\Omega), \tag{8.43}$$

is compact for all $1 \leq q < +\infty$.

Proof. Let $\{u_n\}$ be a bounded sequence in $W^{1,N}(\Omega)$. Since Ω is bounded, Corollary 6.33 implies that the sequence $\{u_n\}$ is bounded on $W^{1,q}(\Omega)$ for $1 \leq q < N$. By Theorem 8.20, there exists a subsequence converging in $L^q(\Omega)$ which gives the result for $1 \leq q < N$.

Let $q \geq N$, $p \in [1, N)$, and $r > q$ be fixed. From the previous argument, there exists a subsequence $\{u_{n_k}\}$ and some u in $L^p(\Omega)$ such that $\{u_{n_k}\}$ converges to u in $L^p(\Omega)$.

On the other hand, from (6.40) we have

$$\begin{aligned}\|u_{n_k} - u_{n_h}\|_{L^q(\Omega)} &\leq \|u_{n_k} - u_{n_h}\|^{\alpha}_{L^p(\Omega)} \|u_{n_k} - u_{n_h}\|^{1-\alpha}_{L^r(\Omega)} \\ &\leq c\|u_{n_k} - u_{n_h}\|^{\alpha}_{L^p(\Omega)},\end{aligned}$$

for some $\alpha \in (0,1)$, $k, h \in \mathbb{N}$, and c independent of h and k.

Consequently, since $\{u_{n_k}\}$ is a Cauchy sequence in $L^p(\Omega)$, it is also a Cauchy sequence in $L^q(\Omega)$, which is complete. Therefore it also converges to u in $L^q(\Omega)$. This concludes the proof. □

Remark 8.22.

(1) It can be shown by counterexamples (see [1]) that the result is false for $q = p^*$, where p^* is defined by (8.5). *The assumptions on Ω are also necessary.*

(2) Let us point out that under the assumptions of Theorem 8.20, one has $H^1(\Omega) \subset L^2(\Omega)$ with compact injection, since $2 < 2^\star = \dfrac{2N}{N-2}$ when $N > 2$. For $N = 1$ and $N = 2$, this result follows from Theorem 8.17 and Theorem 8.21, respectively. By definition, this means that from any bounded sequence in $H^1(\Omega)$, one can extract at least a subsequence strongly convergent in $L^2(\Omega)$. This result, quoted in the literature as the Rellich Theorem, is widely used in PDEs.

The above compact embeddings are, in particular, still true for the space $W_0^{1,p}(\Omega)$, and in this case no assumption on the boundary of the open subset Ω is needed.

Theorem 8.23. *Let Ω be a bounded open subset of $\mathbb{R}^N$ with $N \geq 2$.*
(1) If $N < p \leq +\infty$, the inclusion

$$W_0^{1,p}(\Omega) \subset C^0(\overline{\Omega}), \tag{8.44}$$

is compact.
(2) If $p = N$, the inclusion

$$W_0^{1,N}(\Omega) \dot{\subset} L^q(\Omega), \tag{8.45}$$

is compact for all $1 \leq q < +\infty$.

(3) If $1 \leq p < N$ *and* $\frac{1}{p^*} = \frac{1}{p} - \frac{1}{N}$, *the inclusion*

$$W_0^{1,p}(\Omega) \subset L^q(\Omega), \tag{8.46}$$

is compact for all $1 \leq q < p^*$.

Proof. Let Ω_1 be a bounded open subset of $\mathbb{R}^N$ with a Lipschitz-continuous boundary containing Ω and $p \geq 1$. From Proposition 7.56 it follows that if u is in $W_0^{1,p}(\Omega)$, then its zero extension $\widetilde{u}$ to the set Ω_1 is in $W_0^{1,p}(\Omega_1)$ and

$$\|\widetilde{u}\|_{L^q(\Omega_1)} = \|u\|_{L^q(\Omega)},$$
$$\|\widetilde{u}\|_{W^{1,p}(\Omega_1)} = C\|u\|_{W^{1,p}(\Omega)}.$$

The results follow by applying, respectively, Theorems 8.18, 8.20, and 8.21 in the three cases, for $\Omega = \Omega_1$. □

In particular, from Theorem 8.17 and Theorem 8.23, it follows by duality that the following holds true:

Proposition 8.24. *Let* Ω *be a bounded open subset of* $\mathbb{R}^N$, $N \geq 1$. *The following inclusion is compact:*

$$L^2(\Omega) \subset H^{-1}(\Omega).$$

To end this section, we state without proof a compactness result for functions defined on the boundary (see Definition 7.66).

Theorem 8.25. *Let* $N \geq 2$ *and* Ω *a bounded open subset of* $\mathbb{R}^N$ *with a Lipschitz continuous boundary. The following inclusions are compact:*

$$H^{\frac{1}{2}}(\partial\Omega) \subset L^2(\partial\Omega),$$
$$L^2(\partial\Omega) \subset H^{-\frac{1}{2}}(\partial\Omega).$$

8.3 Some consequences of compactness

We end this chapter with three results which are consequences of the compactness theorems given in the previous section.

The first one is the so-called Poincaré–Wirtinger inequality. Recall that as observed in Remark 7.63, the Poincaré inequality stated in Theorem 7.61 does not hold in $H^1(\Omega)$. Nevertheless, this inequality still holds on the subspace of H^1 consisting of functions with zero average. For simplicity, we state it for the case $p = 2$, but the result is still valid for $p \in (1, +\infty)$, with some modifications.

Theorem 8.26 (Poincaré–Wirtinger Inequality). *Let $N \geq 2$ and suppose that Ω is a connected bounded open subset of $\mathbb{R}^N$ with $\partial\Omega$ Lipschitz continuous. Then there exists a constant c_Ω such that for every $u \in H^1(\Omega)$,*

$$\|u - \mathcal{M}_\Omega(u)\|_{L^2(\Omega)} \leq c_\Omega \|\nabla u\|_{L^2(\Omega)}, \tag{8.47}$$

where $\mathcal{M}_\Omega(u)$ denotes the mean value of u over Ω

$$\mathcal{M}_\Omega(u) = \frac{1}{|\Omega|} \int_\Omega u \, dx. \tag{8.48}$$

Proof. We prove by contradiction. Suppose there exists no constant satisfying (8.47). Then for every $k \in \mathbb{N}$, there exists $u_k \in H^1(\Omega)$ such that

$$\|u_k - \mathcal{M}_\Omega(u_k)\|_{L^2(\Omega)} > k \, \|\nabla u_k\|_{L^2(\Omega)} \, .$$

Define the functions

$$v_k = \frac{u_k - \mathcal{M}_\Omega(u_k)}{\|u_k - \mathcal{M}_\Omega(u_k)\|_{L^2(\Omega)}}, \quad \forall k \in \mathbb{N}.$$

Then $\|v_k\|_{L^2(\Omega)} = 1$, $\mathcal{M}_\Omega(v_k) = 0$, and

$$\|\nabla v_k\|_{L^2(\Omega)} < \frac{1}{k}, \quad \forall k \in \mathbb{N}. \tag{8.49}$$

This implies that the sequence $\{v_k\}$ is bounded on $H^1(\Omega)$. Due to the compactness results proved in the previous section, it has a subsequence, say $\{v_{k_n}\}$, that converges to some v in $L^2(\Omega)$. If $v_{k_n} \to v$, then $\|v\|_{L^2(\Omega)} = 1$ and $\mathcal{M}_\Omega(v) = 0$.

Consequently, using (8.49) we have for every $\varphi \in \mathcal{D}(\Omega)$, $i = 1, \ldots, N$,

$$\int_\Omega v \frac{\partial \varphi}{\partial x_i} \, dx = \lim_{n\to\infty} \int_\Omega v_{k_n} \frac{\partial \varphi}{\partial x_i} \, dx = - \lim_{n\to\infty} \int_\Omega \frac{\partial v_{k_n}}{\partial x_i} \varphi \, dx \quad \to \quad 0.$$

Hence $v \in H^1(\Omega)$ and $\nabla v = 0$ almost everywhere. In view of Theorem 7.40 and the assumptions on Ω, this implies v is constant. Since $\mathcal{M}_\Omega(v) = 0$, this shows that $v = 0$ which contradicts the fact that $\|v\|_{L^2(\Omega)} = 1$. Therefore there must exist a constant c_Ω satisfying (8.47). □

This result allows us to introduce the following space which plays an important role when studying partial differential equations with a Neumann boundary condition. We shall use it in Chapter 9.

Definition 8.27. *Let $N \geq 2$ and Ω a connected bounded open subset of $\mathbb{R}^N$ such that $\partial\Omega$ is Lipschitz continuous. We define the quotient space*

$$W(\Omega) = H^1(\Omega)/\mathbb{R},$$

as the space of equivalence classes with respect to the relation

$$u \simeq v \Longleftrightarrow u - v \text{ is a constant}, \quad \forall u, v \in H^1(\Omega).$$

We denote by $\dot{u}$ the equivalence class represented by u.

The next result is a consequence of the Poincaré–Wirtinger inequality.

Proposition 8.28. *Let $N \geq 2$ and Ω a connected bounded open subset of $\mathbb{R}^N$ such that $\partial\Omega$ is Lipschitz continuous. The following quantity:*

$$\|\dot{u}\|_{W(\Omega)} = \|\nabla u\|_{L^2(\Omega)}, \quad \forall u \in \dot{u},\ \dot{u} \in W(\Omega), \tag{8.50}$$

defines a norm on $W(\Omega)$ for which $W(\Omega)$ is a Banach space.

Moreover, $W(\Omega)$ is a Hilbert space for the scalar product

$$(\dot{v}, \dot{w})_{W(\Omega)} = \sum_{i=1}^{N} \Big(\frac{\partial v}{\partial x_i}, \frac{\partial w}{\partial x_i} \Big)_{L^2(\Omega)},$$

for every $\dot{v}, \dot{w} \in W(\Omega)$ with $v \in \dot{v}$, $w \in \dot{w}$.

Proof. To prove that (8.50) defines a norm, observe that since Ω is connected, Theorem 7.40 implies that if $\|\nabla u\|_{L^2(\Omega)} = 0$, then u is a constant. Hence $u \in \dot{0}$. The other conditions for a norm are obviously satisfied.

We now prove the completeness of $W(\Omega)$. Let $\{\dot{u}_n\}$ be a Cauchy sequence in $W(\Omega)$ and denote by u_n, for every $n \in \mathbb{N}$, the class representative of $\dot{u}_n$ such that $\mathcal{M}_\Omega(u_n) = 0$. From (8.50), it follows that $\{\nabla u_n\}$ is a Cauchy sequence in $L^2(\Omega)$.

By the Poincaré–Wirtinger inequality (8.47), $\{u_n\}$ must be a Cauchy sequence in $H^1(\Omega)$ which is a Banach space by Proposition 7.24. So, $\{u_n\}$ converges to some u in $H^1(\Omega)$ and from (8.50), $\{\dot{u}_n\}$ converges in $W(\Omega)$ to the equivalence class $\dot{u}$ represented by u. This ends the proof. □

The proposition below shows that we can deduce strong convergence in $L^q(\Omega)$ from a weak convergence in $W^{1,p}(\Omega)$ (see Definition 6.66). For simplicity, we treat here only the case $1 \leq p < N$.

Proposition 8.29. *Let $N \geq 2$, $1 \leq p < N$, and Ω a bounded open subset of $\mathbb{R}^N$ with a Lipschitz-continuous boundary. If $\{u_n\}$ is a sequence in $W^{1,p}(\Omega)$ such that*

$$u_n \rightharpoonup u \quad \text{weakly in } W^{1,p}(\Omega),$$

then for any $q \in [1, p^*)$, where $\dfrac{1}{p^*} = \dfrac{1}{p} - \dfrac{1}{N}$,

$$u_n \to u \quad \text{strongly in } L^q(\Omega).$$

Proof. It is clear that if u_n converges weakly to u in $W^{1,p}(\Omega)$, then it also converges weakly to u in $L^p(\Omega)$. Moreover, u_n is bounded on $W^{1,p}(\Omega)$ in view of Theorem 6.69.

Consequently, if $1 \leq q < p^*$, by Theorem 8.21 there exists a subsequence which strongly converges to some u_1 in $L^q(\Omega)$. By Theorem 6.67 (strong convergence implies weak convergence) and from the uniqueness of the weak limit, we deduce that $u = u_1$. Therefore the whole sequence $\{u_n\}$ converges to u in $L^q(\Omega)$ for $1 \leq q < p^*$. □

The third consequence of compactness is a variant of the Poincaré inequality, which is interesting for applications (see Chapter 9).

Proposition 8.30. *Let $N \geq 2$ and Ω a connected bounded open subset of $\mathbb{R}^N$ with a Lipschitz-continuous boundary $\partial\Omega$. Suppose $\partial\Omega = \Gamma_1 \cup \Gamma_2$, with Γ_1 and Γ_2 disjoint and Γ_1 of positive $(n-1)$-dimensional measure. Then there exists a constant C_{Ω,Γ_1} such that*

$$\|u\|_{L^2(\Omega)} \leq C_{\Omega,\Gamma_1} \|\nabla u\|_{L^2(\Omega)}, \tag{8.51}$$

for every $u \in H^1(\Omega)$ with trace $\gamma(u) = 0$ on Γ_1.

Proof. We prove the result by contradiction. Similar to the proof of Theorem 8.26, we can construct a sequence $v_k \in H^1(\Omega)$ such that $\|v_k\|_{L^2(\Omega)} = 1$, $\gamma(v_k) = 0$ on Γ_1, and

$$\|\nabla v_k\|_{L^2(\Omega)} < \frac{1}{k} \quad \text{for any } k \in \mathbb{N}. \tag{8.52}$$

This implies that the sequence $\{v_k\}$ is bounded on $H^1(\Omega)$. By the Eberlein–Šmuljan Theorem (Thm 6.72), it has a subsequence (still denoted $\{v_k\}$) which converges weakly to some v in $H^1(\Omega)$. By Proposition 8.29, it also converges strongly in $L^2(\Omega)$. This implies that $\|v\|_{L^2(\Omega)} = 1$.

Moreover, since the trace operator is a linear and continuous map from $H^1(\Omega)$ to $L^2(\partial\Omega)$ (Theorem 7.49), in view of Theorem 6.71 we have $\gamma(v) = 0$ on Γ_1.

On the other hand, by Proposition 6.74 and (8.52), we have

$$\int_\Omega v \frac{\partial \varphi}{\partial x_i} = \lim_{n\to\infty} \int_\Omega v_{k_n} \frac{\partial \varphi}{\partial x_i} = - \lim_{n\to\infty} \int_\Omega \frac{\partial v_{k_n}}{\partial x_i} \varphi \to 0,$$

for every $\varphi \in \mathcal{D}(\Omega)$, $i = 1, \ldots, N$. It follows that $\nabla v = 0$ almost everywhere. Hence in view of Theorem 7.40 and the assumptions on Ω, v has to be constant. Since $\gamma(v) = 0$ on Γ_1, and as Γ_1 has a positive measure, it follows that $v = 0$. But this contradicts the fact that $\|v\|_{L^2(\Omega)} = 1$. Therefore there exists a constant C_{Ω,Γ_1} satisfying (8.51). □

A direct consequence of this theorem is

Corollary 8.31. *Let $N \geq 2$ and Ω a connected bounded open subset of $\mathbb{R}^N$ with a Lipschitz-continuous boundary $\partial\Omega$. Suppose $\partial\Omega = \Gamma_1 \cup \Gamma_2$ with Γ_1 and Γ_2 disjoint and Γ_1 of positive $(n-1)$-dimensional measure. Let*

$$W_1(\Omega) = \{v \in H^1(\Omega) \big| \gamma(u) = 0 \text{ on } \Gamma_1\}.$$

Then the quantity

$$\|u\|_{W_1(\Omega)} = \|\nabla u\|_{L^2(\Omega)}, \quad \forall u \in W_1(\Omega), \tag{8.53}$$

defines a norm on $W_1(\Omega)$, which is equivalent to that of $H^1(\Omega)$. With this norm, $W_1(\Omega)$ is a Banach space.

Moreover, $W_1(\Omega)$ is a Hilbert space for the scalar product

$$(v, w)_{W_1(\Omega)} = \sum_{i=1}^{N} \left(\frac{\partial v}{\partial x_i}, \frac{\partial w}{\partial x_i}\right)_{L^2(\Omega)},$$

for $v, w \in W_1(\Omega)$.

Remark 8.32. *The same argument as that used in Remark 7.67(3) shows that $L^2(\Omega) \subset (W_1(\Omega))'$. Moreover, for any v is in $L^2(\Omega)$,*

$$\langle v, u\rangle_{(W_1(\Omega))', W_1(\Omega)} = \int_\Omega u\, v\, dx, \quad \forall\, u \in W_1(\Omega),$$

and

$$\|v\|_{(W_1(\Omega))'} \leq \|v\|_{L^2(\Omega)}. \tag{8.54}$$

Chapter 9

Variational Elliptic Problems

We are now ready to study solutions, in the weak sense, of some second order linear elliptic partial differential equations in the divergence form. We consider the equation

$$-\sum_{i,j=1}^{N} \frac{\partial}{\partial x_i}\Big(a_{ij}(x)\frac{\partial u}{\partial x_j}\Big) = f \quad \text{in } \Omega, \tag{9.1}$$

with three types of boundary conditions: Dirichlet, Neumann, and Robin. To simplify the presentation, the Neumann and Robin problems are presented in the case $N \geq 2$. Moreover, the negative sign in (9.1) is adopted for convenience. This is a standard way of formulating variational problems.

We apply the Lax–Milgram Theorem to prove the existence and uniqueness of the weak solutions. We first introduce the notion of continuous and H-elliptic bilinear forms before proceeding to the proof of the Lax–Milgram Theorem and its applications. For more details, we refer the reader to [3], [4], [8], [16], [24], [29], and [41].

9.1 Setting of the problems

Let $A = (a_{ij})_{1\leq i,j\leq N}$ be a matrix vector field on a bounded open subset Ω of $\mathbb{R}^N$. In this section, we treat equation (9.1) with different types of boundary conditions.

Remark 9.1. *The differential operator in (9.1) can be written in the divergence form*

$$-div\,(A(x)\,\nabla u) = -\sum_{i,j=1}^{N} \frac{\partial}{\partial x_i}\Big(a_{ij}(x)\frac{\partial u}{\partial x_j}\Big),$$

so that (9.1) reads

$$-\ div\ (A(x)\ \nabla u) = f \qquad in\ \Omega. \tag{9.2}$$

In particular, if A is the identity matrix in $\mathbb{R}^N$, then this equation reduces to the Laplace equation,

$$-\Delta u = -\sum_{i=1}^{N} \frac{\partial^2 u}{\partial x_i^2} = f.$$

Definition 9.2. *Let $\alpha, \beta \in \mathbb{R}$, such that $0 < \alpha < \beta$. We denote by $M(\alpha, \beta, \Omega)$ the set of $N \times N$ matrix fields $A = (a_{ij})_{1 \leq i,j \leq N} \in (L^\infty(\Omega))^{N^2}$ such that for any $\lambda \in \mathbb{R}^N$*

$$\begin{cases} i) & A(x)\lambda\lambda \geq \alpha|\lambda|^2, \\ ii) & |A(x)\lambda| \leq \beta|\lambda|, \end{cases}$$

almost everywhere on Ω and under the notation $A(x)\lambda\lambda = (A(x)\lambda, \lambda)_{\mathbb{R}^N}$.

In this chapter, we assume that

$$A \in M(\alpha, \beta, \Omega), \tag{9.3}$$

for some α and β such that $0 < \alpha < \beta$.

Remark 9.3.
(1) Observe that if A satisfies (i), the operator $-div\ (A(x)\ \nabla)$ is uniformly elliptic with constant α, in the sense of Definition 3.25.
(2) Condition (ii) clearly implies that $\|A\|_{(L^\infty(\Omega))^{N^2}} \leq \beta$.

For $N \geq 2$ and $x \in \partial\Omega$, we consider the quantity

$$A(x)\nabla u(x) \cdot n(x) = \sum_{i,j=1}^{N} a_{ij}(x)\, \frac{\partial u}{\partial x_j}(x)\, n_i(x)\,, \tag{9.4}$$

where $n = (n_1, ..., n_N)$ denotes the unit outward normal to Ω. This quantity, also denoted by $\dfrac{\partial u}{\partial n_A}$ in the literature, is called the conormal derivative of u (with respect to A) at the point x. When A is the identity matrix, it reduces to the normal derivative already introduced in the first chapter.

We consider some of the boundary conditions on $\partial\Omega$ that have been presented in Section 1.3,

- $u = 0$, (Homogeneous Dirichlet condition)
- $A\nabla u \cdot n = 0$, (Homogeneous Neumann condition)

- $A\nabla u \cdot n = g,$ (Nonhomogeneous Neumann condition)
- $A\nabla u \cdot n + hu = 0,$ (Robin condition)
- $A\nabla u \cdot n + hu = g,$ (Nonhomogeneous Robin condition)

We shall find weak solutions satisfying the variational formulation associated with each problem. As can be seen in the following, to obtain the weak formulation, as a rule we formally multiply the equation by a regular function, then integrate by parts and make use of the boundary conditions.

9.2 Bilinear forms on Hilbert spaces

Definition 9.4. *Let H be a real Hilbert space. A map $a : H \times H \to \mathbb{R}$ is called a bilinear form on H if for any $u \in H$ and $v \in H$, the following maps are linear:*

$$a(u,\cdot) : v \in H \longmapsto a(u,v) \in \mathbb{R},$$
$$a(\cdot,v) : u \in H \longmapsto a(u,v) \in \mathbb{R}.$$

A bilinear form a on H is said to be

(1) symmetric if $a(u,v) = a(v,u), \quad \forall\, u,\, v \in H,$

(2) positive if $a(u,u) \geq 0, \quad \forall\, u \in H,$

(3) bounded if there exists $C > 0$ such that for $u,\, v \in H$

$$|a(u,v)| \leq C \left\|u\right\|_H \left\|v\right\|_H, \tag{9.5}$$

(4) H-elliptic (or coercive on H) with constant α_0 if there exists $\alpha_0 > 0$ such that

$$a(u,u) \geq \alpha_0 \left\|u\right\|^2_H, \quad \forall\, u \in H.$$

In the following, H denotes a real Hilbert space. We also consider the norm on $H \times H$ given by

$$\left\|(u,v)\right\|_{H\times H} = \sqrt{\left\|u\right\|_H^2 + \left\|v\right\|_H^2}, \quad \forall u, v \in H, \tag{9.6}$$

which is equivalent to the norm (6.5) introduced in Examples 6.3. Bilinear forms on Hilbert spaces have the property that boundedness on H is equivalent to continuity on $H \times H$.

Proposition 9.5. *Let $a : H\times H \to \mathbb{R}$ be a bilinear form. Then a is bounded on H if and only if a is continuous on $H \times H$.*

Proof. Suppose a is bounded on H. Then for (u, v), $(u_0, v_0) \in H \times H$, we have by the bilinearity and boundedness of a

$$\begin{aligned}|a(u,v) - a(u_0,v_0)| &= |a(u,v) - a(u,v_0) + a(u,v_0) - a(u_0,v_0)| \\ &\le |a(u, v - v_0)| + |a(u - u_0, v_0)| \\ &\le C\,\|u\|_H \|v - v_0\|_H + C\,\|u - u_0\|_H \|v_0\|_H,\end{aligned}$$

so that $|a(u,v) - a(u_0,v_0)| \to 0$ as $(u,v) \to (u_0,v_0)$. Thus a is continuous at each point $(u_0,v_0) \in H \times H$.

Suppose a is continuous on $H \times H$. In particular, a is continuous at $(0,0)$. For $\varepsilon > 0$, there exists $\delta > 0$ such that if $\|(w,z)\|_{H\times H} < \delta$, then

$$|a(w,z)| < \varepsilon, \tag{9.7}$$

where the norm in $H \times H$ is given by (9.6).

Let u, $v \in H \setminus \{0\}$ and define w and z as

$$w = \frac{\delta}{2}\frac{u}{\|u\|_H}, \qquad z = \frac{\delta}{2}\frac{v}{\|v\|_H}. \tag{9.8}$$

By construction, $\|w\|_H = \|z\|_H = \dfrac{\delta}{2} < \delta$, which implies

$$\|(w,z)\|_{H\times H} = \frac{\delta}{\sqrt{2}} < \delta.$$

Hence, by linearity and using the inequality (9.7), we have

$$\frac{|a(u,v)|}{\|u\|_H\|v\|_H} = \left|a\left(\frac{u}{\|u\|_H}, \frac{v}{\|v\|_H}\right)\right| = \frac{4}{\delta^2}|a(w,z)| \le \frac{4}{\delta^2}\varepsilon. \tag{9.9}$$

This proves that a is bounded as it satisfies (9.5) with $C = \dfrac{4}{\delta^2}\varepsilon$. □

9.3 The Lax–Milgram Theorem

As illustrated in the example given in the beginning of Chapter 7, the problem of finding a solution satisfying a partial differential equation with some boundary conditions can be solved by using a variational formulation. It shall be seen in the next sections that the variational problem is equivalent to the given one when the data are sufficiently smooth.

In this approach, the idea is to write the PDE under consideration in the form

$$\begin{cases} \text{Find } u \in H \text{ such that} \\ a(u,v) = \langle F, v\rangle_{H',H}, \quad \forall v \in H, \end{cases} \tag{9.10}$$

where H is a Hilbert space, a is a bilinear form on H, and $F \in H'$.

This is called a variational formulation of the original PDE and the functions $v \in H$ are usually called test functions. Under suitable assumptions on the form a, a unique u may exist. This is the essence of the Lax–Milgram Theorem.

For the proof of this important theorem, we make use of the classical Banach Fixed Point Theorem which we state here without proof.

Theorem 9.6 (Banach Fixed Point Theorem). *Let X be a Banach space and $f : X \to X$ satisfying*

$$\|f(u) - f(v)\|_X \leq c\|u - v\|_X, \quad \forall u, v \in X \tag{9.11}$$

for some $c \in (0, 1)$. Then there exists a unique point $x \in X$ such that $f(x) = x$.

Observe that the function f in this theorem satisfies a Lipschitz condition on X with a positive constant $c < 1$. In this case, f is called a contraction mapping. The Banach Fixed Point Theorem asserts that a contraction mapping has a unique fixed point.

Theorem 9.7 (Lax–Milgram Theorem). *Let a be a continuous bilinear form on a Hilbert space H and $F \in H'$. If a is H-elliptic with constant α_0, then the variational equation given by (9.10) has a unique solution $u \in H$.*

Moreover, we have the following a priori estimate:

$$\|u\|_H \leq \frac{1}{\alpha_0}\|F\|_{H'}. \tag{9.12}$$

Proof. Let $u \in H$ and consider the map $A_u : H \mapsto \mathbb{R}$ given by

$$A_u(v) = a(u, v). \tag{9.13}$$

We first show that $A_u \in H'$.

Obviously, A_u is a linear functional. Since a is a continuous bilinear form, by Proposition 9.5, it is bounded. Thus there exists some $C > 0$ such that for every u and v in H

$$|\langle A_u, v\rangle| = |a(u, v)| \leq C\|u\|_H\|v\|_H.$$

This implies that A_u is bounded and thus, $A_u \in H'$ with

$$\|A_u\|_{H'} \leq C\|u\|_H, \tag{9.14}$$

where C is independent of u.

We now apply the Riesz Theorem (Thm 6.25) to A_u and $F \in H'$. There exist $\tau(A_u)$ and $\tau(F)$ in H, such that for all $v \in H$,

$$\begin{aligned} \langle A_u, v \rangle_{H',H} &= (\tau(A_u), v)_H, \\ \langle F, v \rangle_{H',H} &= (\tau(F), v)_H. \end{aligned} \tag{9.15}$$

Consequently, recalling (9.13), $u \in H$ satisfies

$$a(u, v) = \langle F, v \rangle_{H',H}, \quad \forall v \in H,$$

if and only if

$$(\tau(A_u) - \tau(F), v)_H = 0, \quad \forall v \in H.$$

This means that problem (9.10) is equivalent to the following one:

$$\begin{cases} \text{Find } u \in H \text{ such that} \\ \tau(A_u) = \tau(F). \end{cases} \tag{9.16}$$

Now define the mapping $\Phi : H \longmapsto H$ as

$$\Phi(v) = v - \rho(\tau(A_v) - \tau(F)), \quad \forall v \in H,$$

where ρ is a constant. Let us show that for a suitable ρ, the mapping Φ is a contraction, that is, there exists a constant $0 < c < 1$ such that

$$\|\Phi(w_1) - \Phi(w_2)\|_H \leq c\|w_1 - w_2\|_H, \quad \forall w_1, w_2 \in H. \tag{9.17}$$

Indeed, the fact that τ is an isometry between H and H', together with (9.14), (9.15) and the H-ellipticity of a, gives

$$\begin{aligned} \|v - \rho\tau(A_v)\|_H^2 &= (v - \rho\tau(A_v), v - \rho\tau(A_v))_H \\ &= \|v\|_H^2 - 2\rho(\tau(A_v), v)_H + \rho^2\|\tau(A_v)\|_H^2 \\ &= \|v\|_H^2 - 2\rho\, a(v, v) + \rho^2\|A_v\|_{H'}^2 \\ &\leq (1 - 2\rho\alpha_0 + \rho^2 C^2)\|v\|_H^2, \end{aligned}$$

for every $v \in H$. If $\rho \in (0, \frac{2\alpha_0}{C^2})$ then $(1 - 2\rho\alpha_0 + \rho^2 C^2) < 1$, which implies

$$\|v - \rho\tau(A_v)\|_H \leq c\|v\|_H, \quad \forall v \in H,$$

for some $c < 1$. Therefore, since

$$\tau(A_{w_1}) - \tau(A_{w_2}) = \tau(A_{w_1 - w_2}),$$

for all $w_1, w_2 \in H$, one has

$$\|\Phi(w_1) - \Phi(w_2)\| = \|w_1 - w_2 - \rho\tau(A_{w_1 - w_2})\| \leq c\|w_1 - w_2\|,$$

which is precisely (9.17). By the Banach Fixed Point Theorem, there exists a unique $u \in H$ such that $\Phi(u) = u$, that is,

$$u - \rho(\tau(A_u) - \tau(F)) = u.$$

So u is the unique solution of (9.16) (and obviously of (9.10)).

To show estimate (9.12), we use the H-ellipticity of a with the constant α_0 to get

$$\alpha_0 \|u\|_H^2 \leq a(u, u) = |\langle F, u\rangle_{H',H}| \leq \|F\|_{H'} \|u\|_H,$$

and this ends the proof of Theorem 9.7. □

Remark 9.8. *If a is symmetric, the theorem is a direct consequence of the Riesz Representation Theorem, since the bilinear form defines an equivalent scalar product on H. Moreover, the solution of (9.10) is the minimum point of a linear functional given in the theorem below.*

Theorem 9.9. *Let a be a continuous bilinear form on a Hilbert space H such that a is positive and symmetric. If $F \in H'$ and J is the functional on H defined by*

$$J(v) = \frac{1}{2} a(v, v) - \langle F, v\rangle_{H',H}, \quad \forall v \in H, \tag{9.18}$$

then u is a solution of the variational equation (9.10) if and only if u is a solution of the following problem:

$$\begin{cases} \textit{Find } u \in H \textit{ such that} \\ J(u) = \inf\limits_{v \in H} J(v). \end{cases} \tag{9.19}$$

Proof. Let u be a solution of (9.10). We show that $J(u) \leq J(v)$ for all $v \in H$. For any $v \in H$, v can be written as $u + w$ for some $w \in H$. We have

$$J(u + w) - J(u) = [a(u, w) - F(w)] + \frac{1}{2} a(w, w) = \frac{1}{2} a(w, w) \geq 0,$$

since a is a positive, symmetric bilinear form and F is linear. Hence u satisfies (9.19).

Conversely, suppose u is a solution of (9.19). For $v \in H$ and $t \in \mathbb{R}$ we have by the assumptions on a and F

$$0 \leq J(u + tv) - J(u) = t\{a(u, v) - F(v)\} + \frac{t^2}{2} a(v, v).$$

The right-hand side, with u and v fixed, is a polynomial of the form $At^2 + Bt$ with $A \geq 0$. Since this polynomial is nonnegative for every t, it follows that $B = 0$, that is $a(u, v) - F(v) = 0$, so u satisfies (9.10). □

Remark 9.10. *If a is a symmetric form which satisfies the hypotheses of Theorem 9.7 and J is defined by (9.19), then Theorem 9.9 implies in particular that problem (9.19) admits a unique solution $u \in H$.*

9.4 Dirichlet boundary conditions

The cases of homogeneous and nonhomogeneous Dirichlet conditions need to be treated separately.

9.4.1 *Homogeneous Dirichlet boundary conditions*

Let us consider the problem

$$\begin{cases} -\text{div } (A \nabla u) = f & \text{in } \Omega, \\ u = 0 & \text{on } \partial\Omega, \end{cases} \tag{9.20}$$

where $A \in M(\alpha, \beta, \Omega)$ and $f \in H^{-1}(\Omega)$.

Its variational formulation is the following:

$$\begin{cases} \text{Find } u \in H_0^1(\Omega) \text{ such that} \\ \displaystyle\int_\Omega A\nabla u \, \nabla v \, dx = \langle f, v \rangle_{H^{-1}(\Omega), H_0^1(\Omega)}, \quad \forall v \in H_0^1(\Omega), \end{cases} \tag{9.21}$$

where

$$\int_\Omega A\nabla u \, \nabla v \, dx = \sum_{i,j=1}^{N} \int_\Omega a_{ij}(x) \frac{\partial u}{\partial x_i} \frac{\partial v}{\partial x_j} \, dx.$$

Formulation (9.21) is obtained by multiplying the first equation in (9.20) (which is exactly equation (9.1)), by a smooth function vanishing on the boundary and then integrating by parts. We prove the existence and uniqueness of the solution by applying the Lax–Milgram Theorem.

A solution of (9.21) is called a weak solution of (9.20) (or simply a solution). As discussed in Sections 7.1 and 9.3, this is justified by the fact that a weak solution is a classical solution when all the data are sufficiently differentiable. Indeed, we have

Proposition 9.11. *Let $\partial\Omega$ be of class C^1. If $A \in (C^1(\overline{\Omega}))^{N^2}$, $f \in C^0(\overline{\Omega})$, and $u \in C^2(\overline{\Omega})$, then u is a (classical) solution of*

$$\begin{cases} -\textit{div } (A(x) \nabla u(x)) = f(x) & \textit{for all } x \textit{ in } \Omega, \\ u(x) = 0 & \textit{for all } x \textit{ in } \partial\Omega, \end{cases} \tag{9.22}$$

if and only if u is a solution of (9.21).

Proof. If u is a solution of (9.22), Theorem 7.49 and Proposition 7.59 imply that u belongs to $H_0^1(\Omega)$. Let v be an arbitrary function in $H_0^1(\Omega)$. By definition, there exists a sequence $\{v_n\} \subset \mathcal{D}(\Omega)$ which converges to v in $H_0^1(\Omega)$. Multiplying the equation in (9.22) by v_n and integrating by parts give

$$\int_\Omega A\nabla u\, \nabla v_n\, dx = \int_\Omega f\, v_n\, dx.$$

In view of Corollary 6.31, passing to the limit as $n \to \infty$ shows that u is a solution of (9.21).

Conversely, if $u \in C^2(\overline{\Omega})$ is solution of (9.21), then

$$\int_\Omega A\nabla u\, \nabla v\, dx = \int_\Omega f\, v\, dx, \quad \forall v \in \mathcal{D}(\Omega).$$

Integrating by parts yields

$$\int_\Omega \left[-\operatorname{div}\,(A\,\nabla u) - f\right] v\, dx = 0, \quad \forall v \in \mathcal{D}(\Omega).$$

By Theorem 6.64, this implies that u is a solution of the equation in (9.22) almost everywhere in Ω. Since both sides of the equation are continuous, this is true for every $x \in \Omega$. By Theorem 7.49 and Proposition 7.59, u satisfies the Dirichlet boundary condition. This completes the proof. □

We shall now apply the Lax–Milgram Theorem in order to prove existence and uniqueness of the solution of problem (9.21).

Theorem 9.12. *Let A be a matrix field in $M(\alpha, \beta, \Omega)$. For any $f \in H^{-1}(\Omega)$, there exists a unique solution $u \in H_0^1(\Omega)$ of problem (9.21). Moreover, u satisfies*

$$\|u\|_{H_0^1(\Omega)} \leq \frac{1}{\alpha}\|f\|_{H^{-1}(\Omega)}, \tag{9.23}$$

where $\|u\|_{H_0^1(\Omega)}$ is given by (7.57).

If $f \in L^2(\Omega)$, then

$$\|u\|_{H_0^1(\Omega)} \leq \frac{C_\Omega}{\alpha}\|f\|_{L^2(\Omega)}, \tag{9.24}$$

where C_Ω is the Poincaré constant given by Theorem 7.61.

Proof. Let a be the form on $H_0^1(\Omega) \times H_0^1(\Omega)$ given by

$$a(u, v) = \int_\Omega A\nabla u\, \nabla v\, dx, \qquad \forall u,\, v \in H_0^1(\Omega). \tag{9.25}$$

We apply the Lax–Milgram Theorem (Theorem 9.7) to a, with $F = f$ and $H = H_0^1(\Omega)$ equipped with the norm (7.57).

Clearly, $a(u, v)$ is a bilinear form by the linearity of the integral. Applying (9.2) and the Cauchy–Schwarz inequality (6.31) gives

$$|a(w,v)| \leq \beta \|\nabla w\|_{L^2(\Omega)} \|\nabla v\|_{L^2(\Omega)} = \beta \|w\|_{H_0^1(\Omega)} \|v\|_{H_0^1(\Omega)}, \tag{9.26}$$

for every w and v in $H_0^1(\Omega)$. In view of Proposition 9.5, this gives the continuity of the form a on $H_0^1(\Omega) \times H_0^1(\Omega)$.

Moreover, from property (i) of Definition 9.2,

$$a(v,v) \geq \alpha \|\nabla v\|^2_{L^2(\Omega)} = \alpha \|v\|^2_{H_0^1(\Omega)}, \quad \forall v \in H_0^1(\Omega), \tag{9.27}$$

which means that a is $H_0^1(\Omega)$-elliptic. Consequently, in view of the Lax–Milgram Theorem we have existence and uniqueness of a solution of (9.21), as well as estimate (9.23). If f is in $L^2(\Omega)$, we use inequality (7.60) to get estimate (9.24). □

Remark 9.13.

(1) Theorem 9.12 shows that the Dirichlet problem (9.20) *is well-posed (in the sense of Definition 1.10) for any of the following choices:*

$$\begin{aligned} &\mathcal{U} = \mathcal{V} = H_0^1(\Omega) \ \ \textit{and} \ \ \mathcal{F} = H^{-1}(\Omega), \\ &\mathcal{U} = \mathcal{V} = H_0^1(\Omega) \ \ \textit{and} \ \ \mathcal{F} = L^2(\Omega). \end{aligned} \tag{9.28}$$

(2) If the matrix A is symmetric, then Theorem 9.9 implies that the solution u given by Theorem 9.12 is the unique minimum point of the functional J defined by

$$J(v) = \frac{1}{2} \int_\Omega A\nabla v \, \nabla v \, dx - \langle f, v \rangle_{H^{-1}(\Omega), H_0^1(\Omega)}, \quad \forall v \in H_0^1(\Omega).$$

9.4.2 *Nonhomogeneous Dirichlet boundary conditions*

Let $N \geq 2$ and assume that Ω is a bounded open set in $\mathbb{R}^N$ with a Lipschitz-continuous boundary $\partial\Omega$. We consider the problem with nonhomogeneus Dirichlet boundary conditions

$$\begin{cases} -\text{div}\,(A\,\nabla u) = f & \text{in } \Omega, \\ u = g & \text{on } \partial\Omega, \end{cases} \tag{9.29}$$

where $A \in M(\alpha, \beta, \Omega)$, $f \in H^{-1}(\Omega)$, and $g \in H^{\frac{1}{2}}(\partial\Omega)$.

We say that u is a (weak) solution of (9.29) if

$$\begin{cases} u \in H^1(\Omega), \\ \displaystyle\int_\Omega A\nabla u \nabla v\, dx = \langle f, v\rangle_{H^{-1}(\Omega), H^1(\Omega)}, \qquad \forall\, v \in \mathcal{D}(\Omega), \\ \gamma(u) = g \quad \text{on } \partial\Omega, \end{cases} \tag{9.30}$$

where γ is the trace operator introduced in Theorem 7.49.

Theorem 9.14. *For $N \geq 2$, let Ω be a bounded open set in $\mathbb{R}^N$ with a Lipschitz-continuous boundary $\partial\Omega$ and A a matrix field in $M(\alpha, \beta, \Omega)$. For any $f \in H^{-1}(\Omega)$ and $g \in H^{\frac{1}{2}}(\partial\Omega)$, there exists a unique solution $u \in H^1(\Omega)$ of problem* (9.30).

Moreover,

$$\|u\|_{H^1(\Omega)} \leq \frac{1}{\alpha}\Big(\|f\|_{H^{-1}(\Omega)} + (\alpha+\beta)\|G\|\|g\|_{H^{\frac{1}{2}}(\partial\Omega)}\Big), \tag{9.31}$$

where G is the lifting operator given by (7.53).

Proof. Let $G \in \mathcal{L}(H^{\frac{1}{2}}(\partial\Omega), H^1(\Omega))$ be the linear continuous lifting operator introduced in Theorem 7.52. By definition, it associates to $g \in H^{\frac{1}{2}}(\partial\Omega)$ a function $u_g = G(g)$ in $H^1(\Omega)$ such that $\gamma(u_g) = g$ on $\partial\Omega$. Thus,

$$\|u_g\|_{H^1(\Omega)} \leq \|G\|\ \|g\|_{H^{\frac{1}{2}}(\partial\Omega)}. \tag{9.32}$$

We introduce the following auxiliary problem:

$$\begin{cases} \text{Find } z \in H_0^1(\Omega) \text{ such that,} \\ \displaystyle\int_\Omega A\nabla z \nabla v\, dx = \langle f, v\rangle_{(H^1(\Omega))', H^1(\Omega)} - \int_\Omega A\nabla u_g \nabla v\, dx \\ \text{for every } v \in H_0^1(\Omega). \end{cases} \tag{9.33}$$

Using (9.32) we have

$$\Big|\int_\Omega A\nabla u_g \nabla v\, dx\Big| \leq \beta\,\|G\|\ \|v\|_{H_0^1(\Omega)} \quad \text{for every } v \in H_0^1(\Omega),$$

so that the map F defined by

$$F(v) = \int_\Omega f\, v\, dx - \int_\Omega A\nabla u_g \nabla v\, dx, \qquad \forall\, v \in H_0^1(\Omega),$$

belongs to $H^{-1}(\Omega)$, with

$$\|F\|_{H^{-1}(\Omega)} \leq \|f\|_{H^{-1}(\Omega)} + \beta\|G\|\|g\|_{H^{\frac{1}{2}}(\partial\Omega)}.$$

Arguing as in the proof of Theorem 9.12, in view of the Lax–Milgram Theorem we deduce that there exists a unique solution z of problem (9.33). Moreover,

$$\|z\|_{H_0^1(\Omega)} \leq \frac{1}{\alpha}\Big(\|f\|_{H^{-1}(\Omega)} + \beta\|G\|\|g\|_{H^{\frac{1}{2}}(\partial\Omega)}\Big). \tag{9.34}$$

Set now $u = z + u_g$, which clearly belongs to $H^1(\Omega)$. Choosing in particular $v \in \mathcal{D}(\Omega)$ as test function in (9.33) we obtain

$$\int_\Omega A\nabla u \nabla v \, dx = \langle f, v\rangle_{H^{-1}(\Omega),H^1(\Omega)}, \qquad \forall\, v \in \mathcal{D}(\Omega).$$

On the other hand, since $z \in H_0^1(\Omega)$, by the linearity of the trace operator and Theorem 7.59,

$$\gamma(u) = \gamma(z) + \gamma(u_g) = \gamma(u_g) = g,$$

which proves that u is a solution of problem (9.30). Its uniqueness follows from that of the corresponding problem with homogeneous Dirichlet conditions.

Finally, the *a priori* estimate (9.31) is a straightforward consequence of (9.32) and (9.34). □

9.5 Neumann boundary conditions

Let $N \geq 2$ and $A \in M(\alpha, \beta, \Omega)$. We consider the problem given by (9.1) with a homogeneous Neumann boundary condition, that is, the problem

$$\begin{cases} -\text{div}\,(A\,\nabla u) = f & \text{in } \Omega, \\ A\,\nabla u \cdot n = 0 & \text{on } \partial\Omega, \end{cases} \tag{9.35}$$

where n is the unit outward normal vector to Ω.

If f is in $(H^1(\Omega))'$, the corresponding variational formulation is

$$\begin{cases} \text{Find } u \in H^1(\Omega) \text{ such that} \\ \displaystyle\int_\Omega A\nabla u\, \nabla v \, dx = \langle f, v\rangle_{(H^1(\Omega))',H^1(\Omega)}, \quad \forall v \in H^1(\Omega). \end{cases} \tag{9.36}$$

As before, the variational formulation is formally obtained by multiplying (9.1) by a smooth function (without any condition on the boundary) and then integrating by parts, which again is justified by the fact that a weak solution is a classical solution when all the data are sufficiently smooth.

However, a difficulty arises in the case of (9.36). Indeed, it is immediately seen that the bilinear form $a(u, v)$ given by

$$a(u,v) = \int_\Omega A\nabla u\, \nabla v \, dx, \qquad \forall u,\, v \in H^1(\Omega),$$

is not coercive on $H^1(\Omega)$, the natural space where solutions are searched for. Indeed, the Poincaré inequality does not hold in this space. As a consequence, we cannot apply the Lax–Milgram Theorem. This can also be seen from the fact that if u is a solution of (9.36), then for every $c \in \mathbb{R}$, $u + c$ is still a solution, so there is no uniqueness of a solution. This means that problem (9.36) is not well-posed (in the sense of Definition 1.10).

We present three situations where the Neumann problem is well-posed and for which the Lax–Milgram Theorem still applies.

In the first situation, we still consider problem (9.35), but we add a condition in the problem in order to have the uniqueness of the solution. Also, we have to prescribe connectedness and some regularity on the open set Ω, and an additional condition on the data f to ensure the existence of a solution of (9.35).

More precisely, we consider the problem

$$\begin{cases} -\text{div}\,(A\nabla u) = f & \text{in } \Omega, \\ A\nabla u \cdot n = 0 & \text{on } \partial\Omega, \\ \mathcal{M}_\Omega(u) = 0, \end{cases} \tag{9.37}$$

where n is the unit outward normal vector to Ω and $\mathcal{M}_\Omega(u)$ denotes the mean value of u over Ω given by (8.48). Clearly, if u is a solution, $u + c$ is not a solution anymore since it does not have a zero mean value.

If f is in $(H^1(\Omega))'$, the corresponding variational formulation is

$$\begin{cases} \text{Find } u \in H^1(\Omega) \text{ such that } \mathcal{M}_\Omega(u) = 0 \text{ and} \\ \displaystyle\int_\Omega A\nabla u\, \nabla v\, dx = \langle f, v\rangle_{(H^1(\Omega))', H^1(\Omega)}, \\ \forall v \in H^1(\Omega). \end{cases} \tag{9.38}$$

To obtain the solution, we use an auxiliary variational formulation posed in the Hilbert space $W(\Omega)$ introduced in Definition 8.27, and then apply the Lax–Milgram Theorem.

Theorem 9.15. *Let $N \geq 2$. Assume that Ω is a connected bounded open set in $\mathbb{R}^N$ with a Lipschitz-continuous boundary $\partial\Omega$ and A is a matrix in $M(\alpha, \beta, \Omega)$. If f is in $(H^1(\Omega))'$ and satisfies the following condition:*

$$\langle f, 1\rangle_{(H^1(\Omega))', H^1(\Omega)} = 0, \tag{9.39}$$

then there exists a unique solution $u \in H^1(\Omega)$ of problem (9.38). *Moreover,*

$$\|\nabla u\|_{L^2(\Omega)} \leq \frac{1 + c_\Omega}{\alpha} \|f\|_{(H^1(\Omega))'}, \tag{9.40}$$

where c_Ω is the constant from the Poincaré–Wirtinger inequality (8.47). *If $f \in L^2(\Omega)$, then*

$$\|\nabla u\|_{L^2(\Omega)} \leq \frac{c_\Omega}{\alpha}\|f\|_{L^2(\Omega)}. \tag{9.41}$$

Proof. Let us first show uniqueness. If u and w are two solutions of (9.38), then

$$\int_\Omega A\nabla(u-w)\,\nabla v\,dx = 0, \quad \forall v \in H^1(\Omega).$$

Since $A \in M(\alpha, \beta, \Omega)$, letting $v = u - w$ gives

$$\alpha\|\nabla v\|^2_{L^2(\Omega)} \leq \int_\Omega A\nabla(u-w)\,\nabla(u-w)\,dx = 0,$$

so $\nabla(u-w) = 0$ almost everywhere in Ω. Due to the assumptions on Ω, Theorem 7.40 implies $u-w$ is constant. Hence $u-w=0$ as $\mathcal{M}_\Omega(u-w)=0$. This shows the uniqueness.

We now prove the existence of a solution. Let $W(\Omega)$ be the Hilbert space given by Definition 8.27, endowed with the norm given by (8.50) and consider the following auxiliary variational formulation:

$$\begin{cases} \text{Find } \dot{u} \in W(\Omega) \text{ such that} \\ \dot{a}(\dot{u}, \dot{v}) = \langle f, v\rangle_{(H^1(\Omega))', H^1(\Omega)}, \\ \forall v \in \dot{v}, \ \forall \dot{v} \in W(\Omega), \end{cases} \tag{9.42}$$

where $\dot{a}$ is defined by

$$\dot{a}(\dot{u}, \dot{v}) = \int_\Omega A\nabla u\,\nabla v\,dx, \quad \forall\, u \in \dot{u}, \forall\, v \in \dot{v}, \tag{9.43}$$

for every $\dot{u}, \dot{v} \in W(\Omega)$. We apply the Lax–Milgram Theorem with $a = \dot{a}$, $H = W(\Omega)$ and F given by

$$F(\dot{v}) = \langle f, v\rangle_{(H^1(\Omega))', H^1(\Omega)}, \qquad \forall\, v \in \dot{v}, \forall\, \dot{v} \in W(\Omega).$$

Let us first show that F is well-defined as an element of $(W(\Omega))'$. Indeed, if w is another representative of $\dot{v}$, then there exists a real constant α such that $w = v + \alpha$. By linearity and (9.39), we have

$$\begin{aligned} \langle f, w\rangle_{(H^1(\Omega))', H^1(\Omega)} &= \langle f, v + c\rangle_{(H^1(\Omega))', H^1(\Omega)} \\ &= \langle f, v\rangle_{(H^1(\Omega))', H^1(\Omega)} + c\,\langle f, 1\rangle_{(H^1(\Omega))', H^1(\Omega)} \\ &= \langle f, v\rangle_{(H^1(\Omega))', H^1(\Omega)}. \end{aligned}$$

This proves that the definition of $F(\dot{v})$ does not depend on the representative of the class $\dot{v}$. Moreover, using the Poincaré–Wirtinger inequality (8.47), we have

$$\begin{aligned}|F(\dot{v})| &= \left|\langle f, v\rangle_{(H^1(\Omega))',H^1(\Omega)}\right| = \left|\langle f, (v - \mathcal{M}_\Omega(v))\rangle_{(H^1(\Omega))',H^1(\Omega)}\right| \\ &\le \|f\|_{(H^1(\Omega))'}\|v - \mathcal{M}_\Omega(v)\|_{H^1(\Omega)} \le (1 + c_\Omega)\|f\|_{(H^1(\Omega))'}\|\nabla v\|_{L^2(\Omega)} \\ &= (1 + c_\Omega)\|f\|_{(H^1(\Omega))'}\|\dot{v}\|_{W(\Omega)}.\end{aligned}$$

Thus F is a linear continuous map on $W(\Omega)$, that is, $F \in (W(\Omega))'$ and

$$\|F\|_{(W(\Omega))'} \le (1 + c_\Omega)\|f\|_{(H^1(\Omega))'}. \tag{9.44}$$

If $f \in L^2(\Omega)$, then by Remark 7.67(1) and using a similar argument yields

$$\begin{aligned}|F(\dot{v})| &= \left|\langle f, (v - \mathcal{M}_\Omega(v))\rangle_{L^2(\Omega),L^2(\Omega)}\right| \\ &\le c_\Omega\|f\|_{L^2(\Omega)}\|\nabla v\|_{L^2(\Omega)} = c_\Omega\|f\|_{L^2(\Omega)}\|\dot{v}\|_{W(\Omega)},\end{aligned}$$

so that in this case

$$\|F\|_{(W(\Omega))'} \le c_\Omega\|f\|_{L^2(\Omega))}. \tag{9.45}$$

On the other hand, $\dot{a}$ is clearly a bilinear form and is continuous on $W(\Omega) \times W(\Omega)$, since from (9.43), (9.3), and the Cauchy–Schwarz inequality (6.31), we have

$$|\dot{a}(\dot{u}, \dot{v})| \le \beta\|\nabla u\|_{L^2(\Omega)}\|\nabla v\|_{L^2(\Omega)} = \beta\|\dot{u}\|_{W(\Omega)}\|\dot{v}\|_{W(\Omega)},$$

for every $u \in \dot{u}$ and $v \in \dot{v}$. Moreover, using again (9.3),

$$a(\dot{u}, \dot{u}) \ge \alpha\|\nabla u\|^2_{L^2(\Omega)} = \alpha\|\dot{u}\|^2_{W(\Omega)},$$

for every $u \in \dot{u}$. This implies that $\dot{a}$ is $W(\Omega)$-elliptic so the Lax–Milgram Theorem provides the existence and uniqueness of a solution $\dot{u}$ of (9.42), which verifies the estimate

$$\|\dot{u}\|_{W(\Omega)} \le \frac{1}{\alpha}\|F\|_{(W(\Omega))'}. \tag{9.46}$$

Hence, if u is the representative of the solution $\dot{u}$ such that $\mathcal{M}_\Omega(u) = 0$, then u is a solution of (9.38). Estimate (9.40) follows from (9.44), (9.46), and the definition of the norm in $W(\Omega)$. Finally, estimate (9.41) follows from (9.45). □

Remark 9.16. *Condition (9.39) on f is called a compatibility condition, which is necessary for the existence of a solution of (9.38), as can be seen by choosing $v = 1$ in the variational formulation. If f is in $L^2(\Omega)$, then the compatibility condition can be written as $\mathcal{M}_\Omega(f) = 0$.*

Let us now discuss the second situation which presents a zero order term in the equation. The problem is

$$\begin{cases} -\text{div}\,(A\,\nabla u) + c\,u \;= f & \text{in } \Omega, \\ A\,\nabla u \cdot n = 0 & \text{on } \partial\Omega, \end{cases} \tag{9.47}$$

where n is the unit outward normal vector to Ω and c is a function in $L^\infty(\Omega)$ satisfying

$$\exists\, c_0 \in \mathbb{R}^+ \quad \text{such that} \quad 0 < c_0 \leq c(x) \quad \text{a.e. in } \Omega. \tag{9.48}$$

If f is given in $(H^1(\Omega))'$, the corresponding variational formulation is

$$\begin{cases} \text{Find } u \in H^1(\Omega) \text{ such that} \\ \displaystyle\int_\Omega A\nabla u\,\nabla v\;dx + \int_\Omega c\,u\,v\;dx = \;\langle f, v\rangle_{(H^1(\Omega))', H^1(\Omega)}, \\ \forall v \in H^1(\Omega). \end{cases} \tag{9.49}$$

Theorem 9.17. *Let $N \geq 2$. If Ω is a bounded open set in $\mathbb{R}^N$, A is a matrix field in $M(\alpha, \beta, \Omega)$, and c is a function in $L^\infty(\Omega)$ satisfying (9.48), then for any $f \in (H^1(\Omega))'$, there exists a unique solution $u \in H^1(\Omega)$ of problem (9.49). Moreover, u satisfies*

$$\|u\|_{H^1(\Omega)} \leq \frac{1}{\min\{\alpha, c_0\}} \|f\|_{(H^1(\Omega))'}. \tag{9.50}$$

If $f \in L^2(\Omega)$, then

$$\|u\|_{H^1(\Omega)} \leq \frac{1}{\min\{\alpha, c_0\}} \|f\|_{L^2(\Omega)}. \tag{9.51}$$

Proof. We apply the Lax–Milgram Theorem (Thm 9.7) with $H = H^1(\Omega)$, $F = f$, and $a(u, v)$ given by

$$a(u, v) = \int_\Omega A\nabla u\,\nabla v\;dx + \int_\Omega c\,u\,v\;dx, \qquad \forall u,\, v \in H^1(\Omega),$$

which is clearly a bilinear form. As $A \in M(\alpha, \beta, \Omega)$, by the Cauchy–Schwarz inequality (6.31) we have

$$\begin{aligned} |a(w, v)| &\leq \beta \|\nabla w\|_{L^2(\Omega)} \|\nabla v\|_{L^2(\Omega)} + \|c\|_{L^\infty(\Omega)} \|w\|_{L^2(\Omega)} \|v\|_{L^2(\Omega)} \\ &\leq \max\left\{\beta,\; \|c\|_{L^\infty(\Omega)}\right\} \|w\|_{H^1(\Omega)} \|v\|_{H^1(\Omega)}, \end{aligned}$$

for every w and v in $H^1(\Omega)$. In view of Proposition 9.5, this gives the continuity of the form a on $H^1(\Omega) \times H^1(\Omega)$.

Moreover, by (9.48), for every $v \in H^1(\Omega)$

$$a(v,v) \geq \alpha \|\nabla v\|^2_{L^2(\Omega)} + c_0 \|v\|^2_{L^2(\Omega)} \geq \min\{\alpha, c_0\} \|v\|^2_{H^1(\Omega)},$$

which implies that a is $H^1(\Omega)$-elliptic. Thus, we can apply the Lax–Milgram Theorem to obtain the existence and uniqueness of a solution of (9.49), as well as estimate (9.50). For f in $L^2(\Omega)$, we use inequality (7.61). □

Remark 9.18. *Let us point out that in Theorem 9.17, regularity and connectedness assumptions on Ω are not necessary.*

The third situation we treat concerns the case where, as in Proposition 8.30, the boundary of the open set Ω consists of two disjoint sets. We prescribe a Dirichlet condition on one part, and a Neumann condition on the other one (without any additional zero order term).

More precisely, suppose that $\partial\Omega = \Gamma_1 \cup \Gamma_2$ where Γ_1 and Γ_2 are disjoint and Γ_1 is of positive $(n-1)$-dimensional measure. Consider the problem

$$\begin{cases} -\text{div}\,(A\nabla u) = f & \text{in } \Omega, \\ u = 0 & \text{on } \Gamma_1, \\ A\nabla u \cdot n = 0 & \text{on } \Gamma_2, \end{cases} \tag{9.52}$$

where n is the unit outward normal vector to Ω. If f is given in $(W_1(\Omega))'$, the corresponding variational formulation is

$$\begin{cases} \text{Find } u \in W_1(\Omega) \text{ such that} \\ \displaystyle\int_\Omega A\nabla u\, \nabla v\, dx = \langle f, v\rangle_{(W_1(\Omega))', W_1(\Omega)}, \\ \forall v \in W_1(\Omega), \end{cases} \tag{9.53}$$

where W_1 is the space introduced in Corollary 8.31, that is,

$$W_1(\Omega) = \{v \in H^1(\Omega) \mid \gamma(u) = 0 \text{ on } \Gamma_1\},$$

endowed with the norm (8.53),

$$\|u\|_{W_1(\Omega)} = \|\nabla u\|_{L^2(\Omega)}, \quad \forall u \in W_1(\Omega).$$

Remark 9.19. *Observe that due to Remark 8.32, if f is in $L^2(\Omega)$, problem (9.53) becomes*

$$\begin{cases} \textit{Find } u \in W_1(\Omega) \textit{ such that} \\ \displaystyle\int_\Omega A\nabla u\, \nabla v\, dx = \int_\Omega f\, v\, dx, \\ \forall v \in W_1(\Omega). \end{cases} \tag{9.54}$$

Theorem 9.20. *Let $N \geq 2$. Suppose Ω is a connected bounded open subset of $\mathbb{R}^N$ with a Lipschitz-continuous boundary such that $\partial\Omega = \Gamma_1 \cup \Gamma_2$, with Γ_1 and Γ_2 disjoint and Γ_1 is of positive $(n-1)$-dimensional measure. If A is a matrix field in $M(\alpha, \beta, \Omega)$, then for any $f \in (H^1(\Omega))'$ there exists a unique solution $u \in H^1(\Omega)$ of problem (9.53). Moreover, u satisfies*

$$\|u\|_{W_1(\Omega)} \leq \frac{1}{\alpha} \|f\|_{(W_1(\Omega))'}. \tag{9.55}$$

If $f \in L^2(\Omega)$, then

$$\|u\|_{W_1(\Omega)} \leq \frac{C_{\Omega,\Gamma_1}}{\alpha} \|f\|_{L^2(\Omega)}, \tag{9.56}$$

where C_{Ω,Γ_1} is the constant given by inequality (8.51) of Proposition 8.30.

Proof. We apply the Lax–Milgram Theorem (Thm 9.7) with $H = W_1(\Omega)$, $F = f$, and $a(u, v)$ as above,

$$a(u, v) = \int_\Omega A\nabla u \, \nabla v \, dx, \qquad \forall u, \, v \in W_1(\Omega).$$

By (8.53) and the Cauchy–Schwarz inequality (6.31), we have

$$|a(w, v)| \leq \beta \|\nabla w\|_{L^2(\Omega)} \|\nabla v\|_{L^2(\Omega)} = \beta \|w\|_{W_1(\Omega)} \|v\|_{W_1(\Omega)},$$

for every w and v in $W_1(\Omega)$. This, in view of Proposition 9.5, gives the continuity of the bilinear form a on $W_1(\Omega) \times W_1(\Omega)$. Moreover, from (9.3)

$$a(v, v) \geq \alpha \|\nabla v\|^2_{L^2(\Omega)} = \alpha \|v\|^2_{W_1(\Omega)}, \quad \forall v \in H^1(\Omega),$$

which means that a is $W_1(\Omega)$-elliptic. So, we can apply the Lax–Milgram Theorem to obtain the existence and uniqueness of a solution of (9.53), as well as estimate (9.55).

Suppose now that f is in $L^2(\Omega)$ and $u \in W_1(\Omega)$ is a solution of (9.53). Using Remark 9.19 and choosing u as the test function in (9.54) give

$$\alpha \|u\|^2_{W_1(\Omega)} \leq a(u, u) = \int_\Omega f \, v \, dx.$$

On the other hand, applying Remark 8.32, the Cauchy–Schwarz inequality, and Proposition 8.30, we obtain

$$\left| \int_\Omega f \, u \, dx \right| \leq \|f\|_{L^2(\Omega)} \|u\|_{L^2(\Omega)} \leq C_{\Omega,\Gamma_1} \|f\|_{L^2(\Omega)} \|u\|_{W_1(\Omega)},$$

which ends the proof. $\square$

9.6 Robin boundary conditions

In this section, we again assume that $N \geq 2$ and that Ω is a bounded open subset of $\mathbb{R}^N$ with a Lipschitz-continuous boundary, so that integrals on the boundary $\partial\Omega$ make sense and the functions in $H^1(\Omega)$ have a well-defined trace on $\partial\Omega$.

We consider the general case of a nonhomogeneous Robin boundary condition of the form

$$\begin{cases} -\operatorname{div}(A\nabla u) + c\,u = f & \text{in } \Omega, \\ A\nabla u \cdot n + h\,u = g & \text{on } \partial\Omega, \end{cases} \tag{9.57}$$

where $A \in M(\alpha, \beta, \Omega)$, n is the unit outward normal vector to Ω, and the data c, h, f and g satisfy the following:

$$\begin{cases} c \in L^\infty(\Omega),\ \ 0 < c_0 \leq c(x) \ \text{ a.e. in } \Omega \ \text{ for some } c_0 \in R_+^*, \\ h \in L^\infty(\partial\Omega),\ h \geq 0 \quad \text{a.e. in } \partial\Omega, \\ f \in (H^1(\Omega))' \ \text{ and } \ g \in L^2(\partial\Omega). \end{cases} \tag{9.58}$$

The corresponding variational formulation is

$$\begin{cases} \text{Find } u \in H^1(\Omega) \text{ such that} \\ \displaystyle\int_\Omega A\nabla u\,\nabla v\,dx + \int_\Omega c\,u\,v\,dx + \int_{\partial\Omega} h\,u\,v\,ds \\ \displaystyle\qquad = \langle f, v\rangle_{(H^1(\Omega))', H^1(\Omega)} + \int_{\partial\Omega} g\,v\,ds, \\ \forall v \in H^1(\Omega). \end{cases} \tag{9.59}$$

Remark 9.21.

(1) Problems with a nonhomogeneous Neumann condition or with a homogeneous Robin condition (they have been mentioned in Section 9.1), are particular cases of problem (9.57).

(2) For simplicity, we assume that $g \in L^2(\partial\Omega)$. *The case* $g \in H^{-\frac{1}{2}}(\partial\Omega)$ *can be treated in a similar way.*

(3) For brevity, we only study here problem (9.57). *One can consider some variants of this problem, adopting the arguments introduced for the Neumann problem in the previous section. In particular, one can consider, as in Theorem 9.20, the case where* $c \geq 0$ *and the boundary of* Ω *consists of two components, with a Dirichlet condition on one component, and a Robin condition on the remaining part.*

Theorem 9.22. *Let $N \geq 2$, Ω a bounded open set in $\mathbb{R}^N$ with $\partial\Omega$ Lipschitz continuous and $A \in M(\alpha, \beta, \Omega)$. If the functions c, h, f, and g satisfy (9.58), then there exists a unique solution $u \in H^1(\Omega)$ of problem (9.59). Moreover, u satisfies the following inequality*

$$\|u\|_{H^1(\Omega)} \leq \frac{1}{\min\{\alpha, c_0\}}\Big(\|f\|_{(H^1(\Omega))'} + \gamma_\Omega \|g\|_{L^2(\partial\Omega)}\Big), \tag{9.60}$$

where γ_Ω is the norm of the trace operator introduced in Theorem 7.49.

If $f \in L^2(\Omega)$, then

$$\|u\|_{H^1(\Omega)} \leq \frac{1}{\min\{\alpha, c_0\}}\Big(\|f\|_{L^2(\Omega)} + \gamma_\Omega \|g\|_{L^2(\partial\Omega)}\Big). \tag{9.61}$$

Proof. We apply the Lax–Milgram Theorem with $H = H^1(\Omega)$, $a(u, v)$ and F given respectively, by

$$a(u, v) = \int_\Omega A\nabla u\, \nabla v\, dx + \int_\Omega c\, u\, v\, dx + \int_{\partial\Omega} h\, u\, v\, ds, \quad \forall u,\, v \in H^1(\Omega),$$

and

$$F(v) = \langle f, v\rangle_{(H^1(\Omega))', H^1(\Omega)} + \int_{\partial\Omega} g\, v\, ds, \qquad \forall u \in H^1(\Omega). \tag{9.62}$$

The map a is obviously a bilinear form. Since $A \in M(\alpha, \beta, \Omega)$, by the Cauchy–Schwarz inequality (6.31) and Theorem 7.49, we have for every w and v in $H^1(\Omega)$,

$$\begin{aligned}
|a(w, v)| &\leq \beta\|\nabla w\|_{L^2(\Omega)}\|\nabla v\|_{L^2(\Omega)} + \|c\|_{L^\infty(\Omega)}\|w\|_{L^2(\Omega)}\|v\|_{L^2(\Omega)} \\
&\quad + \|h\|_{L^\infty(\partial\Omega)}\|w\|_{L^2(\partial\Omega)}\|v\|_{L^2(\partial\Omega)} \\
&\leq \beta\|\nabla w\|_{L^2(\Omega)}\|\nabla v\|_{L^2(\Omega)} + \|c\|_{L^\infty(\Omega)}\|w\|_{L^2(\Omega)}\|v\|_{L^2(\Omega)} \\
&\quad + \gamma_\Omega^2\, \|h\|_{L^\infty(\partial\Omega)}\|w\|_{H^1(\Omega)}\|v\|_{H^1(\Omega)} \\
&\leq \max\left\{\beta, \|c\|_{L^\infty(\Omega)}, \gamma_\Omega^2\, \|h\|_{L^\infty(\partial\Omega)}\right\} \|w\|_{H^1(\Omega)}\|v\|_{H^1(\Omega)},
\end{aligned}$$

where γ_Ω is the norm of the trace operator introduced in Theorem 7.49. In view of Proposition 9.5, this gives the continuity of the form a on $H^1(\Omega) \times H^1(\Omega)$. Moreover, by (9.58),

$$a(v, v) \geq \alpha\|\nabla v\|^2_{L^2(\Omega)} + c_0\|v\|^2_{L^2(\Omega)} \geq \min\left\{\alpha, c_0\right\}\|v\|^2_{H^1(\Omega)}, \quad \forall v \in H^1(\Omega),$$

which means that a is $H^1(\Omega)$-elliptic.

It remains to verify that F (given by (9.62)), defines an element of $(H^1(\Omega)'$. It follows from Proposition 6.11 and Theorem 7.49 that

$$\begin{aligned}|F(v)| &\leq \left|\langle f,v\rangle_{(H^1(\Omega))',H^1(\Omega)}\right| + \|g\|_{L^2(\partial\Omega)}\|v\|_{L^2(\partial\Omega)} \\ &\leq \|f\|_{H^1(\Omega))'}\|v\|_{H^1(\Omega)} + \gamma_\Omega|g\|_{L^2(\partial\Omega)}\|v\|_{H^1(\Omega)},\end{aligned}$$

for every $v \in H^1(\Omega)$. This shows that F is in $(H^1(\Omega)'$ and

$$\|F\|_{(H^1(\Omega)'} \leq \|f\|_{H^1(\Omega))'} + \gamma_\Omega\|g\|_{L^2(\partial\Omega)}.$$

Consequently, we can apply the Lax–Milgram Theorem to obtain the existence and uniqueness of a solution of (9.49), as well as estimate (9.60).

If f is in $L^2(\Omega)$, estimate (9.61) is obtained from (9.60) by using inequality (7.61) from Remark 7.67. The proof is complete. □

9.7 The Dirichlet eigenvalue problem

In this section, we are interested in the eigenvalue problem corresponding to the Dirichlet problem studied in Section 9.4. It consists of finding the set of real numbers λ (called the *spectrum*) such that the following problem has a solution $u \not\equiv 0$,

$$\begin{cases} -\mathrm{div}\,(A\nabla u) = \lambda u & \text{in } \Omega, \\ u = 0 & \text{on } \partial\Omega, \end{cases} \tag{9.63}$$

where, as in Section 9.4, A is assumed to be a symmetric matrix field in the set $M(\alpha, \beta, \Omega)$.

For simplicity, we only discuss here the case of Dirichlet boundary conditions. Nevertheless, in the spirit of the previous sections of this chapter, other boundary conditions can also be considered.

The eigenvalue problem for the Laplace operator was mentioned in Chapter 1 (see (1.18)). We have also seen that the eigenvalue problems play an essential role in the study of the wave equation, as observed in particular in Remark 5.14.

The variational formulation of (9.63) is the following:

$$\begin{cases} \text{Find } \lambda \in \mathbb{R} \text{ such that the problem} \\ \displaystyle\int_\Omega A\nabla u\, \nabla v\, dx = \lambda \int_\Omega u\, v\, dx, \quad \forall v \in H_0^1(\Omega), \\ \text{admits at least a solution } u \not\equiv 0 \text{ in } H_0^1(\Omega). \end{cases} \tag{9.64}$$

Recall that such a λ is called an eigenvalue and a nontrivial solution u is called an eigenfunction corresponding to the eigenvalue λ.

Remark 9.23. *Observe that $\lambda = 0$ is not an eigenvalue, since in this case, in view of Theorem 9.12, the only solution is $u \equiv 0$. Moreover, if λ is an eigenvalue, then it is strictly positive. Indeed, due to the coerciveness of the form associated with the operator A (see (9.27)), if u is a corresponding eigenfunction then*

$$0 < \alpha\|\nabla u\|^2_{L^2(\Omega)} \leq \int_\Omega A\nabla u\, \nabla u\, dx = \lambda \int_\Omega u^2\, dx.$$

To solve problem (9.64), we make use of some classical results concerning the eigenvalues of compact symmetric operators in Hilbert spaces. To this aim, we consider the following abstract eigenvalue problem:

$$\begin{cases} \text{Find } \mu \in \mathbb{R} \text{ such that the equation } B(v) = \mu v \\ \text{admits at least one solution } v \neq 0 \text{ in } H, \end{cases} \tag{9.65}$$

where

$$\begin{cases} H \text{ is a real infinite-dimensional Hilbert space,} \\ B : H \longmapsto H \text{ is a compact linear map,} \\ B \text{ is symmetric, i.e., } \big(B(u), w\big)_H = \big(u, B(w)\big)_H, \ \forall u,\, w \in H. \end{cases} \tag{9.66}$$

A solution μ of (9.65) is also called an eigenvalue and a corresponding $v \not\equiv 0$ an eigenvector. Moreover, the linear space of the eigenvalues corresponding to μ together with the zero vector is called the eigenspace corresponding to the eigenvalue μ.

Let us recall that Hilbert spaces have been introduced in Section 6.2. We refer the reader to Definitions 6.4 and 8.13 for the notions of linear and compact maps, respectively.

In this framework, the following general result holds true (see for instance, [44] or [4] for proofs and details):

Theorem 9.24. *Suppose $B \not\equiv 0$ is a map satisfying (9.66), and such that*

$$B(u) = 0 \Longrightarrow u = 0. \tag{9.67}$$

Under this assumption, the following assertions hold true:

(1) The set of the eigenvalues of problem (9.65) is a countable set of $\mathbb{R}^+$ whose unique accumulation point is zero.

(2) Every eigenvalue is of finite multiplicity, i.e., its corresponding eigenspace is a vector subspace of H of positive finite dimension.

(3) Let $\{\mu_n\}$ be the sequence of the eigenvalues numbered in non-increasing order,

$$\mu_1 \geq \mu_2 \geq \ldots \to 0,$$

where each eigenvalue is repeated as many times as the dimension of its corresponding eigenspace. Then there exists a corresponding sequence of eigenvectors v_n which form a complete orthonormal system in H (see Definition 6.18).

We apply this theorem to problem (9.64) to obtain the following result:

Theorem 9.25. *Let Ω be a bounded open subset of $\mathbb{R}^N$ and A a symmetric matrix field in $M(\alpha, \beta, \Omega)$. Then*

(1) The set of the eigenvalues of problem (9.64) is a countable set whose unique accumulation point is $+\infty$.

(2) Every eigenvalue is of finite multiplicity, that is, its corresponding eigenspace is a vector subspace of $H_0^1(\Omega)$ of positive finite dimension.

(3) Let $\{\lambda_n\}$ the sequence of the eigenvalues numbered in non-decreasing order, that is,

$$0 \leq \lambda_1 \leq \lambda_2 \leq \ldots \to +\infty,$$

where each eigenvalue is repeated as many times as the dimension of its corresponding eigenspace. Then there exists a corresponding sequence of eigenvectors $\{w_n\}$ which forms a complete orthonormal system in $L^2(\Omega)$ (see Definition 6.18).

Proof. Let Φ be the mapping

$$\Phi : f \in L^2(\Omega) \longmapsto u_f \in H_0^1(\Omega),$$

where u_f is the unique solution in $H_0^1(\Omega)$, given by Theorem 9.12, of the problem

$$\begin{cases} -\text{div } (A\nabla u_f) = f & \text{in } \Omega, \\ u_f = 0 & \text{on } \partial\Omega. \end{cases} \tag{9.68}$$

Since from Remark 8.22, the embedding

$$i : w \in H_0^1(\Omega) \longmapsto w \in L^2(\Omega)$$

is compact, the mapping

$$B = i \circ \Phi : f \in L^2(\Omega) \longmapsto u_f \in L^2(\Omega), \tag{9.69}$$

is linear and compact (here, i denotes the identity map). Let us show that B is symmetric, that is,

$$\big(B(g), h\big)_{L^2(\Omega)} = \big(g, B(h)\big)_{L^2(\Omega)} \qquad \text{for any } g,\ h \in L^2(\Omega). \tag{9.70}$$

Let u_g and u_h be the solutions of (9.68) corresponding, to data g and h, respectively. By definition, $B(g) = u_g$ and $B(h) = u_h$. We choose u_h as test function in the variational formulation of (9.68) solved by u_g and choose u_g as test function in that solved by u_h. From (9.68) and the symmetry of A, we have

$$\begin{aligned}\big(B(g), h\big)_{L^2(\Omega)} &= (u_g, h)_{L^2(\Omega)} = \int_\Omega h u_g dx = \int_\Omega A\nabla u_h \nabla u_g \, dx \\ &= \int_\Omega A\nabla u_g \nabla u_h \, dx = \int_\Omega g u_h dx = \big(g, B(h)\big)_{L^2(\Omega)},\end{aligned}$$

which proves (9.70). Consequently, B satisfies (9.66) for $H = L^2(\Omega)$.

On the other hand, B is different from 0, and by Theorem 9.12, it verifies (9.67). Thus Theorem 9.24 applies to B.

At this point, observe that

$$-\text{div}\,(A\nabla u) = \lambda u \iff -\text{div}\,\big(A\nabla(\mu u)\big) = u \iff \mu u = B(u),$$

where $\mu = \lambda^{-1}$. Hence λ is an eigenvalue for problem (9.64) and u a corresponding eigenvector if and only if $\mu = \lambda^{-1}$ is an eigenvalue for problem (9.65), and u is a corresponding eigenvector. This concludes the proof. □

9.8 The eigenvalue problem for the Laplacian

Let us consider the eigenvalue problem (9.63) introduced in Section 9.7, in the particular case where the operator A is the Laplacian Δ,

$$\begin{cases} -\Delta w = \lambda w & \text{in } \Omega, \\ w = 0 & \text{on } \partial\Omega. \end{cases} \tag{9.71}$$

We now prove some properties of a corresponding orthonormal basis which will be used intensively in the next chapter. This basis plays an essential role in the study of variational evolution problems.

Theorem 9.26. *Let Ω be a bounded open subset of $\mathbb{R}^N$ and denote by $\{\lambda_n\}$ the nondecreasing sequence of the eigenvalues of problem* (9.71) *given by Theorem 9.25,*

$$0 \le \lambda_1 \le \lambda_2 \le \ldots \to +\infty,$$

where each eigenvalue is repeated as many times as the dimension of its corresponding eigenspace. If $\{w_n\}$ is a corresponding sequence of eigenvectors forming a complete orthonormal system in $L^2(\Omega)$, then, $\{w_n\}$ is an orthogonal basis of $H_0^1(\Omega)$.

Proof. The variational formulation (9.64) for problem (9.71) implies that for every $j \in \mathbb{N}$,

$$(v, w_j)_{H_0^1(\Omega)} = \lambda_j (v, w_j)_{L^2(\Omega)}, \quad \forall v \in H_0^1(\Omega). \tag{9.72}$$

Since $\{w_n\}$ is orthonormal in $L^2(\Omega)$ and $\lambda_j \neq 0$ for every $j \in \mathbb{N}$, one has

$$\begin{aligned} &(w_i, w_j)_{H_0^1(\Omega)} = 0, \text{ for } i \neq j, \\ &\|w_j\|_{H_0^1(\Omega)} = \sqrt{\lambda_j}, \text{ for every } j \in \mathbb{N}. \end{aligned} \tag{9.73}$$

Consequently

$$v = \sum_{j=1}^{\infty} (v, w_j)_{L^2(\Omega)} w_j = \sum_{j=1}^{\infty} \frac{(v, w_j)_{H_0^1(\Omega)}}{\|w_j\|^2_{H_0^1(\Omega)}}, \quad \forall v \in H_0^1(\Omega). \tag{9.74}$$

This, together with (9.73) and Definition 6.18, prove that $\{w_j\}$ is an orthogonal basis in $H_0^1(\Omega)$. □

We now introduce the following finite-dimensional spaces:

Definition 9.27. *Let $\{w_n\}$ be an orthonormal basis given by Theorem 9.26. For every $n \in \mathbb{N}$, we denote by W_n the n-dimensional subspace of $L^2(\Omega)$ spanned by the set $\{w_1, \dots, w_n\}$.*

Remark 9.28. *Notice that, thanks to Theorem 9.26, W_n is also an n-dimensional subspace of $H_0^1(\Omega)$.*

In the next proposition, we consider the orthogonal projection P_n given by Definition 6.18, from $L^2(\Omega)$ to W_n, and the projection P'_n defined by (6.24) in Corollary 6.23, from $H_0^1(\Omega)$ to W_n.

Proposition 9.29.

(1) Let P_n be the orthogonal projection

$$P_n : v \in L^2(\Omega) \to P_n(v) \doteq \sum_{j=1}^{n} (v, w_j)_{L^2(\Omega)}\, w_j \in W_n. \tag{9.75}$$

Then

$$\|P_n\|_{\mathcal{L}(L^2(\Omega), L^2(\Omega))} = 1. \tag{9.76}$$

(2) The projection P'_n from $H_0^1(\Omega)$ on W_n coincides with the restriction of P_n to $H_0^1(\Omega)$, defined as

$$P_n(v) = \sum_{j=1}^{n} (v, w_j)_{L^2(\Omega)}\, w_j \quad \textit{for all } v \in H_0^1(\Omega). \tag{9.77}$$

It belongs to $\mathcal{L}(H_0^1(\Omega), H_0^1(\Omega))$ *and*

$$\|P_n\|_{\mathcal{L}(H_0^1(\Omega),H_0^1(\Omega))} = 1. \tag{9.78}$$

(3) Moreover, if we extend the operator P_n *to* $H^{-1}(\Omega)$ *by setting*

$$P_n(v) = \sum_{j=1}^{n} \langle v, w_j \rangle_{H^{-1}(\Omega),H_0^1(\Omega)}\, w_j \quad \textit{for all } v \in H^{-1}(\Omega), \tag{9.79}$$

then $P_n \in \mathcal{L}(H^{-1}(\Omega), H^{-1}(\Omega))$ *and*

$$\|P_n\|_{\mathcal{L}(H^{-1}(\Omega),H^{-1}(\Omega))} = 1. \tag{9.80}$$

Proof.
(1) Inequality (6.19) gives

$$\|P_n\|_{\mathcal{L}(L^2(\Omega),L^2(\Omega))} = \sup_{v \in L^2(\Omega)\setminus\{0\}} \frac{\|P_n(v)\|_{L^2(\Omega)}}{\|v\|_L^2(\Omega)} \leq 1.$$

Since (9.75) implies

$$P_n(w_j) = w_j \qquad \text{for } w_j \in W_n, \tag{9.81}$$

it follows that for $v = w_j$ the quotient above is equal to 1. Then the supremum is attained and is equal to 1, which proves (9.76).
(2) Using (9.73), we derive that

$$P_n'(v) = \sum_{j=1}^{n} \frac{(v, w_j)_{H_0^1(\Omega)}}{\|w_j\|^2_{H_0^1(\Omega)}} = \sum_{j=1}^{n} \frac{\lambda_j (v, w_j)_{L^2(\Omega)}}{\lambda_j}\, w_j = P_n(v), \quad \forall v \in H_0^1(\Omega).$$

Moreover, in view of (9.77) and using again (9.73) as well as (9.74),

$$\begin{aligned}
\|P_n(v)\|^2_{H_0^1(\Omega)} &= \Big(\sum_{i=1}^{n} (v, w_i)_{L^2(\Omega)}\, w_i, \sum_{j=1}^{n} (v, w_j)_{L^2(\Omega)}\, w_j\Big)_{H_0^1(\Omega)} \\
&= \sum_{j=1}^{n} (v, w_j)^2_{L^2(\Omega)} \|\nabla w_j\|^2_{L^2(\Omega)} \leq \sum_{j=1}^{\infty} (v, w_j)^2_{L^2(\Omega)} \|\nabla w_j\|^2_{L^2(\Omega)} \\
&= \|\nabla v\|_{L^2(\Omega)} = \|v\|_{H_0^1(\Omega)},
\end{aligned}$$

which together with (9.81) shows (9.78).

(3) Finally, in view of Remark 7.67, observe that (9.79) is an extension of P_n to $H^{-1}(\Omega)$. Let us we apply $P_n(v)$ for $v \in H^{-1}(\Omega))$, to some $u \in H_0^1(\Omega)$,

$$\begin{aligned}
\left|\langle P_n(v), u\rangle_{H^{-1}(\Omega)), H_0^1(\Omega)}\right| &= \left|\sum_{j=1}^{n} \langle v, w_j\rangle_{H^{-1}(\Omega), H_0^1(\Omega)} (u, w_j)_{L^2(\Omega)}\right| \\
&= \left|\left\langle v, \sum_{j=1}^{n} (u, w_j)_{L^2(\Omega)} w_j \right\rangle_{H^{-1}(\Omega)), H_0^1(\Omega)}\right| \\
&= \left|\langle v, P_n(u)\rangle_{H^{-1}(\Omega), H_0^1(\Omega)}\right| \\
&\leq \|v\|_{H^{-1}(\Omega)} \|u\|_{H_0^1(\Omega)},
\end{aligned}$$

where we used inequality (9.78). Hence,

$$\|P_n(v)\|_{H^{-1}(\Omega)} \leq \|v\|_{H^{-1}(\Omega)}.$$

Using again (9.81), this concludes the proof. □

Remark 9.30. *A last important result needed for the applications to evolution problems concerns the following strong convergences:*

$$\begin{aligned}
&P_n v \to v \quad \textit{strongly in } L^2(\Omega) \quad \textit{for all } v \in L^2(\Omega), \\
&P_n v = P_n'(v) \to v \quad \textit{strongly in } H_0^1(\Omega) \quad \textit{for all } v \in H_0^1(\Omega).
\end{aligned} \tag{9.82}$$

These follow directly from Corollary 6.23 and the properties of P_n given in Proposition 9.29.

In the following (in a natural way) we keep the same notation P_n for the projection (9.75), *for its restriction* (9.77) *as well for its extension* (9.79).

9.9 Regularity

In the study of partial differential equations, regularity plays an important role. The question is whether additional assumptions on the data imply additional regularity on weak solutions, which does not necessarily imply that they are classical solutions. There exist many different results in the literature and in general they require difficult and technical proofs, including several preliminary results that exceed the purpose of this book.

To give one idea of the kind of results that can be obtained, we present here three results of varying nature, which for simplicity will be stated in the case of Dirichlet boundary conditions.

The first result is now a classical one (see e.g. [4], [16], or [40]) and provides the L^2-integrability of second or higher order derivatives for the solution.

Theorem 9.31. *Under the assumptions of Theorem 9.12, let* u *be the solution of problem (9.21). Suppose further that for some* $k \in \mathbb{N}$*, the boundary* $\partial\Omega$ *is of class* C^{k+2}*,* $a_{i,j} \in C^{k+1}(\overline{\Omega})$ *and* $f \in H^k(\Omega)$*, where* $H^k(\Omega)$ *is the Sobolev space defined in Remark 7.27. Then,* u *belongs to* $H^{k+2}(\Omega)$ *and there exists a positive constant* C*, depending on* α, β, k*, and* Ω*, such that*

$$\|u\|_{H^{k+2}(\Omega)} \leq C\|f\|_{H^k(\Omega)}.$$

The second result we give here was originally proved in the 60's by N. G. Meyers (see [26]) and states conditions in order to have higher integrability for the gradient of the solution. Let us recall that $W_0^{1,p}(\Omega) \subset H_0^1(\Omega)$ (see Definition 7.64).

Theorem 9.32. *[17], [26] Let* Ω *be a bounded connected open set of* R^N*,* $N \geq 2$*, with a Lipschitz-continuous boundary* $\partial\Omega$ *and let* A *be a matrix field in* $M(\alpha, \beta, \Omega)$*.*

Under these assumptions, there exists a real number $p_0 > 2$*, dependent on* A *and* Ω*, such that for every* $f \in W^{-1,p}(\Omega)$ *with* $p \in [2, p_0)$ *the weak solution* u *of problem (9.21) belongs to* $W_0^{1,p}(\Omega)$*. Further, there exists a constant* c*, depending on* A*,* Ω*, and* p *(and independent of* f*) such that*

$$\|u\|_{W_0^{1,p}(\Omega)} \leq c\|f\|_{W^{-1,p}(\Omega)}.$$

Remark 9.33.

(1) This result was proved in [26] for the case where the boundary $\partial\Omega$ *is of class* C^2*. It has been successively extended in [17] to the case of a Lipschitz-continuous boundary.*

(2) Observe that Theorem 9.32 applies in particular, to the case where f *is of the form*

$$f = g + div\ h,$$

with $g \in L^{\frac{Np}{N+p}}(\Omega)$ *and* $h \in (L^p(\Omega))^N$*, since the data* f *belongs to* $W^{-1,p}(\Omega)$*. Indeed, let* $v \in W_0^{1,p'}(\Omega)$*, where* p' *is the conjugate of* p*. Then in view of the Sobolev continuous embedding given by Theorem 8.4, the function* v *belong to* $L^{(p')^*}(\Omega)$*, where* $\frac{1}{(p')^*} = \frac{1}{p'} - \frac{1}{N}$*. This gives*

$$(p')^* = \frac{Np}{Np - N - p} = \left(\frac{Np}{N+p}\right)',$$

so $(p')^*$ *is exactly the conjugate of* $\frac{Np}{N+p}$. *Consequently,*

$$\left|\langle g+div\ h,\ v\rangle_{W^{-1,p}(\Omega),W_0^{1,p'}(\Omega)}\right| = \left|\int_\Omega g\,v\,dx - \int_\Omega h\nabla v\,dx\right|$$

$$\leq \|g\|_{L^{\frac{Np}{N+p}}(\Omega)}\|v\|_{L^{[\frac{Np}{N+p}]'}(\Omega)} + \|h\|_{(L^p(\Omega))^N}\|\nabla v\|_{L^{p'}(\Omega)}$$

$$\leq c\|g\|_{L^{\frac{Np}{N+p}}(\Omega)}\|\nabla v\|_{L^{p'}(\Omega)} + \|h\|_{(L^p(\Omega))^N}\|\nabla v\|_{L^{p'}(\Omega)},$$

which, using (7.58), implies that $g + div\ h \in W^{-1,p}(\Omega)$.

The third result was proved in the late 50's independently by E. De Giorgi [11] and J. Nash [28] and further extended to more general equations, like for example, quasilinear equations. It is a very difficult and powerful result, which discusses whether solutions in the sense of distributions (hence independent on the boundary conditions) are locally Hölder continuous (see Section 3.6). For a detailed proof of the theorem, see for instance [19], [22] or [42].

Theorem 9.34. *[11], [28] Let* A *be a matrix field in* $M(\alpha,\beta,\Omega)$ *and suppose that for some* $q > N$, $g \in L^{\frac{q}{2}}(\Omega)$ *and* $h \in (L^q(\Omega))^N$. *Let* u *be a solution of*

$$-div\ (A\,\nabla u) = g + div\ h \quad \text{in} \quad \mathcal{D}'(\Omega),$$

in the distribution sense, that is,

$$\int_\Omega A\nabla u\,\nabla\varphi\,dx = \int_\Omega g\varphi\,dx - \int_\Omega h\nabla\varphi\,dx, \quad \forall\varphi \in \mathcal{D}(\Omega).$$

Then $u \in C^{0,\alpha}(\Omega)$ *for some* $\alpha \in (0,1]$.

Chapter 10

Variational Evolution Problems

We now consider time-dependent partial differential equations of second order. We shall prove existence and uniqueness results for two types of linear evolution problems in divergence form, a parabolic problem and a hyperbolic one. To begin with, we shall show how the notions of variational formulation and weak solution can be extended to evolution problems. For simplicity, we only treat the case of Dirichlet boundary conditions. The other boundary conditions, as those considered in Chapter 9 for the elliptic case, can be treated by similar arguments. For general results concerning evolution equations, we refer the reader to [24], [31], [41], and [44] (see also [6], [8], and [16]).

10.1 Setting of the problems

Throughout this chapter, Ω is a bounded open set in $\mathbb{R}^N$, $T > 0$ is a given real number, the variable $x \in \Omega$ represents the space variable, and $t \in (0, T)$ is the time variable. We shall use the classical notation for derivatives of functions depending on x and t, and denote by ∇ the gradient with respect to the space variable and by $'$ the derivative with respect to time.

In Section 10.3.1 we consider the weak formulation of the following parabolic problem:

$$\begin{cases} u'(x,t) - \operatorname{div}\big(A(x)\,\nabla u(x,t)\big) = f(x,t) & \text{in } \Omega \times (0,T), \\ u(x,t) = 0 & \text{on } \partial\Omega \times (0,T), \\ u(x,0) = u^0(x) & \text{in } \Omega, \end{cases} \tag{10.1}$$

where A is a matrix field on Ω.

This equation is the heat equation (1.9) for the general (non isotropic) case with $A(x)$ in place of $k(x)$. As mentioned in Section 1.4, equation

(10.1) models several phenomena, as in particular, the heat diffusion in a body occupying a domain Ω, the solution u representing the temperature. Let us mention that the Black–Scholes equation (see Section 1.4.3), widely used in financial mathematics, is a heat-type equation.

In Section 10.3.1 we consider the weak formulation of the hyperbolic problem (see equation (1.16)),

$$\begin{cases} u''(x,t) - \operatorname{div}\big(A(x)\,\nabla u(x,t)\big) = f(x,t) & \text{in } \Omega \times (0,T), \\ u(x,t) = 0 & \text{on } \partial\Omega \times (0,T), \\ u(x,0) = u^0(x) & \text{in } \Omega, \\ u'(x,0) = u^1(x) & \text{in } \Omega, \end{cases} \tag{10.2}$$

which describes the wave propagation in a body occupying the domain Ω.

To find solutions of the parabolic and hyperbolic problems, we need to define special spaces of functions, called vector-valued functions, depending on the two variables x and t. In the time variable t, these functions have values in a Banach space of functions in x, namely in Sobolev spaces for the applications we have in view. This is the object of Section 10.2 below. As seen in the next sections, this is an appropriate framework for defining variational formulations for problems (10.1) and (10.2) and for their study.

Existence and uniqueness results for both problems (Section 10.4.1 for (10.1) and Section 10.4.2 for (10.2)) are established by using the powerful Galerkin method (also known as Faedo–Galerkin method, see [24]), widely used nowadays for almost all types of PDEs, linear or nonlinear, time-dependent or not. It is an approximation method, nowadays also related to computational methods in PDEs and is actually the basis of the finite elements method.

Suppose we have to solve a PDE in a infinite dimensional function space H. The Galerkin method is based on the idea to approximate H by an increasing sequence of finite dimensional subspaces H_m of H, and to define a sequence of corresponding approximate problems in H_m (much easier to treat in general, since finite dimensional). The method consists of solving these approximate problems and afterwards, passing to the limit as $m \to \infty$. To do so, what is essential is to obtain estimates independent of m for the approximate solutions, allowing to show that they converge to some limit, which actually is proved to be the solution of the original problem.

We start in Section 10.2 by introducing a functional Hilbertian framework, well-adapted to apply the Galerkin method. The properties of separable Hilbert spaces discussed in Section 6.2, namely the fact that they possess orthogonal bases (see Theorem 6.20), will play an essential role in the construction of the approximate problems in finite-dimensional spaces, the first basic ingredient of the Galerkin procedure. This construction is done in the first steps of the existence and uniqueness theorems related to our problems (10.1) and (10.2) (see Sections 10.4.1 and 10.4.2).

Finally, let us point out that there are other techniques to treat evolution problems. For the parabolic case let us mention the method based on the semigroup theory (also called the Hille–Yoshida theory). We refer the reader to [6] and [31] for details.

10.2 Some classes of vector-valued functions

In this section, we introduce some classes of spaces of vector-valued functions and give their main properties that will be needed for our existence and uniqueness results. We only consider here vector-valued functions defined on a bounded interval I of $\mathbb{R}$ of the form $I = (0,T)$ for $T \in \mathbb{R}^+$, which is sufficient for our aim, although one can take any interval $I = (a,b)$ without modifications.

We only state theorems without proofs. For various results and details on vector-valued functions and for the proofs of the results stated below, we refer the reader to [12], [13], and [44] (see also [23], [24], and [35]). Let us begin by recalling some definitions.

Definition 10.1. *Let X be a Banach space. A function $f : (0,T) \to X$ is a step function if f is measurable and takes a finite number of values in X.*

Remark 10.2. *By definition, a step function $f : (0,T) \to X$ can uniquely be written as a finite linear combination of characteristic functions of intervals, that is under the form*

$$f(t) = \sum_{k=1}^{m} \alpha_k \chi_{I_k}(t),$$

with $m \in \mathbb{N}$, $\alpha_k \in X$. Here I_k, $k \in \{1,\dots,m\}$, are pairwise disjoint subintervals of $(0,T)$ and for every k the characteristic function of I_k is defined by

$$\chi_{I_k}(t) = \begin{cases} 1 & \text{if } t \in I_k, \\ 0 & \text{if } t \in \mathbb{R} \setminus I_k. \end{cases} \tag{10.3}$$

The notion of measurability for real-valued functions may be defined in several equivalent ways, which is not true for vector-valued functions (see for instance [13]). For our purpose, we use the following:

Definition 10.3. *Let X be a Banach space. We say that $u : (0,T) \to X$ is a measurable function if u is almost everywhere the limit of step functions.*

With this definition, one can introduce the L^p-spaces for vector-valued functions.

Definition 10.4. *Let X be a Banach space and $p \in [1,\infty]$. We denote by $L^p(0,T;\, X)$ the set of measurable functions $u : t \in (0,T) \to u(t) \in X$ such that $\|u(\cdot)\|_X \in L^p((0,T))$.*

The space $L^p(0,T;\, X)$ is a Banach space. It also has properties similar to the L^p-spaces of real-valued functions.

Proposition 10.5. *Let X be a Banach space and $1 \leq p \leq \infty$. Then $L^p(0,T;\, X)$ is a Banach space for the norm*

$$\|u\|_{L^p(0,T;X)} = \begin{cases} \left[\displaystyle\int_0^T \|u(t)\|_X^p \, dt \right]^{\frac{1}{p}} & \text{if } p < +\infty, \\ \inf \{ \, C \mid \|u(t)\|_X \leq C \;\; \text{a.e. on } (0,T) \} & \text{if } p = +\infty. \end{cases}$$

If X is reflexive and $1 < p < \infty$, the space $L^p(0,T;\, X)$ is also reflexive. If X is separable and $1 \leq p < \infty$, then $L^p(0,T;\, X)$ is separable.

Moreover, if $p \in [1,+\infty)$ and p' is its conjugate, then

$$\left(L^p(0,T;\, X)\right)' = L^{p'}(0,T;\, X')$$

and, if $u \in L^{p'}(0,T;\, X')$ and $v \in L^p(0,T;\, X)$,

$$\langle u, v\rangle_{L^{p'}(0,T;\, X'),L^p(0,T;\, X)} = \int_0^T \langle u(t), v(t)\rangle_{X',X} dt. \tag{10.4}$$

If in particular $p = 2$ and X is a Hilbert space, then $L^2(0,T;\, X)$ is a Hilbert space with respect to the scalar product

$$(u,v)_{L^2(0,T;\, X)} = \int_0^T (u(t), v(t))_X dt, \quad \text{for every } u, v \in L^2(0,T;\, X).$$

Remark 10.6. *From Definition 10.4, it follows that if X and X_1 are two Banach spaces such that $X \subset X_1$ with continuous embedding (see Definition 8.1), then $L^p(0,T;\, X) \subset L^p(0,T;\, X_1)$ is also a continuous embedding. It is easy to verify (by choosing $X = X_1 = \mathbb{R}$) that this property does not extend to compact embeddings. To show that an inclusion is compact, one needs some additional assumptions, for instance, an information on the derivative with respect to time (see Proposition 10.14).*

The notion of distribution can be generalized to vector-valued functions on I as follows:

Definition 10.7. *Let X be a Banach space. A distribution on $(0,T)$ with values in X is a map $T : \mathcal{D}((0,T)) \longmapsto X$ satisfying*
(1) T is linear,
(2) if $\varphi_n \to \varphi$ in $\mathcal{D}((0,T))$, then $T(\varphi_n) \to T(\varphi)$ in X.

We denote by $\mathcal{D}'(0,T;\ X)$ the set of all distributions on $(0,T)$ with values in X. The derivative T' of a distribution T on $(0,T)$ with respect to the time variable t, is defined as

$$T'(\varphi) = - T(\varphi'), \quad \forall \varphi \in \mathcal{D}(a,b). \tag{10.5}$$

In the following, we restrict our discussion to vector-valued functions defined on an open interval $(0,T) \subset \mathbb{R}$, which is what we need in this chapter.

Remark 10.8. *As in Chapter 7 (see Remarks 7.9 and 7.18), we can identify functions in $L^p(0,T;\ X)$ with distributions on $(0,T)$ with values in X. In particular if u is in $L^p(0,T;\ X)$, and $v \in X'$, then the derivative u' with respect to time is a distribution in $\mathcal{D}'(0,T;\ X)$ and verifies*

$$\langle u'(\varphi)\ v\rangle_{X',X} = -\int_0^T \langle u'(\,\cdot\,,t), v\rangle_{X',X}\varphi'\,dt, \quad \forall \varphi \in \mathcal{D}(0,T).$$

Definition 10.9. *The set of continuous functions $u : t \in [0,T] \to u(t) \in X$, when endowed with the norm*

$$\|u\|_{C^0([0,T],\ X)} = \max_{[0,T]} \|u(t)\|_X,$$

is denoted $C^0([0,T],\ X)$.

We now turn our attention to the spaces needed in the evolution problems mentioned at the beginning of this chapter. We will consider the following cases:

$$X = H_0^1(\Omega), \quad X = L^2(\Omega), \quad X = H^{-1}(\Omega).$$

Since X is here a space of functions defined on Ω, an element v of $L^p(0,T;\ X)$ can be regarded as a function of the ordered pair $(x,t) \in \Omega \times (0,T)$ and identified with the function $v_1(x,t)$ defined by

$$v_1(x,t) = v(t)(x) \qquad \text{for a.e. } (x,t) \in \Omega \times (0,T). \tag{10.6}$$

This point of view is well-adapted to evolution problems, where the space and time variables play a different role and the involved functions have different properties with respect to x and t. We shall systematically use this identification throughout this chapter.

In particular, for $p = 2$, $X = L^2(\Omega)$ and v given by (10.6), we have the following result:

Proposition 10.10. *The space* $L^2(0, T; L^2(\Omega))$ *can be identified with the space* $L^2(\Omega \times (0, T))$ *via the map*

$$v \in L^2(0, T; L^2(\Omega)) \longmapsto v_1 \in L^2(\Omega \times (0, T)),$$

with v_1 *given by* (10.6), *which is an isometry.*

Definition 10.11. *The tensor product of the spaces* $H_0^1(\Omega)$ *and* $L^2(0, T)$, *denoted* $H_0^1(\Omega) \otimes L^2(0, T)$, *is the subspace of* $L^2(0, T;\ H_0^1(\Omega))$ *consisting of the functions* $v : \Omega \times (0, T) \to \mathbb{R}$ *of the form*

$$v(x, t) = \sum_{i=1}^{m} v_i(x) h_i(t), \qquad \textit{for a.e. } (x, t) \in \Omega \times (0, T),$$

where $m \in \mathbb{N}$, $v_i \in H_0^1(\Omega)$, *and* $h_i \in L^2(0, T)$ *for any* $i \in \{1, \ldots, m\}$.

Theorem 10.12. *The tensor product* $H_0^1(\Omega) \otimes L^2(0, T)$ *is dense in the space* $L^2(0, T;\ H_0^1(\Omega))$.

Definition 10.13. *We define the spaces* $\mathcal{W}$, $\mathcal{W}_1$ *and* $\mathcal{W}_2$ *as follows:*

$$\begin{aligned}
\mathcal{W} &= \{v \mid v \in L^2(0, T;\ H_0^1(\Omega)),\ v' \in L^2(0, T;\ H^{-1}(\Omega))\}, \\
\mathcal{W}_1 &= \{v \mid v \in L^2(0, T;\ L^2(\Omega)),\ v' \in L^2(0, T;\ H^{-1}(\Omega))\}, \\
\mathcal{W}_2 &= \{v \mid v \in L^2(0, T;\ H_0^1(\Omega)),\ v' \in L^2(0, T;\ L^2(\Omega))\},
\end{aligned}$$

where the time-derivative is taken in the distributional sense according to Definition 10.7.

The next proposition gives the main properties of these spaces, for the proof we refer the reader to [23] and [24].

Proposition 10.14.
(1) The spaces $\mathcal{W}$, $\mathcal{W}_1$ *and* $\mathcal{W}_2$ *are Banach spaces for the norm of the graph defined respectively, by*

$$\begin{aligned}
\|v\|_{\mathcal{W}} &= \|v\|_{L^2(0,T;\ H_0^1(\Omega))} + \|v'\|_{L^2(0,T;\ H^{-1}(\Omega))}, \\
\|v\|_{\mathcal{W}_1} &= \|v\|_{L^2(0,T;L^2(\Omega))} + \|v'\|_{L^2(0,T;H^{-1}(\Omega))}, \\
\|v\|_{\mathcal{W}_2} &= \|v\|_{L^2(0,T;\ H_0^1(\Omega))} + \|v'\|_{L^2(0,T;\ L^2(\Omega))}.
\end{aligned}$$

(2) The following inclusions are continuous:

$$\mathcal{W} \subset C^0([0,T], L^2(\Omega)),$$
$$\mathcal{W}_1 \subset C^0([0,T], H^{-1}(\Omega)).$$

(3) The following inclusions are compact:

$$\mathcal{W} \subset L^2(0,T;\ L^2(\Omega)),$$
$$\mathcal{W}_1 \subset L^2(0,T;\ H^{-1}(\Omega)).$$

(4) For u and v in $\mathcal{W}$, the following differentiation formula holds:

$$\begin{aligned} \frac{d}{dt}\int_\Omega u(x,t)v(x,t)\,dx &= \langle u'(\,\cdot\,,t), v(\,\cdot\,,t)\rangle_{H^{-1}(\Omega),H_0^1(\Omega)} \\ &\quad + \langle v'(\,\cdot\,,t), u(\,\cdot\,,t)\rangle_{H^{-1}(\Omega),H_0^1(\Omega)}. \end{aligned} \tag{10.7}$$

Remark 10.15. *We end this section by giving an interesting property of the weak convergence of the time derivative of a weak converging sequence, which will be used in the proofs of the existence theorems given in the next sections.*

This property states that if $\{u_n\}$ is bounded sequence in $\mathcal{W}$ such that

$$u_n \rightharpoonup u \quad \text{weakly in } L^2(0,T;\ H_0^1(\Omega)) \tag{10.8}$$

(see Definition 6.66), then

$$u_n' \rightharpoonup u' \quad \text{weakly in } L^2(0,T;\ H^{-1}(\Omega)). \tag{10.9}$$

To show this, observe that using the differentiation formulas (10.7) *and* (10.4)*, it follows that*

$$\langle u', v\varphi\rangle_{L^2(0,T;\,H^{-1}(\Omega)),L^2(0,T;\,H_0^1(\Omega))} = -\langle u, v\varphi'\rangle_{L^2(0,T;\,L^2(\Omega)),L^2(0,T;\,L^2(\Omega))},$$

for every $v \in H_0^1(\Omega)$ and $\varphi \in \mathcal{D}(0,T)$, and, by density (see Theorem 6.62), for every $v \in H_0^1(\Omega)$ and $\varphi \in L^2(0,T)$. Then, passing to the limit in the right-hand side of the above equality and using (10.8) *gives:*

$$\langle u_n', v\varphi\rangle_{L^2(0,T;\,H^{-1}(\Omega)),L^2(0,T;\,H_0^1(\Omega))} \to \langle z, v\varphi\rangle_{L^2(0,T;\,H^{-1}(\Omega)),L^2(0,T;\,H_0^1(\Omega))}.$$

By the density result given in Theorem 10.12, this implies that

$$\langle u_n', \psi\rangle_{L^2(0,T;\,H^{-1}(\Omega)),L^2(0,T;\,H_0^1(\Omega))} \to \langle z, \psi\rangle_{L^2(0,T;\,H^{-1}(\Omega)),L^2(0,T;\,H_0^1(\Omega))},$$

for every $\psi \in L^2(0,T;\ H_0^1(\Omega))$, which is exactly (10.9).

By a similar argument, one can show that if $\{u_n\}$ is bounded sequence in $\mathcal{W}_2$ and (10.8) *holds, then the convergence in* (10.8) *takes place in $L^2(0,T;\ L^2(\Omega))$. Moreover,* (10.9) *is still true if $\{u_n\}$ is a bounded sequence in $\mathcal{W}_1$ and the convergence in* (10.8) *takes place in $L^2(0,T;\ L^2(\Omega))$.*

10.3 Variational weak formulations

10.3.1 *Variational formulation of the heat equation*

We define the variational formulation of problem (10.1) as follows:

$$\begin{cases} \text{Find } u \in \mathcal{W} \text{ such that} \\ \langle u'(\cdot\,,t), v\rangle_{H^{-1}(\Omega),H_0^1(\Omega)} + \displaystyle\int_\Omega A(x)\nabla u(x,t)\,\nabla v(x)\,dx \\ \qquad\qquad = \displaystyle\int_\Omega f(x,t)\,v(x)\,dx, \quad \text{a.e. } t\in(0,T), \quad \forall v\in H_0^1(\Omega), \\ u(x,0) = u^0(x) \quad \text{in } \Omega, \end{cases} \tag{10.10}$$

with $\mathcal{W}$ given by Definition 10.13 where we used the identification (10.6).

A solution of (10.10) is called a *weak solution* of (10.1) (or simply a solution). As in the elliptic case, this is justified because one can show that a weak solution is a classical one when all the data are sufficiently smooth.

We shall prove in Section 10.4.1 the existence and uniqueness of a weak solution of problem (10.1) under the following assumptions on the data:

$$\begin{cases} \text{(i)} & A \in M(\alpha,\beta,\Omega), \\ \text{(ii)} & f \in L^2(\Omega\times(0,T)), \\ \text{(iii)} & u^0 \in L^2(\Omega), \end{cases} \tag{10.11}$$

where $0<\alpha<\beta$, and $M(\alpha,\beta,\Omega)$ is given by Definition 9.2.

The equation in the variational formulation (10.10) is formally obtained by multiplying equation (10.1) by a smooth function v depending on x and vanishing on the boundary. When doing that, one gets the equation

$$\begin{aligned} \langle u'(\cdot\,,t), v\rangle_{H^{-1}(\Omega),H_0^1(\Omega)} - \langle \operatorname{div}\big(A\,\nabla u(\cdot\,,t)\big), v\rangle_{H^{-1}(\Omega),H_0^1(\Omega)} \\ = \int_\Omega f(x,t)\,v(x)\,dx. \end{aligned} \tag{10.12}$$

Then (10.10) is derived by observing that, thanks to (10.4), one can write

$$-\langle \operatorname{div}\big(A\,\nabla u(\cdot\,,t)\big), v\rangle_{H^{-1}(\Omega),H_0^1(\Omega)} = \int_\Omega A(x)\nabla u(x,t)\,\nabla v(x)\,dx.$$

With the notation of Chapter 9, we also have

$$\int_\Omega A(x)\nabla u(x,t)\,\nabla v(x)\,dx = a(u(\cdot\,,t),v), \tag{10.13}$$

where the form a was defined as

$$a(w,v) = \int_\Omega A(x)\nabla w\,\nabla v\,dx \quad \text{for} \quad w,v \in H_0^1(\Omega)\times H_0^1(\Omega). \tag{10.14}$$

We showed in the proof of Theorem 9.12 that the map $v \in H_0^1(\Omega) \mapsto a(uv)$ is continuous on $H_0^1(\Omega)$ and that a is a bilinear form (see Definition 9.4) on the product space $H_0^1(\Omega) \times H_0^1(\Omega)$.

The next step is to choose an appropriate function space for the solutions. For evolution problems, this is generally done for any fixed t, the natural choice being a vector-valued function space.

In our case, this space is $\mathcal{W}$ which was introduced in Definition 10.13,

$$\mathcal{W} \ = \big\{v \mid v \in L^2(0,T;\ H_0^1(\Omega)),\ v' \in L^2(0,T;\ H^{-1}(\Omega))\big\}.$$

If u is in $\mathcal{W}$, the duality pairing $\langle u'(\,\cdot\,,t),v\rangle_{H^{-1}(\Omega),H_0^1(\Omega)}$ is well-defined. Moreover, due to Proposition 10.14, any function u in $\mathcal{W}$ belongs to $C^0([0,T],L^2(\Omega))$, and so $u(\,\cdot\,,0)$ is in $L^2(\Omega)$. Due to (10.11)(iii), this gives a meaning to the initial condition u^0 in (10.10) as an equality between functions in $L^2(\Omega)$. In other words, if $u = u(x,t)$ is a solution of (10.10), this can be seen as

$$\lim_{t\to 0} \|u(\,\cdot\,,t)\|_{L^2(\Omega)} = \|u^0\|_{L^2(\Omega)}.$$

One can define the variational formulation of problem (10.1) in several equivalent ways. Definitions 7.5 and 7.10, as well as Theorem 6.64, imply that the equality in the variational formulation above holds for almost every $t \in (0,T)$ if and only if it holds in the sense of distributions, that is, in $\mathcal{D}'(0,T)$. Observe also that if u belongs to $\mathcal{W}$, then both sides of the equation in (10.10) belong to $L^2(0,T)$.

Consequently, problem (10.10) is equivalent to the following one:

$$\begin{cases} \text{Find } u \in \mathcal{W} \text{ such that} \\ \displaystyle\int_0^T \langle u'(\,\cdot\,,t),v\rangle_{H^{-1}(\Omega),H_0^1(\Omega)}\,\varphi(t)\,dt \\ \qquad\qquad\displaystyle + \int_0^T\!\!\int_\Omega A(x)\nabla u(x,t)\,\nabla v(x)\,\varphi(t)\,dx\,dt \\ \displaystyle\quad = \int_0^T\!\!\int_\Omega f(x,t)\,v(x)\,\varphi(t)\,dx\,dt, \\ \forall v \in H_0^1(\Omega) \ \text{ and } \ \forall\,\varphi \in \mathcal{D}(0,T), \\ u(x,0) = u^0(x) \quad \text{in } \Omega, \end{cases} \tag{10.15}$$

which, in view of Theorem 6.62, is also equivalent to the problem

$$
\begin{cases}
\text{Find } u \in \mathcal{W} \text{ such that} \\
\displaystyle\int_0^T \langle u'(\,\cdot\,,t), v\rangle_{H^{-1}(\Omega),H_0^1(\Omega)}\, h(t)\, dt \\
\qquad\qquad\qquad \displaystyle + \int_0^T\!\!\int_\Omega A(x)\nabla u(x,t)\,\nabla v(x)\, h(t)\, dx\, dt \\
\qquad \displaystyle = \int_0^T\!\!\int_\Omega f(x,t)\, v(x)\, h(t)\, dx\, dt, \\
\forall v \in H_0^1(\Omega) \ \text{ and } \ \forall\, h \in L^2(0,T), \\
u(x,0) = u^0(x) \quad \text{in } \Omega.
\end{cases}
\tag{10.16}
$$

Moreover, we have

Proposition 10.16. *Problem* (10.10), *as well as formulations* (10.15) *and* (10.16), *are equivalent to the following one:*

$$
\begin{cases}
\textit{Find } u \in \mathcal{W} \textit{ such that} \\
\displaystyle\int_0^T \langle u'(\,\cdot\,,t), v(\,\cdot\,,t)\rangle_{H^{-1}(\Omega),H_0^1(\Omega)}\, dt \\
\qquad\qquad\qquad \displaystyle + \int_0^T\!\!\int_\Omega A(x)\nabla u(x,t)\,\nabla v(x,t)\, dx\, dt \\
\qquad \displaystyle = \int_0^T\!\!\int_\Omega f(x,t)\, v(x,t)\, dx\, dt, \quad \forall v \in L^2(0,T;\ H_0^1(\Omega)), \\
u(x,0) = u^0(x) \quad \textit{in } \Omega.
\end{cases}
\tag{10.17}
$$

Proof. As observed above, formulations (10.10), (10.15), and (10.16) are equivalent. Hence it suffices to show that (10.17) is equivalent to (10.16).

If $v \in H_0^1(\Omega)$ and $h \in L^2(0,T)$, then $g(x,t) = v(x)h(t)$ clearly belongs to $L^2(0,T;\ H_0^1(\Omega))$, so that a solution of (10.17) is a solution of (10.16).

Suppose that u verifies (10.16). By linearity, this equality is still true by replacing the products $v(x)h(t)$ by finite sums of the form $\sum_{j=1}^m \varphi_j(t)h_j(x)$, with $h_j \in L^2([0,T])$ and $v_j \in H_0^1(\Omega)$ for every $j = 1,\ldots,m$. Thus, since by Theorem 10.12 the tensor product $H_0^1(\Omega) \otimes L^2(0,T)$ is dense in $L^2(0,T); H_0^1(\Omega))$, the function u verifies (10.17). □

Remark 10.17. *Let us point out that if there exists a solution* $u \in \mathcal{W}$ *for this problem, this last formulation* (10.17) *allows us to choose* $v = u$ *as test function. This fact is essential for the proof of the uniqueness of this solution, as can be seen below in Step 6 of the proof of Theorem 10.20.*

10.3.2 *Variational formulation of the wave equation*

We now turn our attention to the hyperbolic equation (10.2) which describes in particular, the wave propagation in a body occupying the domain Ω.

As we proceeded in the preceding section for the heat equation, we start by writing down its weak formulation, that is

$$\begin{cases} \text{Find } u \in \mathcal{W}_2 \text{ such that} \\ \langle u''(\cdot,t), v\rangle_{H^{-1}(\Omega),H_0^1(\Omega)} + \displaystyle\int_\Omega A(x)\nabla u(x,t)\,\nabla v(x)\,dx \\ \qquad = \displaystyle\int_\Omega f(x,t)\,v(x)\,dx, \quad \forall v \in H_0^1(\Omega), \ \text{ a.e. } t \in (0,T), \\ u(x,0) = u^0(x) \qquad \text{in } \Omega, \\ u'(x,0) = u^1(x) \qquad \text{in } \Omega. \end{cases} \tag{10.18}$$

The space $\mathcal{W}_2$ is given by Definition 10.13,

$$\mathcal{W}_2 = \{ w \mid w \in L^2(0,T;\ H_0^1(\Omega)),\ w' \in L^2(0,T;\ L^2(\Omega)) \},$$

where we used again the identification (10.6).

We shall prove in Section 10.4.2 the existence and uniqueness theorem for problem (10.18) under the following assumptions on the data:

$$\begin{cases} \text{(i)} \quad A \in M(\alpha,\beta,\Omega) \ \text{ and } \ a_{ij} = a_{ji} \quad \text{for } \ 1 \leq, i, j, \leq N, \\ \text{(ii)} \quad f \in L^2(\Omega \times (0,T)), \\ \text{(iii)} \quad u^0 \in H_0^1(\Omega), \\ \text{(iv)} \quad u^1 \in L^2(\Omega), \end{cases} \tag{10.19}$$

where $M(\alpha,\beta,\Omega)$ is given by Definition 9.2.

Remark 10.18. *Notice that, in comparison with assumptions* (10.11) *on the matrix A made for the heat equation* (10.10), *we assumed here an additional assumption on the matrix field A, namely that it is symmetric. We will see in the proof of the existence and uniqueness of a solution that this assumption is essential. This is striking, in particular, in Steps 3 and 6 of the proof below.*

Remark 10.19. *For simplicity, we assumed here $f \in L^2(\Omega\times]0,T[)$. As can be seen below, via some straightforward modifications in the proof of Theorem 10.24, one can take also $f \in L^2(0,T;\ H^{-1}(\Omega))$ in* (10.19).

Like in the parabolic case, the variational formulation is formally obtained by multiplying the equation in (10.2), for any fixed t, by a smooth function of x vanishing on the boundary, and then integrating by parts. Here, the suitable functional space for solutions is $\mathcal{W}_2$.

Indeed, if $u \in \mathcal{W}_2$ is a solution of (10.18), then

$$u'' = \operatorname{div}\big(A\,\nabla u\big) + f \quad \text{for a.e. } t \in (0,T),$$

in the sense of distributions. Due to assumptions (10.19), u'' is in $L^2(0,T;\ H^{-1}(\Omega))$ and so the duality pairing $\langle u'(\cdot,t), v\rangle_{H^{-1}(\Omega),H_0^1(\Omega)}$ is well-defined. Moreover, since $u \in \mathcal{W}_2 \subset \mathcal{W}$ and $u' \in \mathcal{W}_1$, Proposition 10.14 implies that

$$u \in C^0([0,T], L^2(\Omega)), \quad u' \in C^0([0,T], H^{-1}(\Omega)).$$

Therefore $u(x,0) \in L^2(\Omega)$ and $u'(x,0) \in H^{-1}(\Omega)$ which, thanks to (10.19), give sense to the initial conditions in (10.18) as an equality between functions in $L^2(\Omega)$ and in $H^{-1}(\Omega)$, respectively.

Following the arguments from Section 10.3.1, it is easily seen that formulation (10.18) is equivalent to the following one:

$$\begin{cases} \text{Find } u \in \mathcal{W}_2 \text{ such that} \\ \displaystyle\int_0^T \langle u''(\,\cdot\,,t), v(\,\cdot\,,t)\rangle_{H^{-1}(\Omega),H_0^1(\Omega)}\,dt \\ \displaystyle\qquad\qquad + \int_0^T\!\!\int_\Omega A(x)\nabla u(x,t)\,\nabla v(x,t)\,dx\,dt \\ \displaystyle\quad = \int_0^T\!\!\int_\Omega f(x,t)\,v(x,t)\,dx\,dt, \quad \forall v \in L^2(0,T;\ H_0^1(\Omega)), \\ u(x,0) = u^0(x) \quad \text{in } \Omega, \\ u'(x,0) = u^1(x) \quad \text{in } \Omega. \end{cases} \tag{10.20}$$

10.4 Existence and uniqueness theorems

The proof of the existence and uniqueness of a solution, either for problem (10.1) or for problem (10.2), makes use of the Galerkin method. To start this procedure, the first step is the construction of a sequence of finite-dimensional approximating problems for each of these equations. To do so, we need to introduce the finite-dimensional spaces where the approximating problems will be defined. For this purpose we shall make use of the special basis $\{w_n\}$, orthonormal in $L^2(\Omega)$ and orthogonal in $H_0^1(\Omega)$, given by Theorem 9.26.

Recall that w_n are the eigenvectors of the Laplacian satisfying (9.72),

$$(v, w_j)_{H_0^1(\Omega)} = \lambda_j (v, w_j)_{L^2(\Omega)}, \quad \forall v \in H_0^1(\Omega),$$

(λ_j being the corresponding eigenvalues). The main properties of these basis have been proved in Section 9.8.

We now consider for every $n \in \mathbb{N}$, the n-dimensional space W_n introduced in Definition 9.27, that is the space spanned by $w_1, \ldots, w_n$. It is in this space that the approximating problems will be formulated.

10.4.1 *The heat equation*

Theorem 10.20. *Under assumptions* (10.11), *problem* (10.10) *has a unique solution* $u \in \mathcal{W}$. *Moreover, there exists a constant* c *depending on* α, β *and* Ω, *such that*

$$\|u\|_{\mathcal{W}} + \|u\|_{L^\infty(0,T;\, L^2(\Omega))} \leq c\big(\|f\|_{L^2(\Omega\times(0,T))} + \|u^0\|_{L^2(\Omega)}\big). \tag{10.21}$$

Proof. The proof consists of seven steps. In the first three, we define the approximating problem and prove *a priori* estimates for the approximate solution. In the fourth and fifth steps we pass to the limit to get equation (10.10). In Step 6 we prove (10.21). The uniqueness is treated in Step 7.

Step 1 (Construction of the approximate problem).
Observe first that since $u^0 \in L^2(\Omega)$, from (9.82),

$$u_n^0 \doteq P_n(u^0) \to u^0 \quad \text{strongly in } L^2(\Omega), \tag{10.22}$$

with P_n given by (9.75).

Using the zero extension of f outside $\Omega \times (0,T)$, together with the density Theorem 6.58 and estimate (6.45) from Chapter 6, we deduce the existence of a sequence $\{f_n\} \subset C^\infty(\overline{\Omega \times (0,T)})$, such that

$$f_n \to f \quad \text{strongly in } \; L^2(\Omega \times (0,T)), \tag{10.23}$$

with

$$\|f_n\|_{L^2(\Omega\times(0,T))} \leq \|f\|_{L^2(\Omega\times(0,T))}. \tag{10.24}$$

For every $n \in \mathbb{N}$, the n-dimensional approximate problem in W_n, is defined as follows:

$$\begin{cases} \text{Find } u_n(\,\cdot\,,t) \in W_n \text{ such that} \\[4pt] \langle u_n'(\,\cdot\,,t), v\rangle_{H^{-1}(\Omega),H_0^1(\Omega)} + \displaystyle\int_\Omega A(x)\nabla u_n(x,t)\,\nabla v(x)\,dx \\[4pt] \quad = \displaystyle\int_\Omega f_n(x,t)\,v(x)\,dx, \quad \forall v \in W_n \;\text{ and } t \in (0,T), \\[4pt] u_n(x,0) = u_n^0(x) \quad \text{in } \Omega. \end{cases} \tag{10.25}$$

In view of the linearity of this problem, we are looking for u_n such that

$$u_n(\,\cdot\,,t) = \sum_{j=1}^{n} g_j^n(t) w_j \in W_n \quad \text{with } g_j \in C([0,T]),\ j = 1,\dots,n, \tag{10.26}$$

and satisfying

$$\begin{cases} \displaystyle\int_\Omega u_n'(x,t)\, w_k(x)\, dx + \int_\Omega A(x)\nabla u_n(x,t)\,\nabla w_k(x)\, dx \\ \displaystyle\quad = \int_\Omega f_n(x,t)\, w_k(x)\, dx \quad \text{for } k = 1,\dots,n \quad \text{and } t \in (0,T), \\ u_n(x,0) = u_n^0(x) \quad \text{in } \Omega. \end{cases} \tag{10.27}$$

Observe now that from (9.75) and (10.22), we have

$$\sum_{j=1}^{n} u_n(\,\cdot\,,0) = g_j^n(0) w_j = u_n^0 = \sum_{j=1}^{n} (u^0, w_j)_{L^2(\Omega)}\, w_j, \tag{10.28}$$

or equivalently, since $w_1, \dots, w_n$ are linearly independent,

$$g_k^n(0) = (u^0, w_k)_{L^2(\Omega)} \quad \text{for } k = 1,\dots,n. \tag{10.29}$$

On the other hand,

$$u_n'(\,\cdot\,,t) = \sum_{j=1}^{n} \frac{dg_j^n}{dt}(t)\, w_j, \tag{10.30}$$

so, from identity (10.29) and using the orthonormality of $\{w_n\}$ in $L^2(\Omega)$, it follows that problem (10.27) is equivalent to the following system of n ordinary differential equations of the first order with n unknowns $g_1^n, \dots, g_n^n$:

$$\begin{cases} \displaystyle\frac{dg_k^n}{dt}(t) + \Big[\sum_{j=1}^{n} \int_\Omega A\nabla w_j\, \nabla w_k\, dx\Big]\, g_j^n(t) = \int_\Omega f_n(x,t)\, w_k\, dx, \\ g_k^n(0) = (u^0, w_k) \quad \text{for } k = 1,\dots,n \quad \text{and } t \in (0,T). \end{cases} \tag{10.31}$$

By classical results (see for instance, [10] Chapter 3), one has a maximal local solution $g_k^n \in C^1([0,T_n))$, $T_n > 0\, (k = 1,\dots,n)$ for $0 < T_n \leq T$.

If $T_n < T$, then necessarily,

$$g_k^n(t) \to +\infty \quad \text{as} \quad t \to T_n \ \ (t < T_n) \quad (k = 1,\dots,n). \tag{10.32}$$

Indeed, if $g_k^n(t) \to l$ as $t \to T_n$, then the extension by l of $g_k^n(t)$ on $[T_n, T]$, would be a continuous solution of (10.31) on $[0,T]$, which contradicts the fact that $[0,T_n)$ is the maximal interval of existence of $g_k^n(t)$. We shall see in the next step that (10.32) is impossible, since it will be shown that g_k^n are bounded independently of k and n, which actually implies

$$T_n = T \quad \text{for } n = 1, 2, \dots. \tag{10.33}$$

so that $g_k^n \in C^1([0, T])$ for $k = 1, \dots, n$.

Step 2 (*A priori* estimates of u_n).

In this step we prove *a priori* estimates (independent of n) of the solution u_n. To do so, for a fixed $t \in [0, T_n[$, we multiply the equation in (10.31) by g_k^n. Summing over k from 1 to n and using (10.26) as well as the orthonormality of the basis $\{w_n\}$ in $L^2(\Omega)$, yield

$$\begin{aligned} \int_\Omega u_n'(x,t)\, u_n(x,t)\, dx &+ \int_\Omega A(x) \nabla u_n(x,t)\, \nabla u_n(x,t)\, dx \\ &= \int_\Omega f_n(x,t)\, u_n(x,t)\, dx. \end{aligned} \tag{10.34}$$

Observe first that

$$\begin{aligned} \int_\Omega u_n'(x,t)\, u_n(x,t)\, dx &= \frac{1}{2} \frac{d}{dt} \int_\Omega u_n(x,t)\, u_n(x,t)\, dx \\ &= \frac{1}{2} \frac{d}{dt} \|u_n(\,\cdot\,, t)\|^2_{L^2(\Omega)}. \end{aligned} \tag{10.35}$$

Using this in (10.34) and recalling that A belongs to $M(\alpha, \beta, \Omega)$ and so satisfies (9.27), then by the Cauchy–Schwarz inequality, the Poincaré inequality (Prop. 7.61) in $H_0^1(\Omega)$, and Young's inequality (6.38) for $p = p' = 2$, we have successively,

$$\begin{aligned} \frac{1}{2} \frac{d}{dt} \|u_n(\,\cdot\,, t)\|^2_{L^2(\Omega)} + \alpha \|u_n(\,\cdot\,, t)\|^2_{H_0^1(\Omega)} &\leq \|f_n(\,\cdot\,, t)\|_{L^2(\Omega)} \|u_n(\,\cdot\,, t)\|_{L^2(\Omega)} \\ &\leq C_\Omega \|f_n(\,\cdot\,, t)\|_{L^2(\Omega)} \|u_n(\,\cdot\,, t)\|_{H_0^1(\Omega)} \\ &= \Big(\frac{C_\Omega}{\sqrt{\alpha}} \|f_n(\,\cdot\,, t)\|_{L^2(\Omega)} \Big) \Big(\sqrt{\alpha} \|u_n(\,\cdot\,, t)\|_{H_0^1(\Omega)} \Big) \\ &\leq \frac{C_\Omega^2}{2\alpha} \|f_n(\,\cdot\,, t)\|^2_{L^2(\Omega)} + \frac{\alpha}{2} \|u_n(\,\cdot\,, t)\|^2_{H_0^1(\Omega)}. \end{aligned}$$

Let us now integrate this last inequality over $[0, t]$. Using (10.22), (10.24), and the properties of P_n from Section 9.8, in particular (9.76), we get

$$\begin{aligned} \|u_n(\,\cdot\,, t)\|^2_{L^2(\Omega)} &+ \alpha \int_0^t \|u_n(\,\cdot\,, \tau)\|^2_{H_0^1(\Omega)}\, d\tau \\ &\leq \|u_n^0\|^2_{L^2(\Omega)} + \frac{C_\Omega^2}{\alpha} \int_0^t \|f_n(\,\cdot\,, \tau)\|^2_{L^2(\Omega)}\, d\tau \\ &\leq \|u^0\|^2_{L^2(\Omega)} + \frac{C_\Omega^2}{\alpha} \|f\|^2_{L^2(\Omega \times (0,T))}. \end{aligned} \tag{10.36}$$

Taking into account the expression of u_n and the properties of the orthonormal basis, this gives

$$\|u_n(\,\cdot\,,t)\|^2_{L^2(\Omega)} = \Big\|\sum_{j=1}^{n} g_j^n(t)w_j\Big\|^2_{L^2(\Omega)} = \sum_{j=1}^{n}(g_j^n(t))^2.$$

This estimate shows that (10.33) holds. Hence, u_n is a global and unique solution on $[0,T]$ of (10.27), and by construction belongs to $C^0([0,T];\,W_n)\cap \mathcal{W}$.

On the other hand, since now (10.36) holds for every $t\in[0,T]$, taking the supremum over $[0,T]$ yields

$$u_n \in L^\infty(0,T;\,L^2(\Omega))\cap L^2(0,T;\,H_0^1(\Omega)),$$

with

$$\begin{aligned}\|u_n\|_{L^\infty(0,T;\,L^2(\Omega))} &+ \|u_n\|_{L^2(0,T;\,H_0^1(\Omega))}\\ &\le C_1\big(\|u^0\|_{L^2(\Omega)} + \|f\|_{L^2(\Omega\times(0,T))}\big),\end{aligned}\tag{10.37}$$

where C_1 only depends on α and Ω. This entails that

$$u_n \text{ is in a bounded set of } L^\infty(0,T;\,L^2(\Omega))\cap L^2(0,T;\,H_0^1(\Omega)).\tag{10.38}$$

Step 3 (*A priori* estimates of u_n').
Let $v\in H_0^1(\Omega)$ and recall that in view of Theorem 9.26, the sequence $\{w_j\}$ is an orthogonal basis in $H_0^1(\Omega)$. Then, using Proposition 6.22, for every $n\in\mathbb{N}$, the function v can be decomposed as

$$v = P_n(v) + v_n \quad \text{with } v_n\in W_n^\perp.\tag{10.39}$$

From the approximate problem (10.25), by using (10.13), we have for t in $[0,T]$ and for any $n\in\mathbb{N}$,

$$\begin{aligned}\big(u_n'(\,\cdot\,,t),P_n(v)\big)_{L^2(\Omega)} &= -a(u_n(\,\cdot\,,t),P_n(v))\\ &\quad + \big(f_n(\,\cdot\,,t),P_n(v)\big)_{L^2(\Omega)}.\end{aligned}\tag{10.40}$$

Note now that from the decomposition of v, it is immediate that

$$\langle u_n'(\,\cdot\,,t),v\rangle_{H^{-1}(\Omega),H_0^1(\Omega)} = \big(u_n'(\,\cdot\,,t),P_n(v)\big)_{L^2(\Omega)}$$

Consequently, working out (10.40), we get successively

$$\begin{aligned}&\big|\langle u_n'(\,\cdot\,,t),v\rangle_{H^{-1}(\Omega),H_0^1(\Omega)}\big|\\ &\quad\le \big|\big(f_n(\,\cdot\,,t),P_n(v)\big)_{L^2(\Omega)}\big| + \big|a(u_n(\,\cdot\,,t),P_n(v))\big|\\ &\quad\le \Big(C_\Omega\|f_n(\,\cdot\,,t)\|_{L^2(\Omega)} + \beta\|u_n(\,\cdot\,,t)\|_{H_0^1(\Omega)}\Big)\|P_n(v)\|_{H_0^1(\Omega)}\\ &\quad\le \Big(C_\Omega\|f_n(\,\cdot\,,t)\|_{L^2(\Omega)} + \beta\|u_n(\,\cdot\,,t)\|_{H_0^1(\Omega)}\Big)\|v\|_{H_0^1(\Omega)},\end{aligned}\tag{10.41}$$

for every $v \in H_0^1(\Omega)$, where we used *a priori* estimate (10.38), Remark 9.3(2) for evaluating $A(u_n)$, and estimate (9.76). Estimate (10.41) implies that

$$\|u_n'(\,\cdot\,,t)\|^2_{H^{-1}(\Omega)} \le C_2\Big(\|f_n(\,\cdot\,,t)\|^2_{L^2(\Omega)} + \|u_n(\,\cdot\,,t)\|^2_{H_0^1(\Omega)}\Big),$$

where the constant C_2 only depends on β and Ω. Integrating the last inequality in t over $(0,T)$ and using (10.37), yield

$$\|u_n'\|_{L^2(0,T;H^{-1}(\Omega))} \le C_3(\|u^0\|_{L^2(\Omega)} + \|f\|_{L^2(\Omega\times(0,T))}), \tag{10.42}$$

where the constant C_3 depends on α, β, and Ω and is independent of n. In conclusion,

$$u_n' \text{ is in a bounded set of } L^2(0,T;\, H^{-1}(\Omega)). \tag{10.43}$$

Step 4 (Passing to the limit).
We will see now that the estimates obtained in the preceding step are sufficient for passing to the limit as $n \to \infty$ in (10.25).

To begin with, note first that Proposition 10.5 implies that the space $L^1(0,T;\, L^2(\Omega))$ is separable, with $\big(L^1(0,T;\, L^2(\Omega))\big)' = L^\infty(0,T;\, L^2(\Omega))$, and that $L^2(0,T;\, H_0^1(\Omega))$ and $L^2(0,T;\; H^{-1}(\Omega))$ are reflexive. Recalling (10.38) and (10.43), then by Theorems 6.72 and 6.79 there exists a subsequence of $\{u_n\}$ (still denoted $\{u_n\}$), and a function $u \in \mathcal{W}$ such that

$$\begin{aligned} u_n &\rightharpoonup u \quad \text{weakly* in } L^\infty(0,T;\, L^2(\Omega)),\\ u_n &\rightharpoonup u \quad \text{weakly in } L^2(0,T;\, H_0^1(\Omega)),\\ u_m' &\rightharpoonup u' \quad \text{weakly in } L^2(0,T;\, H^{-1}(\Omega)), \end{aligned} \tag{10.44}$$

where we used Remark 10.15 for the last convergence. Let us now show that the limit u is solution of the heat equation. To do that, let φ in $L^2(0,T)$ and $v \in H_0^1(\Omega)$. For every $n \in \mathbb{N}$, we have $P_n(v) \in W_n$ and from (9.82),

$$P_n(v) \to v \quad \text{strongly in } H_0^1(\Omega). \tag{10.45}$$

Choose $P_n(v)$ as test function in (10.25), and multiply the result by φ. Integrating with respect to t over $(0,T)$ gives

$$\begin{aligned} &\int_0^T \langle u_n'(\,\cdot\,,t), P_n(v)\rangle_{H^{-1}(\Omega),H_0^1(\Omega)}\varphi(t)\,dx\,dt\\ &\qquad + \int_0^T\!\!\int_\Omega A(x)\,\nabla u_n(x,t)\,\nabla(P_n(v))(x)\varphi(t)\,dx\,dt\\ &\qquad = \int_0^T\!\!\int_\Omega f_n(x,t)\,P_n(v)(x)\varphi(t)\,dx\,dt. \end{aligned} \tag{10.46}$$

Using (10.23), (10.44), (10.45), Remark 7.67 and Proposition 6.74, we can pass to the limit in this equality to obtain

$$\begin{aligned}\int_0^T \langle u'(\,\cdot\,,t), v\rangle_{H^{-1}(\Omega),H_0^1(\Omega)}\varphi(t)\,dx\,dt \\ + \int_0^T\int_\Omega A(x)\,\nabla u(x,t)\,\nabla v(x)\varphi(t)\,dx\,dt \\ = \int_0^T\int_\Omega f(x,t)\,v(x)\varphi(t)\;\,dx\,dt,\end{aligned} \tag{10.47}$$

for every $\varphi \in L^2(0,T)$ and $v \in H_0^1(\Omega)$. This formulation is precisely (10.16) which, as shown in Section 10.3, is one of the equivalent variational formulations of problem (10.1).

It remains to prove that u satisfies the initial condition, namely that $u(\,\cdot\,,0) = u^0$ in Ω, which will imply that u is a solution of (10.10). This is the object of the next step.

Step 5 (Recovery of the initial condition).
Let $v \in H_0^1(\Omega)$. We argue as in the preceding step by taking $P_n(v) \in W_n$ as test functions in (10.25) and multiplying the result by a function φ in $C^\infty([0,T])$ with the additional assumption $\varphi(T) = 0$. Then integrating over $(0,T)$, and recalling the differentiation formula (10.7) from Proposition 10.14, we rewrite (10.46) as follows:

$$\begin{aligned}\int_0^T -\langle P_n(v)(x),\, u(\,\cdot\,,t)\rangle_{H^{-1}(\Omega),H_0^1(\Omega)}\varphi'(t)\,dt \\ + \int_0^T\int_\Omega A(x)\,\nabla u_n(x,t)\,\nabla P_n(v)(x)\varphi(t)\,dx\,dt \\ = \int_0^T\int_\Omega f_n(x,t)\,P_n(v)(x)\varphi(t)\;\,dx\,dt + \big(u_n(\,\cdot\,,0), P_n(v)(x)\varphi(0)\big)_{L^2(\Omega)},\end{aligned}$$

where we can pass to the limit according to convergences (10.22), (10.44), (10.23) and (10.45), to get

$$\begin{aligned}\int_0^T -\langle u(\,\cdot\,,t),\, v\rangle_{H^{-1}(\Omega),H_0^1(\Omega)}\varphi'(t)\,dt \\ + \int_0^T\int_\Omega A(x)\,\nabla u(x,t)\,\nabla v(x)\varphi(t)\,dx\,dt \\ = \int_0^T\int_\Omega f(x,t)\,v(x)\varphi(t)\;\,dx\,dt + \big(u^0, v\,\varphi(0)\big)_{L^2(\Omega)},\end{aligned}$$

for any $v \in H_0^1(\Omega)$ and $\varphi \in C^\infty([0,T])$ with $v(T) = 0$.

On the other hand, using again the differentiation formula (10.7), we rewrite (10.47) as follows:

$$\begin{aligned}\int_0^T -\langle u(\,\cdot\,,t),\, v\rangle_{H^{-1}(\Omega),H_0^1(\Omega)}\,\varphi'(t)\,dt \\ +\int_0^T\int_\Omega A(x)\,\nabla u(x,t)\,\nabla v(x)\varphi(t)\,dx\,dt \\ =\int_0^T\int_\Omega f(x,t)\,\nabla v(x)\varphi(t)\;\;dx\;dt+\big(u(0),v\,\varphi(0)\big)_{L^2(\Omega)},\end{aligned}\tag{10.48}$$

for any $v\in H_0^1(\Omega)$ and $\varphi\in C^\infty([0,T])$ with $\varphi(T)=0$. Comparing with the previous equality, the arbitrariness of v and φ (actually of $\varphi(0)$), implies $u(0)=u^0$.

Step 6 (*A priori* estimate (10.21)).

To obtain *a priori* estimate (10.21), we use the *a priori* estimates of u_n obtained in the first two steps. From convergences (10.44), and the lower semi-continuity of the norm with respect to the weak and weak* convergences (stated in Propositions 6.69 and 6.78), we have

$$\begin{aligned}\|u\|_{L^\infty(0,T;\,L^2(\Omega))} &\le \liminf_{n\to\infty}\|u_n\|_{L^\infty(0,T;\,L^2(\Omega))}\,,\\ \|u\|_{L^2(0,T;\,H_0^1(\Omega))} &\le \liminf_{n\to\infty}\|u_n\|_{L^2(0,T;\,H_0^1(\Omega))}\,,\\ \|u'\|_{L^2(0,T;\;H^{-1}(\Omega))} &\le \liminf_{n\to\infty}\|u_n'\|_{L^2(0,T;\,H^{-1}(\Omega))}\,.\end{aligned}$$

Since $\sum\{\liminf\}\le\liminf\{\sum\}$, by using estimates (10.37) and (10.42) we deduce (10.21).

Step 7 (Uniqueness).

To complete the proof of Theorem 10.20, it remains to show that (10.10) has a unique solution. Suppose that u_1 and u_2 are two different solutions of (10.10) corresponding to the same data.

In view of Proposition 10.16 the difference $w=u_1-u_2$ satisfies

$$\begin{cases}\displaystyle\int_0^T\langle w'(\,\cdot\,,t),v(\,\cdot\,,t)\rangle_{H^{-1}(\Omega),H_0^1(\Omega)}\,dt \\ \qquad\qquad\displaystyle+\int_0^T\int_\Omega A(x)\nabla w(x,t)\,\nabla v(x,t)\,dx\,dt=0,\\ \forall v\in L^2(0,T;\;H_0^1(\Omega)),\\ w(x,0)=0\quad\text{in }\Omega.\end{cases}$$

Choosing $v=w$, using formula (10.35) and the ellipticity of A, by Proposition 10.14(4) it follows that

$$\frac{1}{2}\frac{d}{dt}\|w(\,\cdot\,,t)\|^2_{L^2(\Omega)} + \alpha\|w(\,\cdot\,,t)\|^2_{H^1_0(\Omega)} \le 0,$$

which integrated over $(0,t)$ with $t \in (0,T)$, gives

$$\|w(\,\cdot\,,t)\|^2_{L^2(\Omega)} + 2\alpha\int_0^t \|w(\,\cdot\,,\tau)\|^2_{H^1_0(\Omega)}\,d\tau \le 0 \qquad \text{for a.e. } t \in (0,T).$$

This implies that $w \equiv 0$ and hence $u_1 = u_2$ and this ends the proof. □

Remark 10.21. *It should be pointed out that the Galerkin method is actually the version for infinite-dimensional spaces of the method of separation of variables that we used in Chapters 4 and 5 for the one, two, or three dimensional heat and wave equations, respectively.*

Remark 10.22. *In the context of classical PDEs, we listed in Chapter 4 several properties characteristic of parabolic equations. Among them, we mentioned the maximum principle in Theorem 4.1 and the regularizing effect in Remark 4.8. It can be proven that these properties are still true for the general case of the heat equation treated in this chapter. The maximum principle states that the solution u of problem* (10.10) *satisfies*

$$\min\{0, \inf_\Omega u_0\} \le u(x,t) \le \max\{0, \sup_\Omega u_0\}, \quad \forall (x,t) \in \Omega \times (0,T).$$

Concerning the regularizing property, one proves that the solution u is not only in $C^0([0,T], L^2(\Omega))$, a consequence of the fact that $u \in \mathcal{W}$ (see Proposition 10.14), but much more regular. Indeed, for every $\varepsilon > 0$,

$$u \in C^\infty\big(\overline{\Omega} \times [\varepsilon, T)\big).$$

So, with a data u_0 only in $L^2(\Omega)$, hence possibly discontinuous, u is C^∞ with respect to x for any $t > 0$. This implies in particular, that the heat equation is not time reversible. For proofs and related comments we refer the reader, for instance, to [4], [16] and [19].

10.4.2 *The wave equation*

In this section we turn our attention to the hyperbolic partial differential equation (10.2) that we recall here,

$$\begin{cases} u''(x,t) - \operatorname{div}\big(A(x)\,\nabla u(x,t)\big) = f(x,t) & \text{in } \Omega \times (0,T),\\ u(x,t) = 0 & \text{on } \partial\Omega \times (0,T),\\ u(x,0) = u^0(x) & \text{in } \Omega,\\ u'(x,0) = u^1(x) & \text{in } \Omega, \end{cases}$$

which describes the wave propagation in a body occupying the domain Ω. Its weak formulation (10.18) was given in Section 10.3.2, that is

$$\begin{cases} \text{Find } u \in \mathcal{W}_2 \text{ such that} \\ \langle u''(\cdot,t), v\rangle_{H^{-1}(\Omega),H_0^1(\Omega)} + \displaystyle\int_\Omega A(x)\nabla u(x,t)\,\nabla v(x)\,dx \\ \qquad = \displaystyle\int_\Omega f(x,t)\,v(x)\,dx, \quad \forall v \in H_0^1(\Omega), \ \text{a.e. } t \in (0,T), \\ u(x,0) = u^0(x) \qquad \text{in } \Omega, \\ u'(x,0) = u^1(x) \qquad \text{in } \Omega. \end{cases}$$

The space $\mathcal{W}_2$ is given by Definition 10.13,

$$\mathcal{W}_2 = \{w \mid w \in L^2(0,T;\ H_0^1(\Omega)),\ w' \in L^2(0,T;\ L^2(\Omega))\},$$

where we used again the identification (10.6).

In the proof of the existence theorem we will need a simple version of the classical Gronwall Lemma, a tool very useful when treating evolution problems. Let us recall it before going further.

Lemma 10.23. *Let h be a function in $C([0,T])$. Assume that there exists $\eta \in \mathbb{R}^+$ such that*

$$w(t) \le \eta + \int_0^t w(\tau)\,d\tau, \quad \forall\, t \in [0,T]. \tag{10.49}$$

Then the following estimate holds true:

$$w(t) \le \eta e^t \quad \textit{for all } t \in [0,T]. \tag{10.50}$$

Proof. A simple computation shows that inequality (10.49) is equivalent to the following one:

$$\frac{d}{dt}\Big(e^{-t}\int_0^t w(\tau)\,d\tau\Big) \le \eta\, e^{-t}.$$

Integrating from 0 to t gives

$$e^{-t}\int_0^t w(\tau)\,d\tau \le \eta(-e^{-t}+1),$$

which, used in (10.49) yields (10.50). □

Theorem 10.24. *Under assumptions* (10.19), *problem* (10.18) *has a unique solution $u \in \mathcal{W}_2$. Moreover,*

$$\begin{aligned} u\ &\in L^\infty(0,T;\ H_0^1(\Omega)), \\ u'\ &\in L^\infty(0,T;\ L^2(\Omega)), \\ u''\ &\in L^\infty(0,T;\ H^{-1}(\Omega)), \end{aligned}$$

and there exists a constant c depending on α, β, Ω, and T such that

$$\begin{aligned}\|u\|_{L^\infty(0,T;\,H_0^1(\Omega))} + \|u'\|_{L^\infty(0,T;\,L^2(\Omega))} + \|u''\|_{L^\infty(0,T;\,H^{-1}(\Omega))}\\ \leq c\big(\|f\|_{L^2(\Omega\times(0,T))} + \|u^0\|_{H_0^1(\Omega)} + \|u^1\|_{L^2(\Omega)}\big).\end{aligned} \tag{10.51}$$

Proof. Many arguments in the proof of this result are similar to those used in the proof of Theorem 10.20. However, for the reader's convenience we will give below the proof of Theorem 10.24, pointing out the main differences with the heat equation. We will skip the details which are just the same as those in the case of the heat equation.

Step 1 (Construction of the approximate problem). As for the case of the heat equation, let $\{w_n\}$ be a orthonormal basis of eigenvectors related to problem (9.71) and introduced in Section 9.8 (see Theorem 9.26). Denote W_n the space spanned by $\{w_i\}_{1\leq i\leq n}$.

The n-dimensional approximate problem in W_n is defined as follows:

$$\begin{cases}\text{Find } u_n(\,\cdot\,,t) = \displaystyle\sum_{j=1}^{n} g_j^n(t)w_j \in W_n \text{ such that}\\ \displaystyle\int_\Omega u_n''(x,t)w_k(x)\,dx + \int_\Omega A(x)\nabla u_n(x,t)\,\nabla w_k(x)\,dx\\ \qquad = \displaystyle\int_\Omega f_n(x,t)\,w_k(x)\,dx,\ \ \forall k=1,\dots,n,\quad \text{a.e. } t\in(0,T),\\ u_n(x,0) = u_n^0(x) \quad \text{in } \Omega,\\ u_n'(x,0) = u_n^1(x) \quad \text{in } \Omega,\end{cases} \tag{10.52}$$

where the sequence $\{f_n\}$ belongs to $C^\infty(\overline{\Omega\times(0,T)})$ and satisfies (10.23) and (10.24), introduced in Step 1 of the proof of Theorem 10.20. For the initial conditions, taking into account (10.19), we set

$$u_n^0 \doteq P_n(u^0) = \sum_{j=1}^{n}(u^0\,,w_j)w_j,\quad u_n^1 \doteq P_n(u^1) = \sum_{j=1}^{n}(u^1\,,w_j)w_j. \tag{10.53}$$

Observe that as in (10.29),

$$\sum_{j=1}^{n} g_j^n(0)w_j = u_n(\,\cdot\,,0) = u_n^0, \tag{10.54}$$

which, since $w_1,\dots,w_n$ are linearly independent, means that

$$g_k^n(0) = (u^0, w_k),\qquad \forall k=1,\dots,n. \tag{10.55}$$

Similarly,

$$\sum_{j=1}^{n}(g_j^n)'(0)w_j = u_n'(\,\cdot\,,0) = u_n^1, \tag{10.56}$$

so that

$$(g_k^n)'(0) = (u^1, w_k), \qquad \forall k = 1,\ldots,n. \tag{10.57}$$

Thus, problem (10.52) is equivalent to the following system of n linear ordinary differential equations of the second order with unknowns $g_1^n,\ldots,g_n^n$:

$$\begin{cases} \dfrac{d^2 g_k^n}{dt^2}(t) + \displaystyle\sum_{j=1}^{n} g_j^n(t)\int_\Omega A(x)\nabla w_j\,\nabla w_k\,dx = \int_\Omega f_n(x,t)\,w_k\,dx, \\ g_k^n(0) = (u^0, w_k), \qquad (g_k^n)'(0) = (u^1, w_k), \qquad \forall k = 1,\ldots,n. \end{cases} \tag{10.58}$$

Classical results imply the existence of a solution $(g_1^n,\ldots,g_n^n)$ with g_k^n are in $C^1([0,T_n))$. Using the argument from Step 1 of the proof of Theorem 10.20 (see (10.32)), it follows that if $T_n < T$, then necessarily $g_k^n(t) \to +\infty$ as $t \to T_n$ $(t < T_n)$.

The *a priori* estimates that will be proven in the next step show that this is impossible since g_k^n are bounded independently of k and n, so that we can take $T_n = T$. In particular, this means that (10.52) has a unique solution $(g_1^n,\ldots,g_n^n)$, with $g_k^n \in C^1([0,T])$ for all $k = 1,\ldots,n$.

Step 2 (*A priori* estimates for u_n and u_n'). Multiply the equation in (10.58) by $(g_k^n)'$ and sum over k from 1 to n, to get

$$\begin{aligned} \int_\Omega u_n''(x,t)\,u_n'(x,t)\,dx &+ \int_\Omega A(x)\nabla u_n(x,t)\,\nabla u_n'(x,t)\,dx \\ &= \int_\Omega f(x,t)\,u_n'(x,t)\,dx. \end{aligned} \tag{10.59}$$

By the same computation which led to (10.35), it is easily seen that

$$\int_\Omega u_n''(x,t)\,u_n'(x,t)\,dx = \frac{1}{2}\frac{d}{dt}\|u'(\,\cdot\,,t)\|^2_{L^2(\Omega)}. \tag{10.60}$$

Let us now go to the second integral in (10.59). It is at this point that the hypothesis on the symmetry of matrix A plays an essential role. It allows us to show that

$$\begin{aligned} &\int_\Omega A(x)\nabla u_n(x,t)\,\nabla u_n'(x,t)\,dx \\ &\qquad = \frac{1}{2}\frac{d}{dt}\int_\Omega A(x)\nabla u_n(x,t)\,\nabla u_n(x,t)\,dx. \end{aligned} \tag{10.61}$$

Indeed,

$$\frac{d}{dt}\int_\Omega A(x)\nabla u_n(x,t)\,\nabla u_n(x,t)\,dx = \frac{d}{dt}\sum_{i,j=1}^N \int_\Omega a_{ij}(x)\frac{\partial u_n(x,t)}{\partial x_i}\,\frac{\partial u_n(x,t)}{\partial x_j}\,dx$$

$$= \sum_{i,j=1}^N\left[\int_\Omega a_{ij}(x)\frac{\partial u_n'(x,t)}{\partial x_i}\,\frac{\partial u_n(x,t)}{\partial x_j}\,dx + \int_\Omega a_{ij}(x)\frac{\partial u_n(x,t)}{\partial x_i}\,\frac{\partial u_n'(x,t)}{\partial x_j}\,dx\right]$$

$$= \sum_{i,j=1}^N\left[\int_\Omega a_{ji}(x)\frac{\partial u_n'(x,t)}{\partial x_i}\,\frac{\partial u_n(x,t)}{\partial x_j}\,dx + \int_\Omega a_{ij}(x)\frac{\partial u_n(x,t)}{\partial x_i}\,\frac{\partial u_n'(x,t)}{\partial x_j}\,dx\right],$$

and (10.61) follows simply by interchanging in the last integral the summation indices i and j. With (10.60) and (10.61), from (10.59) we have successively,

$$\begin{aligned}\frac{d}{dt}\Big(\|u_n'(\cdot,t)\|^2_{L^2(\Omega)} + \int_\Omega A(x)\nabla u_n(x,t)\,\nabla u_n(x,t)\,dx\Big)&\\ \le 2\|f\|_{L^2(\Omega)}\|u_n'(\cdot,t)\|_{L^2(\Omega)}&\\ \le \|f(\cdot,t)\|^2_{L^2(\Omega)} + \|u_n'(\cdot,t)\|^2_{L^2(\Omega)},&\end{aligned} \tag{10.62}$$

which we integrate for $t \le T$ on the interval $(0,t)$. Using Definition 9.2 of the set $M(\alpha,\beta,\Omega)$ to bound the integral in the left-hand side, yields

$$\begin{aligned}&\|u_n'(\cdot,t)\|^2_{L^2(\Omega)} + \alpha\|u_n(\cdot,t)\|^2_{H_0^1(\Omega)}\\ &\quad\le \|u_n^1\|^2_{L^2(\Omega)} + \int_\Omega A(x)\nabla u_n^0(x)\,\nabla u_n^0(x)\,dx\\ &\qquad + \int_0^T \|f(\cdot,\tau)\|^2_{L^2(\Omega)}\,dt + \int_0^t \|u_n'(\cdot,\tau)\|^2_{L^2(\Omega)}\,d\tau\\ &\quad\le \|u_n^1\|^2_{L^2(\Omega)} + \beta\|u_n^0\|^2_{H_0^1(\Omega)}\\ &\qquad + \|f\|^2_{L^2(\Omega\times(0,T))} + \int_0^t \|u_n'(\cdot,\tau)\|^2_{L^2(\Omega)}\,d\tau.\end{aligned}$$

In view of definition (10.53) of u_n^0 and u_n^1, one has

$$\begin{aligned}&\|u_n'(\cdot,t)\|^2_{L^2(\Omega)} + \alpha\|u_n(\cdot,t)\|^2_{H_0^1(\Omega)}\\ &\quad\le \|u^1\|^2_{L^2(\Omega)} + \beta\|u^0\|^2_{H_0^1(\Omega)} + \|f\|^2_{L^2(\Omega\times(0.T))}\\ &\qquad + \int_0^t \Big[\|u_n'(\cdot,\tau)\|^2_{L^2(\Omega)} + \alpha\|u_n(\cdot,\tau)\|^2_{H_0^1(\Omega)}\Big]\,d\tau.\end{aligned} \tag{10.63}$$

We are now in position to apply Lemma 10.23 with

$$w(t) = \|u_n'(\cdot,t)\|^2_{L^2(\Omega)} + \alpha\|u_n(\cdot,t)\|^2_{H_0^1(\Omega)},$$

$$\eta = \|u^1\|^2_{L^2(\Omega)} + \beta\|u^0\|^2_{H_0^1(\Omega)} + \|f\|^2_{L^2(\Omega\times(0.T))},$$

to get the estimate

$$\begin{aligned}
&\|u_n\|_{L^\infty(0,T;\, H_0^1(\Omega))} + \|u_n'\|_{L^\infty(0,T;\, L^2(\Omega))} \\
&\qquad \leq c_1\big(\|f\|_{L^2(\Omega\times]0,T[)} + \|u^0\|_{L^2(\Omega)} + \|u^1\|_{H_0^1(\Omega)}\big),
\end{aligned} \tag{10.64}$$

where c_1 depends only on α, β, Ω and T. Consequently,

$$\begin{aligned}
&u_n \text{ is in a bounded set of } L^\infty(0,T;\, H_0^1(\Omega)),\\
&u_n' \text{ is in a bounded set of } L^\infty(0,T;\, L^2(\Omega)).
\end{aligned} \tag{10.65}$$

To proceed further, we need to obtain an *a priori* estimates for u_n''.

Step 3 (*A priori* estimates for u_n''). The procedure is exactly the same as that of Step 3 of the proof of Theorem 10.20. With notation (10.14), for $v \in H_0^1(\Omega)$ decomposed as $v = P_n(v) + v_n$ with $v_n \in W_n^\perp$, from (10.52) we get

$$\begin{aligned}
(u_n''(\cdot\,,t), P_n(v)) &= \langle u_n'', v\rangle_{H^{-1}(\Omega),H_0^1(\Omega)}\\
&= -a(u_n(\cdot\,,t), P_n(v)) + (f(\cdot\,,t), P_n(v)).
\end{aligned}$$

Thanks to inequality (9.80) one immediately has

$$\|u_n''(\cdot\,,t)\|_{H^{-1}(\Omega)} \leq C\big(\|f(\cdot\,,t)\|_{L^2(\Omega} + \|u_n(\cdot\,,t)\|_{H_0^1(\Omega)}\big) \text{ a.e. for } t \in [0,T].$$

Integrating in time over $(0,T)$, and using the former *a priori* estimate (10.64), yield the required estimate for u_n'' and ends the proof of (10.51).

As a conclusion

$$u_n'' \text{ is in a bounded set of } L^\infty(0,T;\, H^{-1}(\Omega)). \tag{10.66}$$

Step 4 (Passing to the limit). Due to (10.65) and (10.66), using Remark 10.15, one may extract from the approximate sequence $\{u_n\}$ a subsequence still denoted $\{u_n\}$, such that,

$$\begin{aligned}
u_n &\rightharpoonup u \quad \text{weakly* in } L^\infty(0,T;\, H_0^1(\Omega)),\\
u_n' &\rightharpoonup u' \quad \text{weakly* in } L^\infty(0,T;\, L^2(\Omega)),\\
u_n'' &\rightharpoonup u'' \quad \text{weakly in } L^2(0,T;\, H^{-1}(\Omega)).
\end{aligned} \tag{10.67}$$

Observe that the first convergences entail that $u \in \mathcal{W}_2$, the space where the solutions of the weak wave equation (10.18) are searched.

The rest of the proof is the same as that of Step 4 of the proof of Theorem 10.20, expression (10.46) being replaced by the following one:

$$
\begin{aligned}
\int_0^T \langle u_n''(\cdot, t), P_n(v)\rangle_{H^{-1}(\Omega), H_0^1(\Omega)}\varphi(t)\, dx\, dt \\
+ \int_0^T \int_\Omega A(x)\, \nabla u_n(x,t)\, \nabla P_n(v)(x)\varphi(t)\, dx\, dt \qquad (10.68) \\
= \int_0^T \int_\Omega f_n(x,t)\, P_n(v)(x)\varphi(t)\, dx\, dt,
\end{aligned}
$$

where the passage to the limit is straightforward. The density argument used at the end of Step 4 of the proof of Theorem 10.20 allows us to obtain precisely the variational formulation (10.20), that is

$$
\begin{aligned}
\int_0^T \langle u''(\cdot, t), v\rangle_{H^{-1}(\Omega), H_0^1(\Omega)}\, dx\, dt \\
+ \int_0^T \int_\Omega A(x)\, \nabla u(x,t)\, \nabla v(x,t)\, dx\, dt \\
= \int_0^T \int_\Omega f(x,t)\, v(x,t)\, dx\, dt, \quad \forall v \in L^2(0,T); H_0^1(\Omega)),
\end{aligned}
$$

Step 5 (Recovery of the initial data). One has to show that the initial conditions $u(x,0) = u^0(x)$ and $u'(x,0) = u^1(x)$ are satisfied. Here again, the proof is modelled on that of Theorem 10.20. We choose a test function $v_n \in W_n$ in (10.52) satisfying (10.45), and a function $\varphi \in C^\infty([0,T])$ such that $\varphi(T) = \varphi'(T) = 0$. Then, following along the lines of Step 5 of the proof of Theorem 10.20, the result follows easily.

Step 6 (*A priori* estimates on u). The *a priori* estimate on the solution u stated in (10.51) can be derived from the corresponding estimates on u_n following the arguments of the corresponding step for the heat equation.

Step 7 (Uniqueness). Let u_1 and u_2 be two solutions corresponding to the same data. Their difference $w = u_1 - u_2$ satisfies

$$
\begin{cases}
\langle w''(t), v\rangle_{H^{-1}(\Omega), H_0^1(\Omega)} + \displaystyle\int_\Omega A(x)\nabla w(x,t)\, \nabla v(x) = 0, \\
\qquad\qquad\qquad \text{in } \mathcal{D}'(0,T), \quad \forall v \in H_0^1(\Omega), \\
w(x,0) = 0 \quad \text{in } \Omega, \\
w'(x,0) = 0 \quad \text{in } \Omega.
\end{cases}
\qquad (10.69)
$$

Here appears one of the major differences with the heat equation. As noticed in Remark 10.17, the solution of the heat equation (10.1) belongs to $\mathcal{W}$, so it could be taken as test function in this equation. Indeed, since

$u \in L^2(0,T;\ H_0^1(\Omega))$ and $u' \in L^2(0,T;\ H^{-1}(\Omega))$, the two spaces being in duality, the term $\langle u'(t), u\rangle_{H^{-1}(\Omega),H_0^1(\Omega)}$ is well-defined. This is not the case for the wave equation. To continue the proof, we would like to use w' in the equation from (10.69) above. But this is not possible because of the term $\langle w''(t), v\rangle_{H^{-1}(\Omega),H_0^1(\Omega)}$.

Indeed, we know from the *a priori* estimates that w' is only in $L^\infty(0,T;\ L^2(\Omega))$, so the duality pairing does not make sense for w in place of v.

This difficulty is solved by introducing a special test function φ, defined on $[0,T]$, and given by

$$\varphi(x,t) = \begin{cases} -\displaystyle\int_t^s w(x,\tau)\,d\tau & \text{for } 0 \le t \le s, \\ 0 & \text{for } s \le t \le T, \end{cases}$$

for s fixed in $(0,T)$.

Obviously, $\varphi(\,\cdot\,,t)$ belongs to $H_0^1(\Omega)$ for $t \in [0,T]$. Observe also that $\varphi'(\,\cdot\,,t) \in L^2(\Omega)$ for $t \in [0,T]$. Indeed,

$$\varphi'(x,t) = \begin{cases} -w(x,t) & \text{for } 0 \le t \le s, \\ 0 & \text{for } s \le t \le T. \end{cases} \tag{10.70}$$

Let us now take φ as test function in the variational formulation (10.69) and integrate over $(0,T)$, to get

$$\begin{aligned}\int_0^s \langle w''(\,\cdot\,,t), \varphi(\,\cdot\,,t)\rangle_{H^{-1}(\Omega),H_0^1(\Omega)}dt \\ + \int_0^s\int_\Omega A(x)\nabla w(x,t)\,\nabla\varphi(x,t)\,dxdt = 0.\end{aligned} \tag{10.71}$$

For the first integral term, taking into account the properties of w (see (10.69)) and of φ (in particular the fact that by construction, $\varphi(x,s) = 0$), and using also (10.35), we have successively,

$$\begin{aligned}\int_0^s\langle w''(\,\cdot\,,t),\varphi(\,\cdot\,,t)\rangle_{H^{-1}(\Omega),H_0^1(\Omega)}\,dt &= -\int_0^s\langle w'(\,\cdot\,,t), \varphi'(\,\cdot\,,t)\rangle_{H^{-1}(\Omega),H_0^1(\Omega)}\,dt \\ &\quad + \int_\Omega w'(x,s)\,\varphi(x,s)\,dx - \int_\Omega w'(x,0)\,\varphi(x,0)\,dx \\ &= -\int_0^s\int_\Omega \varphi'(x,t)\,w'(x,t)\,dx\,dt = \int_0^s\int_\Omega w(x,t)\,w'(x,t)\,dx\,dt \\ &= \frac{1}{2}\int_0^s \frac{d}{dt}\|w(\,\cdot\,,t)\|_{L^2(\Omega)}^2dt.\end{aligned}$$

Recalling (10.61) and (10.70), the second term in (10.71) is given by

$$\int_0^s\int_\Omega A(x)\nabla w(x,t)\,\nabla\varphi(x,t)\,dxdt = -\int_0^s\int_\Omega A(x)\nabla\varphi'(x,t)\,\nabla\varphi(x,t)\,dxdt$$
$$= -\frac{1}{2}\int_0^s\frac{d}{dt}\int_\Omega A(x)\nabla\varphi(x,t)\,\nabla\varphi(x,t)\,dxdt.$$

Consequently, (10.71) reads

$$\int_0^s\frac{d}{dt}\Big(\|w(\,\cdot\,,t)\|^2_{L^2(\Omega)} - \int_\Omega A(x)\nabla\varphi(x,t)\,\nabla\varphi(x,t)\,dx\Big)dt = 0,$$

so that

$$\|w(\,\cdot\,,s)\|^2_{L^2(\Omega)} + \int_\Omega A(x)\nabla\varphi(x,0)\,\nabla\varphi(x,0)\,dx = 0.$$

Since by definition

$$\nabla\varphi(x,s) = 0 \quad \text{a.e. on } \Omega,$$

by the ellipticity of A, this implies

$$\|w(\,\cdot\,,s)\|^2_{L^2(\Omega)} + \alpha\|\nabla\varphi(x,0)\|^2_{L^2(\Omega)} \leq 0,$$

and therefore,

$$w(s) = 0.$$

The arbitrariness of s in $(0,T)$ implies that $w \equiv 0$. □

Bibliography

[1] R.A. ADAMS, *Sobolev Spaces*, Academic Press, New York, 1965.

[2] F. BLACK AND M. SCHOLES, The pricing of options and corporate liabilities, Journal of Political Economy, 81 (3) 1973, 637-654.

[3] H. BREZIS, *Analyse fonctionelle: Théorie et Applications*, Masson, Paris, 1987.

[4] H. BREZIS, *Functional Analysis, Sobolev Spaces and Partial Differential Equations*, Universitext, Springer Verlag, 2011.

[5] A.P. CALDERÓN, Lebesgue spaces of differentiable functions and distributions, in Proc. Symp. Pure Math., IV, 1961, 33-49.

[6] T. CAZENAVE AND A. HARAUX, *An Introduction to Semilinear Evolution Equations*, Oxford Lecture in Mathematics and its Applications n. 13, Clarendon Press, Oxford, 1998.

[7] D. CHENAIS, On the existence of a solution in a domain identification problem, Journal Math. Anal. and Appl. (52) 1975, 189-219.

[8] D. CIORANESCU AND P. DONATO, *An Introduction to Homogenization*, Oxford Lecture Series in Mathematics and its Applications, vol. 17, Oxford University Press, 1999.

[9] D. CIORANESCU, V. GIRAULT AND K. RAJAGOPAL *Mechanics and Mathematics of Fluids of the Differential Type*, Springer, 2016.

[10] E. CODDINGTON AND N. LEVINSON, *Theory of Ordinary Differential Equations*, Mc Graw Hill, New York, 1955.

[11] E. DE GIORGI, Sulla differenziabilità e l'analiticità delle estremali degli Integrali multipli regolari, Mem. Accad. Sci. Torino Cl. Sci. Fis. Mat. Natur. (3) 25-43, 1957.

[12] J. DIESTEL AND J.J. UHL, *Vector Measures*, Mathematical Surveys, 15, American Mathematical Society, Providence, 1977.

[13] N. DINCULEANU, *Vector Measures*, Pergamon Press, New York, 1967.

[14] N. DUNFORD AND J.T. SCHWARTZ, *Linear Operators*, Interscience Publishers, New York, 1958.

[15] YU. V. EGOROV AND M. A. SHUBIN, *Foundations of the Classical Theory of Partial Differential Equations*, Springer, Berlin Heidelberg, 1998.

[16] L. C. EVANS, *Partial Differential Equations*, Graduate Studies in Mathe-

matics Vol. 19, American Mathematical Society, 1998.

[17] T. Gallouët and A. Monier, *On the regularity of Solutions to Elliptic Equations*, Rendiconti di Matematica, Serie VII, Volume 19, Roma, 1999, 471-488.

[18] P. R. Garabedian, *Partial Differential Equations*, Wiley, New York, 1964.

[19] D. Gilbarg and N. Trudinger, *Elliptic Partial Differential Equations of Second Order*, Classics in Mathematics, Springer Verlag, Berlin Heidelberg New York 1983.

[20] E. Kreyszig, *Advanced Engineering Mathematics*, 7th edition, John Wiley and Sons, New York, 1993.

[21] S. Kesavan, *Topics in Functional Analysis and Applications*, John Wiley & Sons, Inc., New York, 1989.

[22] O. A. Ladyzenskaya and N.N. Ural'tseva, *Linear and Quasilinear Elliptic Equations*, English Translation: Academic Press, New York, 1968.

[23] J.-L. Lions, *Quelques méthodes de résolution des problèmes aux limites non linéaires*, Dunod, Paris, 1969.

[24] J.-L. Lions and E. Magenes, *Problèmes aux limites non homogènes et applications*, Vol. 1 and Vol. 2, Dunod, Paris, 1968.

[25] R. Merton, Theory of rational option pricing, Bell J. Economics and Management Sci. 4 (1) 1973, 141-183. doi:10.2307/3003143.

[26] N. G. Meyers, An L^p-estimate for the gradient of solutions of second order elliptic divergence equations, Ann. Sc. Norm. Sup. Pisa, 17 1963, 189-206.

[27] V. P. Mikhailov, *Partial Differential Equations*, Mir, Moscow, 1976.

[28] J. Nash, Continuity of solutions of parabolic and elliptic equations. Amer. J. Math. 1958(80), 931- 954.

[29] J. Nečas, *Les méthodes directes en théorie des équations elliptiques*, Masson, Paris, 1967.

[30] J. Ockendon, S. Howinson, A. Lacey and A. Movchan, *Applied Partial Differential Equations*, Oxford University Press, 2006.

[31] A. Pazy, *Semigroups of Linear Operators and Applications to Partial Differential Equations*, Lecture Notes, University of Maryland, 1974.

[32] R. Precup, *Ecuaţii cu derivate parţiale* (in romanian), Transilvania Press, Cluj, 1997.

[33] H. L. Royden, *Real Analysis*, Macmillan, 1963.

[34] W. Rudin, *Real and Complex Analysis*, McGraw Hill, New York, 1966.

[35] L. Schwartz, *Théorie des distributions*, Hermann, Paris, 1965.

[36] G. Stampacchia, Équations elliptiques du second ordre à coefficients discontinus, Séminaire sur les équations aux dérivées partielles, Collège de France, 1964.

[37] G. Stampacchia, Le problème de Dirichlet pour les équations elliptiques du second ordre à coefficients discontinus, Ann. Inst. Fourier (Grenoble) (15) fasc. 1, 1965, 189-258.

[38] E. M. Stein, *Singular Integrals and Differentiability Properties of Functions*, Princeton University Press, Princeton, New Jersey, 1970.

[39] M. E. Taylor, *Partial Differential Equations: Basic Theory*, Springer, New York, 1996.

[40] G. Troianiello, *Elliptic Differential Equations and Obstacle Problems*, The University Series in Mathematics, Plenum Press, New York, 1987.
[41] J. Wloka, *Partial Differential Equations*, Cambridge University Press, 1987.
[42] Z. Wu, J. Yin and C. Wang, *Elliptic and Parabolic Equations*, World Scientific, 2006.
[43] K. Yosida, *Functional Analysis*, Springer-Verlag, Berlin, 1964.
[44] E. Zeidler, *Nonlinear Functional Analysis and its Applications* (Part I and Part II), Springer-Verlag, Berlin, 1980.

Index

Printed in the United States
By Bookmasters